Management of Energy and Manufacturing System

Management of Energy and Manufacturing System

Editors

Tangbin Xia
Ershun Pan
Rongxi Wang
Yupeng Li
Xi Gu

MDPI • Basel • Beijing • Wuhan • Barcelona • Belgrade • Manchester • Tokyo • Cluj • Tianjin

Editors

Tangbin Xia
Shanghai Jiao Tong
University
China

Ershun Pan
Shanghai Jiao Tong
University
China

Rongxi Wang
Xi'an Jiaotong University
China

Yupeng Li
China University of Mining
and Technology
China

Xi Gu
The State University of New
Jersey
USA

Editorial Office
MDPI
St. Alban-Anlage 66
4052 Basel, Switzerland

This is a reprint of articles from the Special Issue published online in the open access journal *Energies* (ISSN 1996-1073) (available at: https://www.mdpi.com/journal/energies/special_issues/Energy_Manufacturing_System).

For citation purposes, cite each article independently as indicated on the article page online and as indicated below:

LastName, A.A.; LastName, B.B.; LastName, C.C. Article Title. *Journal Name* **Year**, *Volume Number*, Page Range.

ISBN 978-3-0365-7604-6 (Hbk)
ISBN 978-3-0365-7605-3 (PDF)

Contents

About the Editors

Tangbin Xia

Tangbin Xia received a Ph.D. degree in MECHANICAL ENGINEERING (INDUSTRIAL ENGINEERING) from Shanghai Jiao Tong University in 2014. He was a Visiting Scholar with WuMRC, University of Michigan and ISyE, Georgia Institute of Technology. Currently, he is an Associate Professor (Tenured) and serves as Deputy Director of the Department of Industrial Engineering at Shanghai Jiao Tong University, State Key Laboratory of Mechanical System and Vibration. He has published 160 papers. His major research interests include intelligent maintenance systems, prognostics & health management, and advanced manufacturing. He is a member of IEEE, ASME, IISE and INFORMS.

Ershun Pan

Ershun Pan received a Ph.D. degree in mechanical engineering from Shanghai Jiao Tong University in 2000. Currently, he is a professor and serves as director of the Department of Industrial Engineering at Shanghai Jiao Tong University. He is also Vice President of the China Institute of Quality Research. He previously worked at the University of Michigan as a visiting scholar and published more than 200 papers in various journals. His research and teaching interests are in the areas of the theory and methods of quality control, reliability engineering and maintenance strategy, and lean manufacturing technology.

Rongxi Wang

Rongxi Wang is an Associate Professor in State Key Laboratory for Manufacturing Systems Engineering at Xi'an Jiaotong University. He received his Ph.D degree in mechanical engineering from Xi'an Jiaotong University, Xi'an, China. His current research interests include reliability and maintenance, quality control and performance condition evaluation of distributed and complex electromechanical systems and the theory of complex network. During the past 5 years, about 20 articles have been published in top journals, and some work has been applied in actual applications which play an effective role in ensuring the reliability of complex electromechanical systems.

Yupeng Li

Yupeng Li received a B.S. degree in Industrial Engineering, the M.S. degree in Resource Development and Planning from China University of Mining and Technology, Xuzhou, in 2006 and 2009, respectively, and the Ph. D degree in Mechanical Engineering from Shanghai Jiao Tong University in 2015. He is currently Associate Professor of Mechanical Engineering with School of Mines, China University of Mining and Technology. His research interests include product evolution control and operation management of manufacturing and service systems.

Xi Gu

Xi Gu obtained a B.S. degree in mechanical engineering from Zhejiang University, Hangzhou, Zhejiang, China, an M.S. in industrial and operations engineering and a Ph.D. degree in mechanical engineering, both from the University of Michigan, Ann Arbor, MI, USA. He is currently an Assistant Teaching Professor with the Department of Mechanical and Aerospace Engineering at Rutgers, The State University of New Jersey. His research interests include smart manufacturing, design and operation of reconfigurable manufacturing systems, maintenance decision-making in manufacturing systems, and engineering education. He was the recipient of the 2020 Outstanding Young Manufacturing Engineer Award from the Society of Manufacturing Engineers (SME), and the 2014 International Conference on Frontiers of Design and Manufacturing best paper award.

Preface to "Management of Energy and Manufacturing System"

This reprint aims to publish recent advances and technical challenges in, as well as novel design methodologies for, energy manufacturing systems. Key to the present conception of sustainability is the capacity to meet current needs without sacrificing the future ability to do so. Maintaining this principle, which until now has been a key sustainability concept, is becoming more challenging than ever, with an increasing population rate, energy poverty, global warming, and surging demand for products and services each presenting their own unique challenges. Manufacturing is in a prime position to address this challenge, making a significant economic contribution to the global GDP and having a high influence over the environment and humanity.

These published papers can provide references for engineers, scholars, and business managers in the field of energy optimization.

Tangbin Xia, Ershun Pan, Rongxi Wang, Yupeng Li, and Xi Gu
Editors

Article

Research on Machining Workshop Batch Scheduling Incorporating the Completion Time and Non-Processing Energy Consumption Considering Product Structure

Nailiang Li * and Caihong Feng

Department of Industrial Engineering, School of Mines, China University of Mining and Technology, Xuzhou 221116, China; TS19020220A31@cumt.edu.cn
* Correspondence: cumt_lnl@126.com

Abstract: Energy-saving scheduling is a well-known issue in the manufacturing system. The flexibility of the workshop increases the difficulty of scheduling. In the workshop schedule, considering the collaborative optimization of multi-level structure product production and energy consumption has certain practical significance. The process sequence of parts and components should be consistent with the assembly sequence. Additionally, the non-production energy consumption (NPEC) (such as the energy consumption of workpiece handling, equipment standby, and workpiece conversion) generated by the auxiliary machining operations, which make up the majority of the total energy consumption, should not be ignored. A sub-batch priority is set according to the upper and lower coupling relationship in the product structure. A bi-objective batch scheduling model that minimizes the total energy consumption and the total completion time is developed, and the multi-objective gray wolf optimizer (MOGWO) is employed as the solution to obtain the optimal schedule scheme. A case study is performed to demonstrate the potential possibilities concerning NPEC in regard to reducing the total energy consumption and to show the effectiveness of the algorithm. Compared with the traditional optimization model, the joint optimization of NPEC and PEC can reduce the energy consumption of standby and handling by 9.95% and 22.28%, respectively.

Keywords: multi-level structure; non-production energy consumption (NPEC); sub-batch priority; multi-objective gray wolf optimizer (MOGWO)

Citation: Li, N.; Feng, C. Research on Machining Workshop Batch Scheduling Incorporating the Completion Time and Non-Processing Energy Consumption Considering Product Structure. *Energies* **2021**, *14*, 6079. https://doi.org/10.3390/en14196079

Received: 3 August 2021
Accepted: 18 September 2021
Published: 24 September 2021

Publisher's Note: MDPI stays neutral with regard to jurisdictional claims in published maps and institutional affiliations.

1. Introduction

Scheduling has been certified to be critical in manufacturing for improving the productivity of the manufacturing system and utilization of equipment, as well as shorting the manufacturing cycle [1]. Traditional machining scheduling typically envisions product processing phases where jobs are independent of each other and are not sequence constrained. In the conventional processing scheduling method, the two processes of processing and assembly are separated from each other, which will lead to the destruction of the original parallel relationship between processing and assembly [2]. Generating a processing sequence on the base of the assembly process would reduce the frequency of this phenomenon. Nevertheless, because the process is complicated (such as product hierarchy, number of workpieces, and processing steps), the resolution of these problems considered processing sequence in machining systems is more difficult than traditional machining scheduling. Moreover, it is an NP-hard problem [3].

Temporally, the completion time of workpieces is affected by the hierarchy of the product tree. For example, a product that includes 10 levels is processed from the bottom up. If each level is processed when the previous level is finished, without a doubt, it will prolong the entire production cycle. Therefore, batching the workpieces and making the scheduling scheme in the production of the multi-level structures have practical significance for reducing the completion time.

The manufacturing industry has consumed large amounts of energy in the process of transforming resources into products or services, leading to many environmental problems [4]. The realization of low-carbon manufacturing is extremely important for improving the sustainability of the manufacturing industry [5]. Energy-efficient scheduling, which can be approved in the manufacturing industry, has achieved energy conservation and emissions reduction [6]. Therefore, the scheduling scheme should not only consider the rationality of the process sequence for the workpieces but also reduce the energy consumption of the production system.

The processing energy consumption (PEC) and non-processing energy consumption (NPEC) of the machine are the two components of the production energy consumption of the workshop. PEC stands for the energy consumption of the machine at the processing stage, which is related to the processing power and processing time of the machine. NPEC is the sum of standby energy consumption, conversion energy consumption, and handling energy consumption. Compared with the energy consumed by the machine in other operating phases, the equipment consumes less energy when processing workpieces, especially in mass production, which generally only accounts for approximately 10% of the total energy consumption [7].

Most of the energy consumption in production is generated by auxiliary operations. Here, auxiliary operations are defined as operations that are not directly involved in processing but indispensable in the production process, such as those for equipment standby, state conversion, and workpiece handling. Compared with the energy consumption generated by the processing phase, the energy consumption generated by auxiliary operations can be large. Therefore, if the focus of energy conservation is on developments in processing and energy-saving equipment [8], the considerable energy-saving potential of auxiliary operations will be ignored. In addition, relative to changing a processing technology or researching and developing more energy-saving processing, an optimized workshop scheduling scheme can provide good application value with a low investment [9]. Therefore, in a production system based on processing sequences of workpieces, comprehensively considering the energy consumption composition in the production process, optimizing the allocation of workshop resources, and formulating reasonable scheduling arrangements will be more conducive to reducing energy consumption and improving efficiency in manufacturing enterprises.

Gray wolf optimization [10] (GWO) is a new intelligent optimization algorithm proposed in recent years. Compared with the genetic algorithm (GA) and particle swarm optimization (PSO), GWO algorithm results are more competitive [11]. At present, the gray wolf algorithm has been widely applied in thermodynamics [12], power systems [13], energy and fuels [14], cloud technology [15], and workshop scheduling [16–18]. Lu [16] embedded genetic operators into the multi-objective GWO to enhance the searchability of the algorithm. Qin [17] used the improved multi-objective gray wolf algorithm to solve the casting shop scheduling to minimize the production cycle, total production cost, and total delivery delay. Lu [18] added a random search model based on traditional GWO search to enhance global search capability. Although GWO has been successfully used in many different types of production environments, there is limited literature on GWO to solve energy-saving scheduling problems in a machine-shop, especially to optimize auxiliary production energy consumption. Therefore, we extended the single-objective GWO to the multi-objective GWO to consider completion time and total energy consumption minimization.

Reducing energy consumption through NPCE optimization and minimum completion time are major design goals. The research motivation and research problem will be clearer in the discussion of background research, and then the related mathematical model will be introduced, followed by the MOGWO algorithm and how it is applied to optimize the bi-objective scheduling problem; finally, a presentation of a case study will demonstrate the model and algorithm in the case of two different energy consumption optimization objectives and different algorithms.

In the rest of this paper, the current research progress will be introduced in Section 2 and a bi-objective model based on the research problem is established in Section 3. In Section 4, the working procedure of MOGWO of this optimization problem is described. Section 5 conducts case analysis and comparison. Finally, Section 6 summarizes this article.

2. Literature Review

The scheduling research of multi-level structure products was first studied for an assembly workshop [19,20]. Li et al. [21] developed four batch strategies to solve three different multi-level product assembly problems. Lu et al. [22] studied tree-like products scheduling problems in an assembly workshop, aiming to minimize the assembly completion time. Wan et al. [23] proposed a visual modeling and scheduling model for assembly processing based on a workflow for the assembly process of complex products and designed heuristic scheduling rules. Suharyanti et al. [24] investigated the optimal lot size of complex products in the job shop. Batching can effectively reduce product production cycles. However, the scheduling research on multi-level structure products is not enough to solve the problem of collaboration production because sequence constraint scheduling should be compatible with the assembly sequence and has higher complexity.

A scheduling problem that comprehensively considers the optimization of the energy consumption with the traditional targets (completion time/production cost) is complex, and it is particularly important to carry out in-depth research on this [25]. At present, a large amount of energy waste occurring in processes that have nothing to do with equipment processing operations has been found. Dahmus and Gutowski [7] analyzed the energy consumption of machining and proved that the actual processing operations only accounted for a small part of the total energy consumption, whereas auxiliary operation energy consumption accounted for 30–50%. The idle time of the machine occupied 16% of the total completion time [26]. Thus, the NPEC is large. Based on the knowledge regarding total energy consumption, it has become a trend to decompose the total energy consumption and reduce energy waste through optimization of the scheduling scheme. Wang [27] simulated a processing process and classification of energy consumption by product quality. In general, the total energy consumption can divide into PEC and NPEC, where NPEC, as indicated above, refers to the energy consumption generated by auxiliary operations such as equipment start-up, shutdown, and idling.

In recent years, more and more research of NPEC has been conducted thoroughly. Luan et al. [28] studied the energy consumption of non-cutting status and established an accurate power model to accurately predict the power of the feed motion. Liu et al. [29] improved the machine utilization rate by 8.2% by optimizing the processing sequence for the workpieces. Peng et al. [30] considered standby energy consumption. Wu et al. [31] studied a renewable energy scheduling problem of a flow shop and established a multi-objective renewable energy power supply model, intending to reduce the processing and idle energy consumption during processing. Gilles et al. [32] investigated the impacts of batch production on energy consumption and order completion time. It was considered that batching could effectively reduce the number of conversions and equipment standby time, thereby reducing the conversion energy consumption and standby energy consumption. Che et al. [25] used a clustering algorithm to determine whether a shutdown operation was required between two tasks to reduce the standby energy consumption and/or optimally sort the processing tasks. Liu et al. [29] integrated scattered short standby periods into a long standby time and judged whether off or assigned other tasks. Wang et al. [33] considered the power changes in the standby state and processing states of the machine. However, the above research mostly focused on the conversion and standby energy consumption separately. In the research of NPEC, they should be considered more systematically.

In the research articles above, most of the influences of workpiece handling on the energy consumption of the workshop were ignored. However, it is more realistic to consider the scheduling and optimization of such handling. The impact of handling on energy consumption has generally not been considered [34–36]. The handling of workpieces between machines is also part of auxiliary processing, and its energy consumption should belong to the NPEC in the system. Different from previous studies, the NPEC would be determined based on the energy consumption of the equipment standby, conversion, and workpiece handling. In addition, a machine may perform many processing tasks, and the frequent shutdown and start-up of the equipment will increase energy consumption. Therefore, this part of the energy consumption is generally not considered.

Based on the above discussion, the research on multi-level product production and energy-conscious scheduling in the machining workshop is limited. The existing workshop scheduling methods cannot meet the requirements of production and energy saving because each mode has its characteristics. The research on NPEC should be considered when proposing a scheduling method that is different from the traditional optimal energy consumption scheduling. Therefore, the scheduling model of the machining workshop is more complicated than that of the traditional workshop. To bridge this gap, a scheduling method was introduced specifically for machining workshops to minimize completion time and energy consumption.

3. Problem Statement

3.1. Related Product Structure

In real-world production, a product is a combination of a group of parts/components with order constraints and can be represented using a tree structure diagram. The processing and assembly of products are conducted according to the tree structure. The edges in the tree represent the assembly constraint relationships, the leaf nodes are the parts/components, and the root node is the final product. Each level's leaf node can be called a child node of its upper-level node, and the processing starts from the lowest-level leaf node. It is composed of n parts/components with order constraints. The production of components includes several processes, each process is handled by a machine (in M), and the alternative processing equipment for different processes can be identical.

According to the structural characteristics and commonalities between products, the literature [37] has generally summarized product structures into three types: flat, tall, and complex. The corresponding product structure trees are shown in Figure 1. The flat type is a single-layer product and is directly assembled from first-level parts into products. The tall type has multiple levels of parts/components, and each sub-workpiece contains at most two nodes. The complex type is a multi-layer composite of flat- and tall-type tree structures, in which at least one parent node contains more than two child nodes, as shown in Figure 1. The nodes of each tree are arranged hierarchically, where level 0 represents the complete product, and level 1 is the hierarchical arrangement of parts. For example, the structure of the B product is divided into 1–3 levels, and the branch nodes under it are called components, such as P_1 and P_2. The production sequence is as follows: first produce the workpieces J_3 and J_4, and then the superior P_2 and J_2 for production, and so on.

Based on the coupling relationship(s) between the parts/components in the product structure, batch production and handling are conducted, in which each type of workpiece is divided into equal batches, and the numbers and size of the sub-batch of each product are determined, as well as the sub-batch production and handling sequence. A batch scheduling problem based on the product structure will face difficulties caused by the coordination of the processing times between the workpieces. The arrangement of the production sequence of each sub-batch to meet the coupling sequence of the products is very important. For example, in the mass production of the workpieces in a machining workshop, each type of workpiece can be divided into an equal number of sub-batches. According to the coupling relationship between the workpieces and the production time of the workpiece, the lower-level workpieces are produced first, and then the start processing time of the upper-level workpieces will be later than the next level of workpieces, minimizing machine standby while optimizing handling equipment to reduce NPEC.

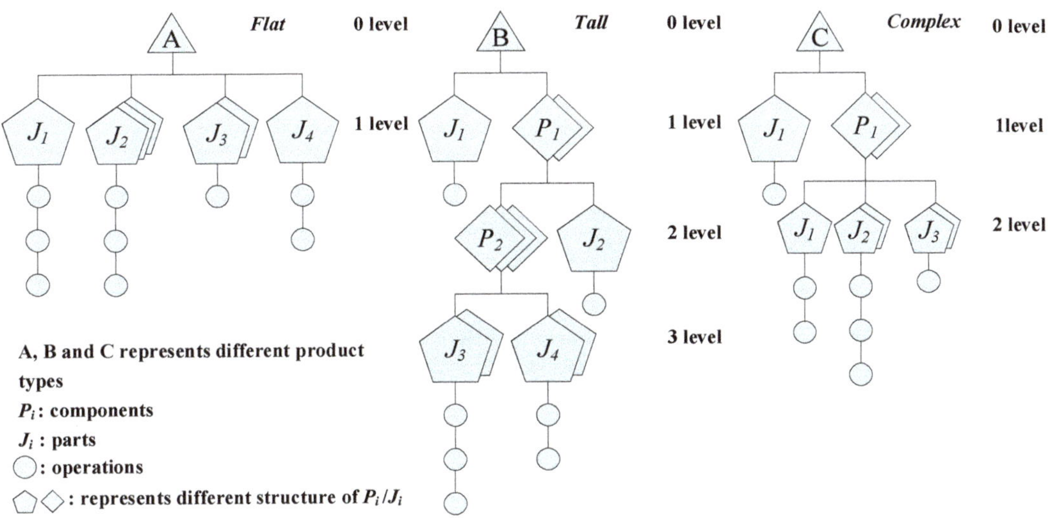

Figure 1. Three types of product structure.

3.2. Problem Definition and Assumptions

The problem can be expressed as follows. Related sets and decision variables are shown in the Appendix A.

1. The product contains n types of workpieces $J = \{J_1, J_2, J_3 \ldots J_n\}$, the number of workpieces J_j is Q_j, and each workpiece contains O_j processes.

2. There are h types of handling equipment in the workshop. A specific piece of handling equipment is expressed as H_h. The handling speed and power of the same type of handling equipment are the same; the speed of H_h is V_h, the power when handling parts/components j is P_j^h, and the rated capacity of the workpiece j on the handling equipment is S_{jh}.

3. After a certain process of the workpiece is processed on equipment m, it needs to be transported to the selected equipment m' of the next process. The locations of all equipment in the workshop are fixed, and the distance between equipment m and equipment m' is $d_{mm'}$. After the last process of a batch of workpieces is processed on equipment m, they are transported to assembly workshop P for assembly. The distance between equipment m and assembly workshop P is d_{mp}.

The following assumptions are used in the scheduling.

- Alternating machines for different processes can be the same.
- Each type of sub-batch of workpieces can only be transported to the next process processing equipment for processing/waiting after the previous process is completed according to the process sequence.
- At most, one workpiece is processed on each machine at a time, and one workpiece is processed on at most one piece of equipment at any time.
- The processing and handling equipment are available at the initial moment.
- A process is not interrupted once it starts processing.
- The equipment requires preparation time before processing different types of workpieces successively; in contrast, processing the same types of parts does not require preparation time.
- The number of pieces of handling equipment is limited. If the number of sub-batches of workpieces is greater than the rated capacity of the handling equipment, multiple pieces of equipment must be moved simultaneously or multiple times by one piece of equipment.

- The time required for loading and unloading workpieces is ignored.

3.3. Problem Formulation

3.3.1. Objectives

During workpieces processing, there are three main states of equipment: the processing state, conversion state, and standby state. The equipment power and state vary with time [25], as shown in Figure 2. S_m^i and C_m^i represent the start and completion times of the ith operation on device m, respectively, and $O_{j(1)i(1)}^m$ represents the O_{ji} processing on equipment m.

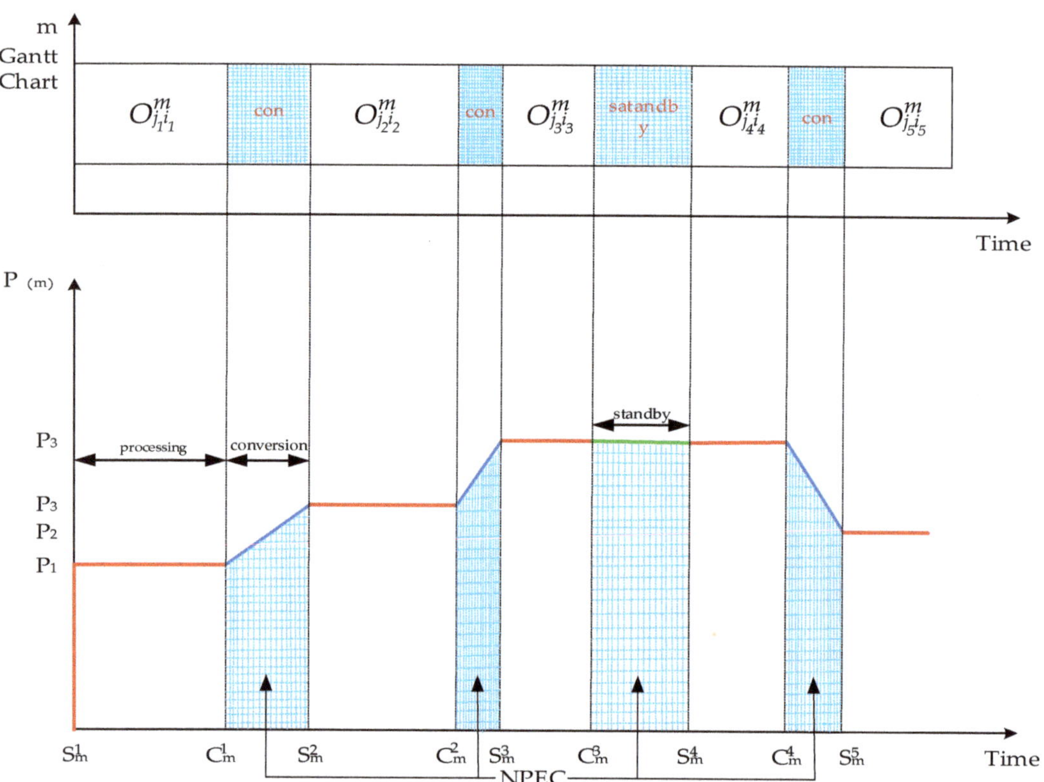

Figure 2. Gantt chart of *m* and the distribution of corresponding power.

The total energy consumption of the production system is composed of the PEC (E_p) and NPEC (E_n) during the processing of components and parts. Of these, the NPEC consists of the state transition energy consumption (E_s), the standby energy consumption (E_w), and the handing energy consumption (E_d) (including the energy consumption of the moving parts/components between machines and moving to the assembly workshop when finished).

- The energy consumption criterion E

It consists of the PEC and NPEC. Specifically, this energy consumption metric is given as follows:

$$E = E_p + E_n \tag{1}$$

The PEC consumed by all workpieces owing to the processing, as shown in Equation (2), as follows

$$E_p = \sum_{j=1}^{n} \sum_{k=1}^{B_j} \sum_{i=1}^{O_j} \sum_{m=1}^{w} \alpha_j TB_{jkim} \tag{2}$$

The NPEC consists of the equipment standby energy consumption (E_w), conversion energy consumption (E_s), and handling energy consumption (E_d). The equation is given by Equation (3), as follows:

$$E_n = E_w + E_s + E_d \tag{3}$$

Among them, E_w refers to the idling state where the equipment is on non-stop, and no parts/components are processing on the machine. The formula for the energy consumption during the standby period is shown as follows:

$$E_w = \sum_{k=1}^{B_j} \sum_{j=1}^{n} \sum_{m=1}^{w} \left(SB_{j(k+1)im} - CB_{jkim} \right) P_m^w \tag{4}$$

In the above, $SB_{j(k+1)m}$ represents the start time of the $(k+1)$th sub-batch of j on the processing equipment m, represents the kth sub-batch on equipment m completion time, and P_m^w is the standby power of the machine.

E_s represents the energy consumption of the equipment state transition. In Equation (5), $\left| \alpha_j - \alpha_{j'} \right|$ is the absolute value of the difference in energy consumption involved in switching power owing to processing different types of workpieces ($j \rightarrow j'$); $R_{jj'm}$ is a 0–1 variable. If the processing of the workpiece on equipment m is different from the workpieces to be processed, $R_{jj'm} = 1$; otherwise, $R_{jj'm} = 0$.

$$E_s = \sum_{j=1}^{n} \sum_{k=1}^{B_j} \sum_{i=1}^{O_j} \sum_{m=1}^{M} R_{jkim} R_{jj'm} \left| \alpha_j - \alpha_{j'} \right| \tag{5}$$

The workshop handling energy consumption is related to the sub-batch quantity B_{pj} and sub-batch Q_{pjk} of j, selected handling equipment H_h, distance $d_{mm'}$ between the equipment, and distance d_{mP} between the equipment and assembly workshop P. E_d represents the energy consumption during handling. The transportation of workpieces includes two parts: one part comprises transporting the current sub-batch process to the next processing machine after the completion of the current sub-batch process, and the other comprises transporting it to assembly workshop P after the last process of the sub-batch process is completed. The two parts of energy consumption are described in detail as follows. This can be expressed using Equation (6).

$$E_d = E_{jkimm'}^h + E_{jkmP}^h \tag{6}$$

After the process O_{ji} is processed on equipment m, the workpiece will be transported to the next process $O_{j(i+1)}$. The energy consumption $E_{jkimm'}^h$ of the transportation equipment H_h at the selected equipment m' is shown in Equation (7).

$$
\begin{aligned}
E_{jkimm'}^h &= \sum_{j=1}^{n} \sum_{k=1}^{B_j} \sum_{i=1}^{O_j} \sum_{m=1}^{w} \sum_{h=1}^{H} S_{jkih}^1 H_{jkimh}^1 n_{jk} P_j^h t_{jkimm'}^h \\
&= \sum_{j=1}^{n} \sum_{k=1}^{B_j} \sum_{i=1}^{O_j} \sum_{m=1}^{w} \sum_{h=1}^{H} S_{jkih}^1 H_{jiamh}^1 \left\lceil \frac{Q_{pjk}}{s_{jh}} \right\rceil P_j^h \frac{d_{mm'}}{V_h}
\end{aligned} \tag{7}
$$

Here, n_{jk} represents the number of pieces of handling equipment required for the kth sub-batch of j, and $t_{jkimm'}^h$ indicates the time that H_h moves j from process O_{ji} of equipment m to m'. S_{jkih}^β is a 0–1 variable. If H_h was selected to handle the ith process of the kth sub-batch of j, then $S_{jkih}^\beta = 1$; otherwise, $S_{jkih}^\beta = 0$. H_{jkimh}^β is also a 0–1 variable. β can take two values, 1 and 2; $\beta = 1$ indicates that the workpieces are transported between equipment; $\beta = 2$ indicates that the workpieces are transported from the last piece of equipment m to assembly shop P. If the ith process of the kth sub-batch of j is transported by H_h, then $H_{jkimh}^\beta = 1$; otherwise, $H_{jkimh}^\beta = 0$.

After the workpieces j are processed, they are transported from equipment m to assembly workshop P by H_h. The energy consumption E^h_{jkmP} of Hh is given by Equation (8), as follows:

$$E^h_{jkmP} = \sum_{j=1}^{n} \sum_{k=1}^{B_j} \sum_{h=1}^{H} X_{jkO_jm} S^2_{jkh} H^2_{jkimh} n_{jk} P^h_j t^h_{jkmP} = \sum_{j=1}^{n} \sum_{k=1}^{B_j} \sum_{h=1}^{H} X_{jkO_jm} S^2_{jkh} H^2_{jlimh} \left\lceil \frac{Q_{pjk}}{S_{jh}} \right\rceil P^h_j \frac{d_{mP}}{V_h} \tag{8}$$

In the above, t^h_{jkmP} represents the time taken by H_h to transport the kth sub-batch of j from m where the last process is located to the assembly workshop P. X_{jkO_jm} is a 0–1 variable. If the last process of the kth sub-batch of j is completed on equipment m, then $X_{jkO_jm} = 1$; otherwise, $X_{jkO_jm} = 0$.

- The makespan criterion C

The makespan criterion is defined as the maximum time for the last sub-batch for the equipment in the workshop to be processed and transported to the assembly workshop; $CB_{jB_jO_jm}$ represents the completion time of the last process O_j of the last sub-batch B_j of the workpiece j on the equipment m. In addition, $X_{jkO_jm} S^{2l}_{jB_jO_jh} H^2_{jB_jO_jh}$ represents the completion of the processing of the part j after choosing the handling equipment h and transporting it to the assembly workshop. The calculation is shown in Equation (9), as follows.

$$C = \sum_{j=1}^{n} \sum_{m=1}^{M} CB_{jB_jO_jm} + \sum_{j=1}^{n} \sum_{m=1}^{M} \sum_{h=1}^{H} X_{jkO_jm} S^2_{jB_jO_jh} H^2_{jB_jO_jh} \frac{d_{mP}}{V_h} \tag{9}$$

The multi-objective model is shown in Equations (10) and (11).

$$f_1 = min\ C_{max} \tag{10}$$

$$f_2 = min\ E \tag{11}$$

3.3.2. Constraints

Two issues should be considered in the scheduling:

- The influences of the division of the workpieces into sub-batches, process equipment selection, equipment standby, and state transition in the processing process on the energy consumption and completion time must be considered.
- The number of handling equipment types is limited, and a type of handling equipment needs to be selected during the handling process. If a sub-batch of workpieces is larger than the rated capacity of the handling equipment, multiple pieces of handling equipment must be selected for simultaneous or multiple handling.

The batches of workpieces should satisfy the condition that the sum of the divided sub-batch batches is equal to the processing quantity of the workpieces; moreover, the number of divided sub-batches should not exceed the total quantity of workpieces.

$$\begin{cases} Q_j = \sum_{k=1}^{B_j} Q_{jk} \\ 2 \le B_j \le Q_{jk} \end{cases} \tag{12}$$

The workpieces are split into equal batches. If the number of batches is not an integer, it is rounded down, and the remaining workpieces comprise a single batch.

$$Q_{jk} = \begin{cases} \lfloor Q_j \div B_j \rfloor\ k \le B_{j-1} \\ Q_j - \lfloor Q_j \div B_j \rfloor \times (B_j - 1)\ k = B_j \end{cases} \tag{13}$$

If the processing type of the current workpieces is the same as that of the workpieces being processed, there is no need to switch the state of the machine; otherwise, the state needs to be switched.

$$R_{jj'm} = \begin{cases} 0, P_{j'k'i'm} = NP_{jkim} \\ 1, P_{j'k'i'm} \neq NP_{jkim} \end{cases} \tag{14}$$

The scheduling is performed according to the sub-batch priority relationships of the parts/components in the product structure tree. The process starts processing time at the completion time of the last sub-batch process in the selected equipment, and the sub-batch workpieces after the previous process are completed and transported according to the maximum value between the device moments.

$$\begin{cases} SB_{jkO_{j1}m} = CB_{jko_jm} = 0 \\ SB_{jkim} = max\{CB_m^{n-1}, CB_{jk(i-1)m'} + t_{jkimm'}^h\} \end{cases} \tag{15}$$

The production process of the same sub-batch should not be interrupted.

$$CB_{jkim} = SB_{jkim} + TB_{jkim} \tag{16}$$

The part production sequence should meet the requirement that the production time of the lower-level parts/components is earlier than the start-up time of the upper-level parts/components; that is, the start-up time of the first process in the n-level parts/components sub-batch should be later than the $(n + 1)$ level parts/components.

$$S_{jn1} \geq S_{j(n+1)1} \tag{17}$$

Among them, S_{jn1} is the start time of the first process of the n-level parts/components, and $S_{j(n+1)1}$ is the start time of the first process of the first sub-batch of the $(n + 1)$ parts/components j.

For two adjacent processes for the same workpiece, the processing sequence constraints between the processes need to be met, and the next process can only be conducted after the previous process is completed and the workpiece is transported to the selected equipment m' for the start of the next process.

$$SB_{jk(i+1)m'} \geq R_{jj'm}R_{jkim} + TB_{pjko_{ji}} + t_{jkimm'}^h \tag{18}$$

In each process, O_{ji} can only select one piece of machine for processing.

$$\sum_{m=1}^{w} MP_{jim} = 1 \tag{19}$$

One type of handling equipment is selected for each handling instance.

$$\sum_{h=1}^{H} H_{jkimh}^{\beta} = 1 \tag{20}$$

4. Multi-Objective Gray Wolf Optimization Algorithm

4.1. Basic Gray Wolf Optimization Algorithm

Mirjalili [10] proposed the gray wolf optimizer (GWO) in 2014. The core of the algorithm is to manage an optimization problem by imitating the hunting process of a gray wolf population. Owing to its balance of local and global search capabilities, convergence speed, and depth balancing, it has attracted widespread attention since its proposal.

The basic idea is that α wolf was chosen to be the most suitable plan, and β wolf and δ wolf were the second and third optimal plans. The rest are ω. α and β are the guiders of hunting, followed by δ and ω wolf. The equation for simulating the corresponding behaviors is defined as follows.

$$\vec{D} = \left| \vec{C} \cdot \vec{X_p}(t) - \vec{X}(t) \right| \tag{21}$$

Among them, $\vec{D}$ represents the distance to the prey, t indicates the current iteration, $\vec{X_p}$ represents the prey's position vector, and $\vec{X}$ represents the wolf's position vector. The coefficient vectors are represented by $\vec{A}$ and $\vec{C}$, and the formula is:

$$\vec{A} = 2\vec{a} \cdot \vec{r_1} - \vec{a} \tag{22}$$

$$\vec{C} = 2\vec{r_2} \tag{23}$$

In the search process, $\vec{a}$ linearly decreases from 2 to 0 and is used to emphasize detection and discovery of prey. $\vec{R_1}$ and $\vec{r_2}$ are selected in the range [0,1] randomly. α, β are the guider in hunting, and δ wolves also can join the hunting. The location of the prey (optimal) is unknown. Simulating the hunting behavior of gray wolves, α, β, and δ wolves are assumed to be more familiar with the potential location of their prey. In each hunt for prey, the three best solutions represented by α, β and δ wolves will be saved and used in each search, guiding other wolves to the possible position of the prey. The hunting formula is given by Equations (24)–(26).

$$\vec{D_\alpha} = \left| \vec{C_1} \cdot \vec{X_\alpha} - \vec{X} \right|, \vec{D_\beta} = \left| \vec{C_2} \cdot \vec{X_\beta} - \vec{X} \right|, \vec{D_\alpha} = \left| \vec{C_1} \cdot \vec{X_\alpha} - \vec{X} \right| \tag{24}$$

$$\vec{X_1} = \vec{X_\alpha} - \vec{A_1} \cdot \vec{D_\alpha}, \vec{X_2} = \vec{X_\beta} - \vec{A_2} \cdot \vec{D_\beta}, \vec{X_3} = \vec{X_\delta} - \vec{A_3} \cdot \vec{D_\delta} \tag{25}$$

$$\vec{X}(t+1) = \frac{\vec{X_1} + \vec{X_2} + \vec{X_3}}{3} \tag{26}$$

All in all, *GWO* starts with guiding the search process by α, β, and δ. When $\left| \vec{A} \right| > 1$, they diverge and look for prey; otherwise, they find and attack the prey. Finally, if the stopping criterion is met, the optimal solution (i.e., prey) is output. In brief, in each iteration of the algorithm, individuals in the population were divided into α, β, δ and ω. The first three belong to the individuals at the decision-making level, representing the historical solution of optimal, suboptimal, and third optimal. ω corresponds to the other individuals. In the algorithm iterations, α, β and δ are locating prey and guiding ω to update its position, completing a sequence of actions including approaching, surrounding, and attacking the prey.

4.2. Application of Multi-Objective Gray Wolf Algorithm

The multi-objective gray wolf algorithm (MOGWO) [38] added two new components based on the gray wolf algorithm by Mirjalili in 2016. The first component is the archive, which served to store the currently acquired non-dominant Pareto optimal solutions. Then comes the leader selection strategy, which helps decision-makers to choose α, β and δ as the leader of the search process from the archived results. The basic flow chart is shown in Figure 3.

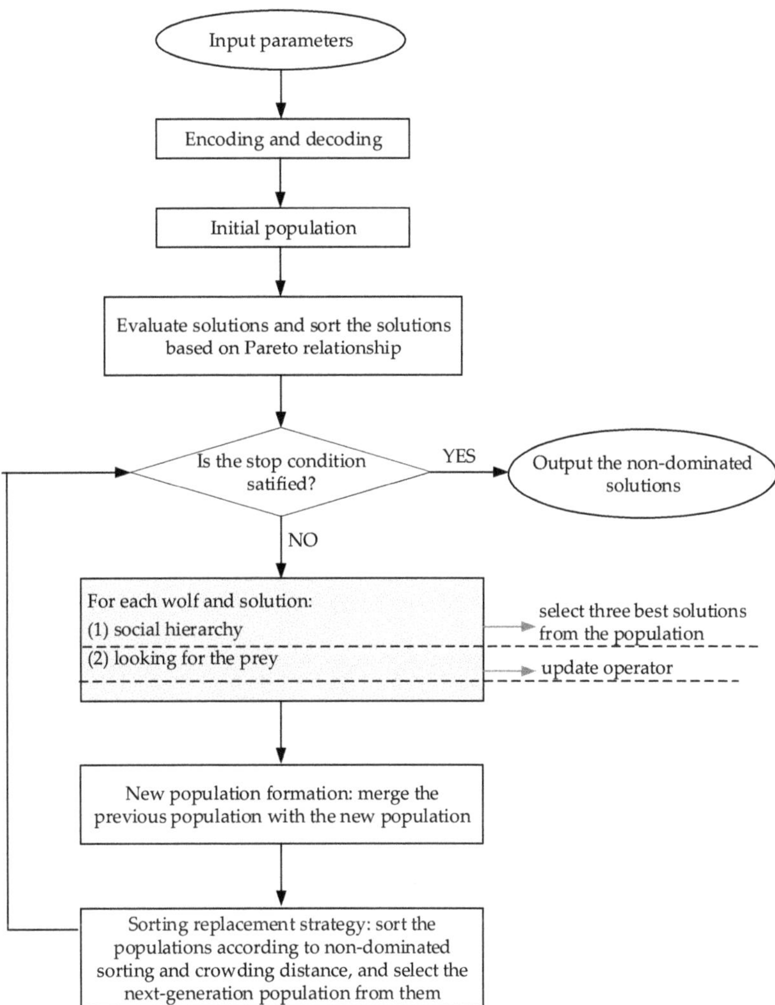

Figure 3. Multi-objective gray wolf algorithm flow chart.

4.2.1. Encoding and Decoding Mechanism

Before applying the MOGWO algorithm to a specific problem, we designed an encoding and decoding scheme. It connects the solution space of the problem with the search space under consideration. Hence, designing the correct codec scheme is an important issue that affects the performance of the algorithm.

Code design is based on the types of parts/components and the number of batches. In each chromosome, a gene is represented by three or four numbers. For example, "301" represents the first sub-batch of the third parts/components, and "1003" represents the third sub-batch of the tenth parts/components. In addition, the number sequence in the chromosome represents the processing operations of each sub-batch of parts/components. As shown in Figure 4, each type of part/component needs to go through multiple processing operations. The first occurrence of "101" represents the first processing operation of the first sub-batch of parts/components J_1, and the second occurrence represents the second processing operation, and so on. Taking an example for illustration, the optional processing equipment of three types of workpieces and the processing time for each process are shown in Table 1.

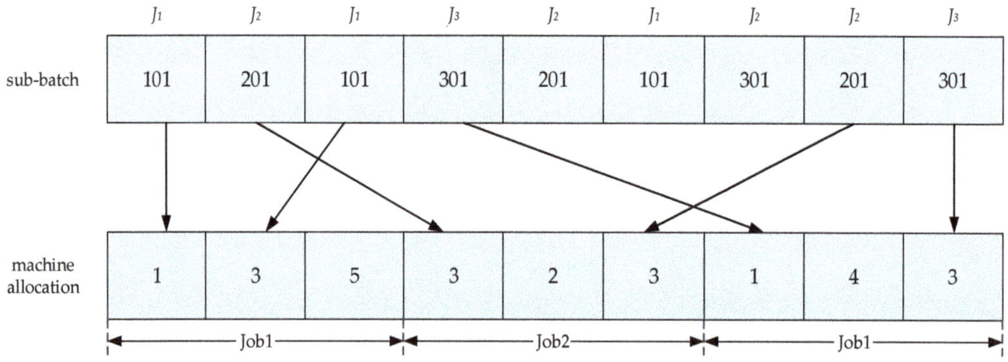

	J_1	J_2	J_1	J_3	J_2	J_1	J_2	J_2	J_3
sub-batch	101	201	101	301	201	101	301	201	301
machine allocation	1	3	5	3	2	3	1	4	3

Figure 4. Chromosome coding, for example.

Table 1. Equipment scheduling problem, for example (unit: min).

Sub-Batch	Quantity	Operations	M1	M2	M3	M4	M5
		O_{11}	2	-	5	-	6
101	30	O_{12}	-	6	5	6	-
		O_{13}	4	3	-	-	8
		O_{21}	5	-	4	-	-
201	40	O_{22}	9	5	-	6	-
		O_{23}	-	5	4	7	-
		O_{31}	6	-	9	10	-
301	30	O_{32}	5	7	-	6	-
		O_{33}	-	8	6	-	7

According to the coding method of the part/component arrangement, it can be assumed that the position of a gray wolf individual in this problem is [1–3], that is, the processing order of the parts/components is 1-3-2. Then, the part/component placement is decoded into a viable scheduling scheme. A Gantt chart corresponding to the first sub-batch of parts/components is shown in Figure 5. Through the decoding process, a suitable machine is selected for each process in each station for processing, and the order of each part and the start time are determined to obtain the objective function value. In Figure 5, initially, all the processes of the first sub-batch of J_1 are arranged on the machine that can process it earliest, and then the other sub-batches of other parts/components are scheduled. The various processes of the parts/components are arranged on the machine that can complete its processing earliest; if the processing completion time on the allocated machine is less than the earliest processing start time of the scheduled parts/components, it will be arranged before the scheduled parts/components (and so on for the remaining parts/components). Each sub-batch of parts is arranged before the position of each machine; otherwise, it is arranged behind the arranged parts/components.

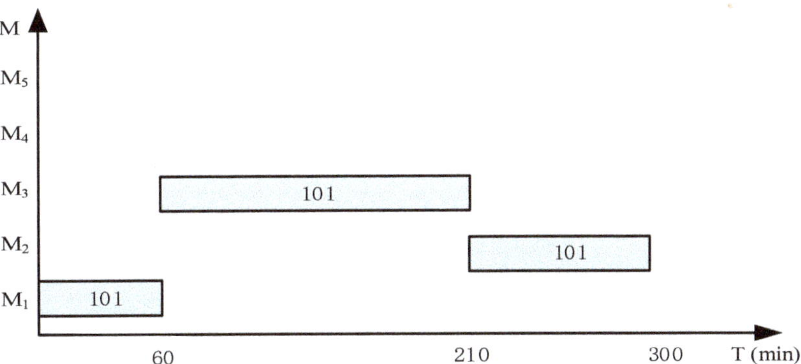

Figure 5. Gantt chart of first sub-batch.

4.2.2. Initialization

Initial solutions can affect the obtaining of the optimal solution to a degree. In our study, the problem can be divided into two sub-problems: equipment selection and process sequencing. Therefore, in the initialization phase, the most suitable processing equipment with the lowest processing energy consumption and processing time is selected based on the above encoding and decoding scheme. Then, the sorting plan is obtained according to the processing priority rules of the workpiece and the remaining load at most.

4.2.3. Roulette Selection

In a multi-object search space, comparing solutions is usually not easy; thus, the leader selection mechanism has been designed to solve this problem. The leader provides the α, β, and δ wolves in the least crowded search places. The selection is performed by using the roulette method. The probability of each hypercube is calculated as follows: two individuals are randomly selected from N individuals each time, and the individuals with the lower ranking levels are selected first. If the ranking levels are the same, the crowding degree is the first large individual to generate a population of $N/2$ individuals. The process merges the two generated populations into a new progeny population (population size is N).

4.2.4. Social Hierarchy

Due to the Pareto advantage of multi-objective planning, the optimal result is usually not a single, which is called a "trade-off solution" in multi-objective planning. According to the Pareto dominance relationship, a population can be divided into several levels. The first-level solution (non-dominant solution or compromise solution) can be expressed as a solution. If there are more than three levels in the whole, β and δ are the second- and third-level solutions. In this study, social stratification was conducted by assuming these three situations:

- Select α, β and δ randomly from the non-dominated level or the first level.
- Select α and β from the current two levels.
- Select α, β, and δ wolves from the first three levels, respectively (only have two levels).

4.2.5. Update Operator

In this study, individuals are no longer updated according to the decision level, and a hybrid search method combining local search and global search is adopted. The wolf pack generated by each iteration of the algorithm is divided into two parts: the search and tracking operations are carried out, respectively. Then, in the process of searching, the number in each group is dynamically adjusted to achieve the purpose of the individual update.

4.2.6. Sorting Replacement Strategy

Different from the basic GWO algorithm, the newly generated solution is evaluated based on two fitness values: maximum time to completion and total energy consumption. The parent population and progeny individuals produced by global search and local search operations combined to a large new species, then using the non-dominated sorting method and crowded degree to sort the new species. Among them, based on the classification of the solutions, the sorting method reduces the computing complexity, and the crowded degree calculation can save with low levels of similar solution and keep the diversity of the solution space. At the same time, the distribution of individuals on the current Pareto frontier should be as broad and uniform as possible, and the introduction of an elite retention mechanism is conducive to maintaining excellent individuals and improving the overall evolution level of the population.

5. A Case Study

5.1. Data Preparation

Taking into account the following 10×10 production workshop example based on a realistic situation, it comprehensively considers the quantity required for each type of part or component, its handling energy consumption, and its level in the product structure.

For the three different types of products, their complex structures were different, resulting in different processing priorities and handling complexities. As described above, the flat product structure is a single layer, that is, all components have the same priority, and the processing and handling scheduling are relatively simple; in contrast, tall and complex products contain multi-layered parts/components structures. Taking a typical tall product as an example, the specific product structure tree is shown in Figure 6. In the figure, B represents the corresponding product, and the arrow in the figure represents the position of part J_i. The part P_i in the product structure was divided into different levels; each component or part had a certain demand; they were mixed in batches, and the production was completed. Subsequently, it was transported to the assembly workshop by a transport vehicle for assembly.

Table 2 displays the detail settings of the components, parts, and equipment. The parts and the components are set to 6 and 4, respectively. The rated capacity of the workpiece on the handling equipment is listed in Table 3. In batch scheduling, the production batch had a U-shaped relationship with the production cycle. In general, production batches that are excessively large or small will lead to a longer production cycle. In this study, each workpiece was divided into 2–3 batches, as shown in Table 4. There were three types of handling vehicles, each of which has three available equipment. The power of the handling vehicle was 20 kW, the speed was 30 m/min, and the distance between the assembly workshop and production workshop was 200 m. The information of each type of part and component and the required power is shown in Table 4. In the workshop, the distance between adjacent equipment is 5 m. All cases were simulated in MATLAB R2016b and were tested many times. The algorithm was programmed using Matlab2016b on a personal computer with an Intel(R) Core (TM) i5-930M CPU @ 2.50 GHz. The values of the algorithm parameters were determined by preliminary experiments, and the specific parameters were determined by comprehensive experiments as follows: number of iterations: 250; the number of grids per dimension:15, and population size: 20.

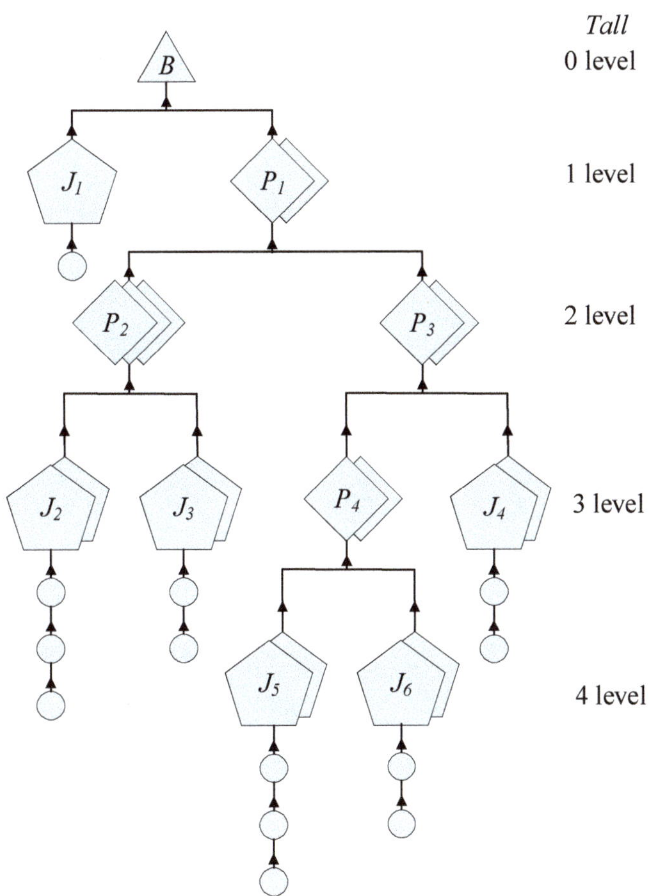

Figure 6. Product structure tree of tall.

Table 2. Parts/components information (10×10).

| Parts/Components | Equipment (Preparation Time/Processing Time) (min) | | | | | | | | | |
	O_{ji}	M_1	M_2	M_3	M_4	M_5	M_6	M_7	M_8	M_9	M_{10}
J_1	O_{11}	[2,18]	[3,15]	—	[3,20]	—	—	—	—	—	—
	O_{12}	—	—	—	—	—	—	—	[3,20]	[2,14]	—
	O_{13}	[3,17]	[2,20]	—	[3,21]	—	—	—	—	—	—
P_1	O_{21}	—	[2,12]	[3,20]	—	—	—	[2,15]	—	—	—
	O_{22}	[2,15]	[2,14]	—	—	[3,18]	—	—	—	—	—
	O_{23}	—	—	—	—	—	—	—	[3,17]	[2,20]	—
P_2	O_{31}	—	[3,20]	[4,20]	—	—	—	[4,18]	—	—	—
	O_{32}	—	—	—	—	—	—	—	—	[2,18]	[3,15]
P_3	O_{41}	[1,15]	[1,10]	—	[2,10]	—	—	—	—	—	—
	O_{42}	—	—	—	—	—	—	—	[2,16]	[2,14]	—
	O_{43}	[2,15]	[2.5,17]	—	[4,18]	—	—	—	—	—	—
	O_{44}	[2.5,16]	[2,15]	—	—	[3,15]	—	—	—	—	—

Table 2. *Cont.*

Parts/Components	Equipment (Preparation Time/Processing Time) (min)										
	O_{ji}	M_1	M_2	M_3	M_4	M_5	M_6	M_7	M_8	M_9	M_{10}
J_2	O_{51}	[2,20]	[3,25]	—	[2.5,16]	—	—	—	—	—	—
	O_{52}	—	—	—	—	—	—	—	—	[3,20]	[4,20]
	O_{53}	—	[3,28]	[3,30]	—	—	—	[5,25]	—	—	—
J_3	O_{61}	—	—	—	[1,15]	[2,25]	[2,30]	—	—	—	—
	O_{62}	—	—	—	[2,22]	[4,20]	—	—	—	—	—
	O_{63}	[2,20]	[1,10]	—	—	—	—	[1,15]	—	—	—
P_4	O_{71}	—	—	—	[1,10]	[2,25]	[2,30]	—	—	—	—
	O_{72}	—	—	—	[2,15]	[3,10]	[2,20]	—	—	—	—
	O_{73}	[3,15]	[3,18]	[4,30]	—	—	—	—	—	—	—
J_4	O_{81}	—	—	—	[1,20]	[3,25]	—	—	—	—	—
	O_{82}	[3,15]	[3,25]	—	—	—	—	[4,30]	—	—	—
	O_{83}	—	—	[2,10]	—	[3,15]	[2,16]	—	—	—	—
J_5	O_{91}	—	—	[1,10]	—	[2,15]	[1,20]	—	—	—	—
	O_{92}	—	—	[3,15]	—	[2,20]	—	—	—	—	—
	O_{93}	—	[1,5]	[2,5]	—	—	—	[2,7]	—	—	—
J_6	O_{101}	—	—	[2,10]	[3,15]	—	[3,20]	—	—	—	—
	O_{102}	—	—	[1,15]	[2,10]	—	[2,18]	—	—	—	—
	O_{103}	[1,10]	[3,15]	—	—	—	—	—	—	—	—

Table 3. The rated capacity of the workpiece on the handling equipment.

H \ Workpiece	J_1	P_1	P_2	P_3	J_2	J_3	P_4	J_4	J_5	J_6
H_1	60	50	40	70	65	50	90	68	70	70
H_2	40	60	80	85	70	70	110	80	60	70
H_3	30	60	70	60	80	60	100	60	60	83

Table 4. Other information about workpieces.

Workpiece	J_1	P_1	P_2	P_3	J_2	J_3	P_4	J_4	J_5	J_6
Types (part/com)	part	com	com	com	part	part	com	part	part	part
level	1	1	2	2	3	3	3	3	4	4
Quantity	300	300	600	300	600	300	900	300	600	300
α_j (kW)	20	15	20	25	22	25	30	20	15	20
Symbols in Gantt chart	Job1	Job2	Job3	Job4	Job5	Job6	Job7	Job8	Job9	Job10
Quantity of sub-lots in 5.2	2	2	3	2	3	2	2	2	2	2
Quantity of sub-lots in 5.3.1	2	3	3	2	3	2	3	2	3	2

5.2. Result Analysis

In the experiment, taking into account the total energy consumption and completion time as the objective function and comprehensively considering all the energy consumption involved in the production process, the MOGWO algorithm was used to solve the problem. As shown in Figure 7, the values of the two optimization objectives stabilize at the 189th iteration. Figure 8a shows the population distribution results after 200 iterations. There are eight solutions set in Figure 8b. The point with lower energy consumption and early completion time than others was a selection in Figure 8b. The Gantt chart of this scheme is shown in Figure 9. As the total energy consumption optimization reduces the equipment conversion and waiting time, the completion time is 38,400.86 min, and the total energy consumption is 356,750 kWh. Table 5 displays the handling process of the first sub-batch of each workpiece between the machines and the workshop, where J01 (J = 1, 2, 3, . . . ,10) represents the first sub-batch of the *J*th workpiece.

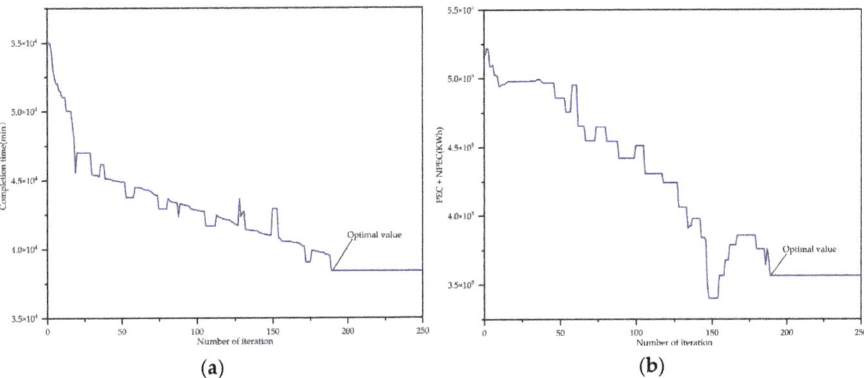

Figure 7. Convergence process of two objectives: (**a**) convergence process of C objective; (**b**) convergence process of E objective.

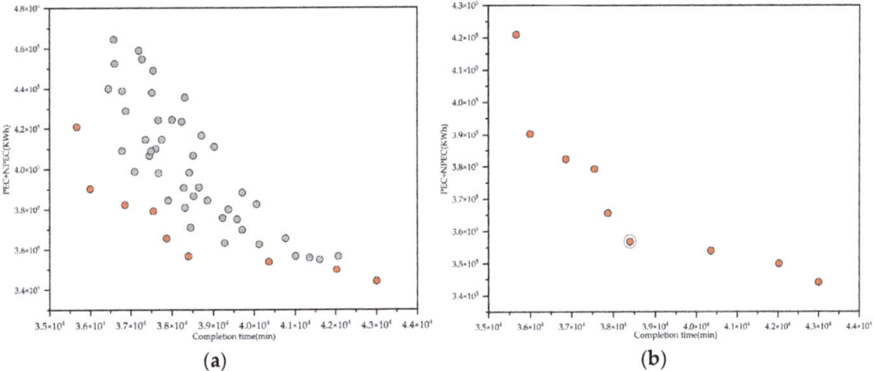

Figure 8. Results obtained considering PEC+NPEC: (**a**) final population distribution; (**b**) Pareto frontier distribution.

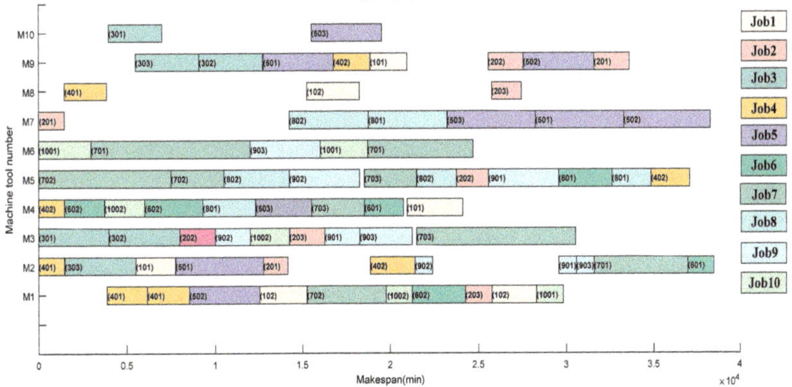

Figure 9. Gantt chart of the optimized schedule scheme.

Table 5. One sub-batch handling process for each kind of workpiece.

J01	Quantity	Operations	M	H	Start Time (min)	End Time (min)	Starting Location	Arrival Location
(J_1) 101	150	O_{11}	M_2	H_1	7756	7756.5	M_2	M_9
		O_{12}	M_9	H_1	20,907.5	20,907.8	M_9	M_4
		O_{13}	M_4	H_2	24,058	24,064.5	M_4	P
(P_1) 201	100	O_{21}	M_7	H_2	2250	2250.5	M_7	M_2
		O_{22}	M_2	H_2	14,861	14,861.5	M_2	M_9
		O_{23}	M_9	H_3	33,914	33,920.5	M_9	P
(P_2) 301	200	O_{31}	M_3	H_2	4004	4004.5	M_3	M_{10}
		O32	M_{10}	H_3	7007.5	7014	M_{10}	P
(P_3) 401	150	O_{41}	M_2	H_2	1500	1500.5	M_2	M_8
		O_{42}	M_8	H_2	3100.5	3101	M_8	M_1
		O_{43}	M_1	_	4601	4601	M_1	M_1
		O_{44}	M_1	H_1	6201	6207.5	M_1	P
($J2$) 501	200	O_{51}	M_2	H_3	12,759	12,759.5	M_2	M_9
		O_{52}	M_9	H_3	16,759.5	16,759.7	M_9	M_7
		O_{53}	M_7	H_2	33,118	33,124.5	M_7	P
(J_3) 601	150	O_{61}	M_4	H_1	21,115	21,115.1	M_4	M_5
		O_{62}	M_5	H_2	32,564	32,564.4	M_5	M_2
		O_{63}	M_2	H_1	38,965	38,971.5	M_2	P
(P_4) 701	300	O_{71}	M_6	-	12,005	12,005	M_6	M_6
		O_{72}	M_6	H_2	24,710	24,710.5	M_6	M_2
		O_{73}	M_2	H_3	37,464	37,470.5	M_2	P
(J_4) 801	150	O_{81}	M_4	H_1	11,751	11,751.3	M_4	M_7
		O_{82}	M_7	H_1	23,224	23,224.2	M_7	M_5
		O_{83}	M_5	H_2	34,817	34,823.5	M_5	P
(J_5) 901	200	O_{91}	M_3	H_2	18,865	18,865.3	M_3	M_5
		O_{92}	M_5	H_2	29,560	29,560.4	M_5	M_2
		O_{93}	M_2	H_2	30,560.4	30,566.9	M_2	P
(J_6) 1001	150	O_{101}	M_6	-	3003	3003	M_6	M_6
		O_{102}	M_6	H_2	18,708	18,708.5	M_6	M_1
		O_{103}	M_1	H_2	29,572	29,579.5	M_1	P

5.3. Comparison Analysis

5.3.1. Comparison with the Traditional Model without Considering the NPEC

Different from the experiment above, only PEC is considered in the energy consumption model in this section. As shown in Figure 10, the values of the two optimization objectives stabilize at the 158th iteration. Figure 11a shows the population distribution results after 200 iterations. There are 10 solutions set in Figure 11b. The point with lower energy consumption and earlier completion time than others is selected in Figure 11b. The Gantt chart of this scheme is as shown in Figure 12, the total completion time is 40,900.43 min, and the PEC is 227,000 kWh.

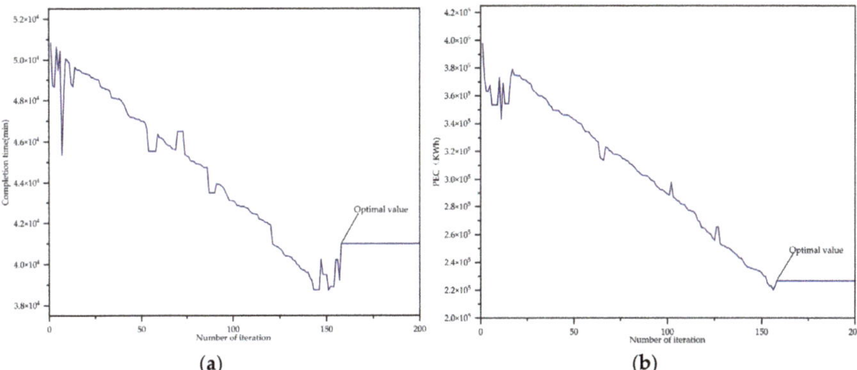

Figure 10. Convergence process of two objectives: (**a**) convergence process of C objective; (**b**) convergence process of PEC objective.

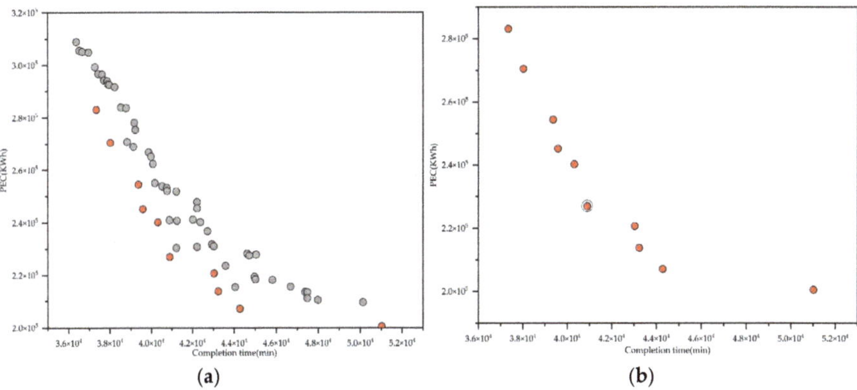

Figure 11. Results obtained in comparison experiment: (**a**) final population distribution; (**b**) Pareto frontier distribution.

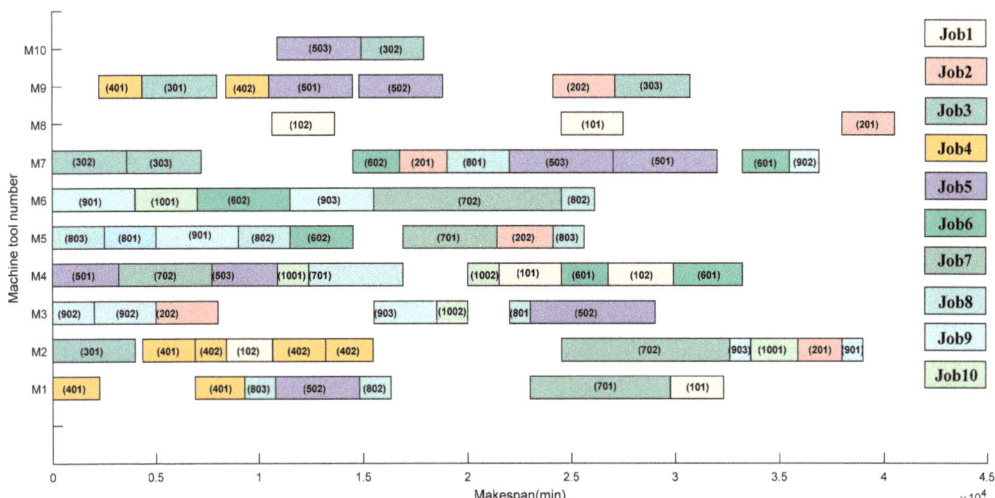

Figure 12. Gantt chart of the optimized scheme from the traditional model.

Table 6 displays the energy consumption values and completion times for the two optimization schemes. As could be seen from the table, owing to the reduced standby and conversion time, the total completion time is better than that of the comparative experiment; in terms of energy consumption, the PEC values of the two schemes are very close; the slight difference in processing times may be owing to the different preparation times required on different equipment. The optimization of the handling equipment is not considered in the comparative experiment. The handling energy consumption in the comparative experiment is estimated through the historical handling time, and the energy consumption in our experiment is optimized. As could be seen in Table 6, the optimization of the handling energy consumption can significantly reduce the NPEC, thereby reducing the total energy consumption. Compared to the comparative experiment, the standby energy consumption and handling energy consumption are reduced by 9.95% and 22.28%, respectively. The utilization rates of each machine in the two experiments are shown in Figure 13. It can be seen from the figure that the machine utilization is mostly higher than that in the comparative experiment.

Table 6. Comparison of energy consumption and time of two experiments. (unit: kWh).

	Optimize (PEC + NPEC)	Optimize PEC
E_p	2.33×10^5	2.27×10^5
E_w	4.75×10^4	5.33×10^4
E_s	550	588
E_d	1.57×10^4	2.02×10^4
C	38,400.86	40,900.43

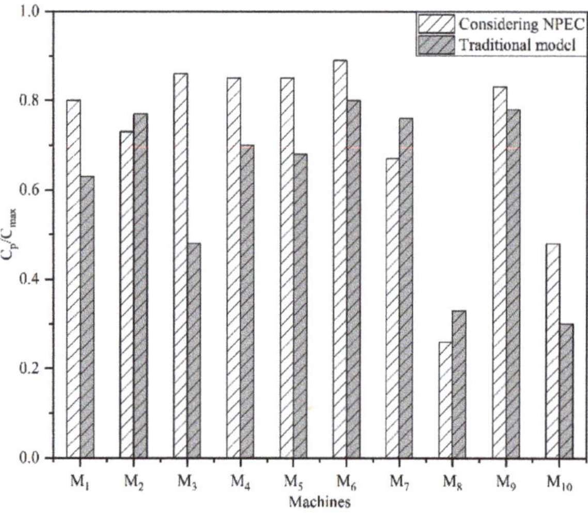

Figure 13. Comparison in C_p/C for machines.

5.3.2. Algorithm Comparison with the NSGA-II

In this section, the most popular multi-objective heuristic algorithm NSGA-II [39] was selected in the experiment for comparative analysis. NSGA-II is an improvement of the NSGA algorithm. It is one of the most outstanding evolutionary multi-objective optimization algorithms so far. The parameters of the algorithm are set as follows: the population size is 100, the maximum number of iterations is set to $G = 200$, crossover probability $Pc = 0.9$, and the mutation probability $Pm = 0.2$. The parameters of MOGWO are set in Section 5.1. They use the same initialization strategy and encoding scheme.

Comparing the performance of multi-objective optimization algorithms is more based on the following criteria: (a) the degree of similarity between the solution set obtained by the operation result and the real Pareto solution set, that is, convergence; (b) the uniformity of the solution set on the Pareto frontier, namely diversity; (c) comprehensively measuring convergence and diversity. According to references [40,41], the following two indicators are used for algorithm performance evaluation: Δ metric and inverted generational distance. The real Pareto frontier of the research in this paper is the set of non-dominated solutions in the final solution set. However, the real Pareto frontier is not known. The final solution set is obtained by the calculation example through multiple independent operations of the algorithm. The specific introduction and calculation formula of these two indicators are given below.

Δ metric: Describe the uniformity of the Pareto front obtained by the algorithm. The calculation method is as follows. The smaller Δ, the more even the solution is, and the better the performance of the algorithm. When Δ is equal to 0, it indicates that the solution obtained by the algorithm is uniformly distributed in the solution set space, generally only appearing under ideal circumstances.

$$\Delta = \frac{d_f + d_t + \sum_{i=1}^{n-1}\left| d_i - \bar{d}\right|}{d_f + d_t + (n-1)\bar{d}} \tag{27}$$

Here, d_f and d_t are the distances between the boundary point of the Pareto frontier obtained by the algorithm and the actual Pareto frontier boundary point of the problem to be solved; n represents the number of solutions in the Pareto frontier obtained by the algorithm operation, and d_i represents the value obtained by the algorithm operation. $\bar{d}$ represents the average value of all d_t.

Inverted generational distance: A variant of iterative distance not only reflects the convergence of the algorithm but also shows a diversity index, which is a comprehensive evaluation index. The formula of *IGD* is as follows:

$$IGD(S, \, P^*) = \frac{\sum_{x\in P^*} dist(x, S)}{|P^*|} \tag{28}$$

Among them, $dist(x, \, S)$ represents the individual $x \in P^*$ to the nearest Euclidean distance on S, and $|P^*|$ is the cardinality of the set P^*. The smaller the value of *IGD*, the more it can approach the entire PF. In addition, when $IGD(S, \, P^*) = 0$, it means S is a subset of P^*.

In this paper, two intelligent optimization algorithms are selected for comparison. The results are shown Table 7. The table counts the minimum, average, and standard deviation of the indicators.

Table 7. Comparison of two algorithms.

Evaluation Index	MOGWO			NSGA-II		
	Min	Agv	Sd	Min	Agv	Sd
Spread	0.343	0.553	0.107	0.522	0.677	0.115
Inverted generational distance	0.074	0.089	0.021	0.090	0.125	0.029

From Table 7, we can draw the following two points:

- According to the spread value (Δ) in the table, MOGWO is better than the NSGA-II algorithm, and the MOGWO algorithm is more evenly distributed than the solution set obtained by NSGA-II. It is due to the neighborhood search mechanism of the MOGWO, which can increase the probability of obtaining the optimal solution, thereby improving the uniformity of the solution set distribution, and the algorithm has better optimization capabilities.

- According to the inverted generational distance value in the table, the solution obtained by MOGWO has better convergence and distribution than the NSGA-II algorithm. This is because of the unique hierarchical system of the MOGWO algorithm, which can be selected from different dominance levels. The optimal solution to improve the convergence and distribution of the algorithm was chosen.

The Pareto frontiers obtained by the two algorithms are shown in Figure 14. Observation shows that the solution obtained by using the MOGWO algorithm is numerically smaller than the other solution set, that is, it can dominate the solutions obtained by NSGA-II. Additionally, MOGWO proves to be effective in reducing the total energy consumption of scheduling plans. Therefore, the selected algorithm has the best solution effect.

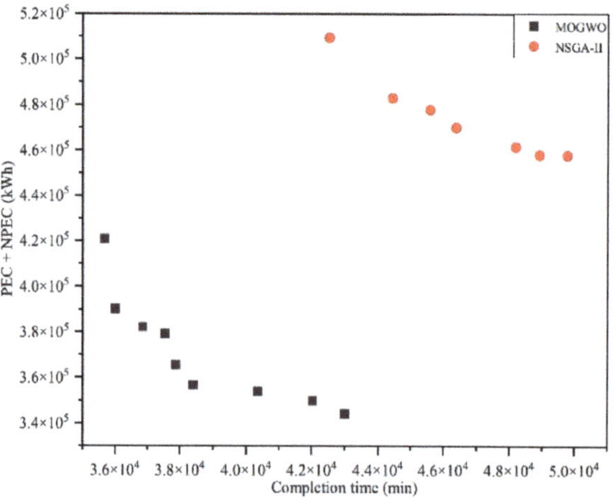

Figure 14. Comparison of two algorithms.

5.4. Sensitivity Analysis

Sensitivity analysis was used to compare the influence of energy consumption of different auxiliary production operations in this section. If the objective function value and fitness value of different optimization operations change greatly, the sensitivity coefficient and the corresponding auxiliary production operation will be large and sensitive.

Sensitivity analysis is the importance of factor variables of the model to the value of the optimization objective function. The calculation is shown in Equation (29), as follows.

$$S_{fA} = \frac{\Delta f / f}{\Delta A / A} \tag{29}$$

Here, S_{fA} represents the sensitivity of target function value f to parameter A, and $\Delta A / A$ means the rate of change of a parameter; $\Delta f / f$ represents the change rate of the target function value caused by the change of factor variable ΔA. The sensitivity of total energy consumption to E_W, E_s, and E_d will be analyzed in our experiment. In this analysis, $\Delta A / A$ represents the rate of change relative to historical observations when different auxiliary operation optimizations are considered. The change rate of total energy consumption due to different auxiliary operations of optimization is expressed by $\Delta f / f$. Table 8 shows the sensitivity coefficients under different optimization schemes; the larger the sensitivity coefficient, the higher the sensitivity of the target to the variation of parameters.

Table 8. Sensitivity coefficients under different optimization combinations.

Schemes	MOGWO			$\Delta A/A$	$\Delta f/f$	S_{fA}
	E_w	E_s	E_d			
A	5.33×10^4	588	2.02×10^4	-	-	-
A_1	4.63×10^4	-	-	13.13%	2.32%	0.18
A_2	-	523	-	11.05%	0.22%	0.02
A_3	-	-	1.49×10^4	26.24%	1.76%	0.07

Table 8 compares the changes in total energy consumption when considering different combinations of auxiliary optimization operations. Among them, A represents the scheduling scheme without considering auxiliary operation optimization, where E_w, E_s, and E_d are historical forecast values. The standby time, conversion operation, and handling operation are considered separately, and the scheme is expressed by A_1, A_2, and A_3, respectively. Data indicate the three sensitivity coefficients are all positive and the target function value has a certain sensitivity to parameter changes, proving that the optimization model proposed in our research can obtain satisfactory results in reducing energy consumption, but there have been slightly different changes in the target function to the parameter between three schemes. From Table 8, energy consumption is significantly reduced by optimizing standby time, and the target quantity changes significantly to E_w; second is handling energy consumption optimization, followed by conversion energy consumption. In this research, due to the large standby time and handling times, the energy consumption optimization effect is relatively significant, especially in the case of large total energy consumption. Through the sensitivity analysis of the above optimization scheme, it is reasonable and effective to comprehensively consider the NPEC in the optimization model.

6. Conclusions and Future Works

This study researched energy consumption optimization in the production process of a machining workshop, starting from the assembly relationship between the parts/ components of the multi-level product structure. By analyzing the existing energy consumption research, we conducted a systematic study on the NPEC of the workshop. The research was mainly conducted from the following two aspects.

- The production of a multi-level product structure is combined with energy consumption optimization. The start processing times of different levels of workpieces are set, and the characteristics of the PEC and NPEC in the production system are considered. With the goal of minimum completion time and energy consumption, equipment standby, workpiece conversion, and handling constraints are established, and the MOGWO is adopted to solve the problem.
- The total energy consumption optimization results are compared with those of an optimization plan considering only the PEC. The results show that after considering the NPEC optimization as proposed here, the standby energy consumption and handling energy consumption are reduced by 9.95% and 22.28%, respectively. This provides a feasible research direction for the study of energy-saving scheduling in workshops.

Production scheduling considering energy-saving measures is of great significance for the realization of energy savings and emissions reduction. This study has a set of limitations: for example, we suppose the same power during the standby time and there is no interruption in the processing that limit the versatility of our method. For future works, other possibilities, such as equipment failures and emergency order insertions, will be further integrated into the optimization model.

Author Contributions: Conceptualization, N.L. and C.F.; methodology, N.L. and C.F.; software, C.F.; formal analysis, N.L.; investigation, C.F.; resources, N.L.; data curation, N.L. and C.F.; writing—original draft preparation, C.F.; writing—review and editing, N.L. and C.F; visualization, N.L. and

C.F.; supervision, N.L.; funding acquisition, N.L. All authors have read and agreed to the published version of the manuscript.

Funding: This research received no external funding.

Institutional Review Board Statement: Not applicable.

Data Availability Statement: Data is contained within the article.

Acknowledgments: The authors would like to thank the researchers in the Department of Industrial Engineering, China University of Mining and Technology, for their inputs.

Conflicts of Interest: The authors declare no conflict of interest.

Nomenclature

C	completion time
T_P	processing time
E_p	energy consumption of processing
E_w	energy consumption of stand by state
E_d	energy consumption of handling
E^h_{jkmP}	energy consumption of the kth sub-batch of j transported from m to P
B_{jk}	the kth sub-batch of workpiece j
$d_{mm'}$	distance between machines
S^i_m	start time of the ith operation on the machine m
C^i_m	completion time of the ith process on the machine m
CB^n_m	completion time of the nth process on machine m
O_{ji}	the jth operation of the workpiece j
w	each type of workpiece is processed on w sets of equipment
$t^h_{jkimm'}$	transportation time of from m to m'
SB_{jkim}	start time of the ith operation of the kth sub-batch about workpiece j on machine m
CB_{jkim}	completion time of the ith process of the kth sub-batch about workpiece j on machine m
$P_{j'k'i'm}$	kth sub-batch ith operation of workpiece j is processed on equipment m
S_{jh}	rated capacity of work j on equipment h
$\alpha_j, \alpha_{j'}$	the processing power of workpiece j, j'
E_{max}	total energy consumption
T_R	setup time
E_n	non-processing energy consumption
E_s	conversion energy consumption
$E^h_{jkimm'}$	energy consumption of the workpiece j transported from m to m'
Q_j	quantity of workpiece j
Q_{jk}	quantity of j of B_{jk}
d_{mP}	distance between processing workshop and P
P^h_j	power of handling H_h
SB^n_m	start time of the nth process on the machine m
$O^m_{j(1)i(1)}$	ith operation of workpiece j processed on machine m
n_{jk}	the number of h required for the kth sub-batch of workpiece j
P^w_m	standby power of machine
S_{jn1}	start time of the first process of n-level parts/components
TB_{jkim}	processing time of the ith operation of the kth sub-batch about workpiece j on equipment m
R_{jkim}	the set-up time of the ith operation of the kth sub-batch about workpiece j on equipment m
NP_{jkim}	the kth sub-batch and the ith operation of the jth workpiece are being processed on the m
V_h	speed of H_h
MP_{jim}	each process O_{ji} can only choose one piece of machine for processing
t^h_{jkmP}	the time it takes for H_h to transport the kth sub-batch of the workpiece j from the equipment m where the last process to the assembly workshop P

Appendix A

Table A1. Sets.

Nomenclature	Description
J	Parts set; $J = \{J_1, J_2, J_3 \ldots J_n\} \cup \{P_1, P_2, P_3, \ldots P_n\}$, $j \in J$
M	a finite set of M machines; $m = 1,2,3 \ldots M$
H	a finite set of H handle equipment; $H_h \in H$; $h = 1,2,3$
O_j	Process set; $O_{ji} \in O_j$, $i = 1,2,3 \ldots O_j$
B_j	Number of sub-batch of j; $j = 1, 2, \ldots n$

Table A2. Decision variables.

Nomenclature	Description
$R_{jj'm}$	If the currently processed part/component j of the processing equipment m is different from the part/component j' to be processed, then $R_{jj'm} = 1$, otherwise, $R_{jj'm} = 0$.
S_{jkih}^{β}	If the equipment H_h is handling the ith process of the kth sub-batch of parts/components, then $S_{jkih}^{\beta} = 1$, otherwise, $S_{jkih}^{\beta} = 0$.
H_{jkimh}^{β}	If the ith process of the kth sub-batch of the part/component j is carried by the equipment H_h, then $H_{jkimh}^{\beta} = 1$, otherwise, $H_{jkimh}^{\beta} = 0$.
X_{jkO_jm}	If the last process of the kth sub-batch of j is completed on machine m, $X_{jkO_jm} = 1$, otherwise, $X_{jkO_jm} = 0$.

References

1. Pinedo, M.L. *Scheduling: Theory, Algorithms, and Systems*; Springer: New York, NY, USA, 2016.
2. Shi, F.; Zhao, S.K.; Meng, Y. Hybrid algorithm based on improved extended shifting bottleneck procedure and GA for assembly job shop scheduling problem. *Int. J. Prod. Res.* **2019**, *1*, 89–100. [CrossRef]
3. Pereira, M.T.; Santoro, M.C. An integrative heuristic method for detailed operations scheduling in assembly job shop systems. *Int. J. Prod. Res.* **2011**, *49*, 6089–6105. [CrossRef]
4. Kumar, M.L.; Kaur, J.; Singh, J. Dynamic and static energy efficient scheduling of task graphs on multiprocessors: A Heuristic. *IEEE Access* **2020**, *8*, 176351–176362. [CrossRef]
5. Jiang, Z.P.; Gao, D.; Lu, Y.; Kong, L.H.; Shang, Z.D. Electrical energy consumption of CNC machine tools based on empirical modeling. *Int. J. Adv. Manuf. Technol.* **2019**, *100*, 2255–2267. [CrossRef]
6. Li, K.Q. Energy and time constrained scheduling for optimized quality of service. *Sust. Comput.* **2019**, *22*, 134–138. [CrossRef]
7. Dahmus, J.B.; Gutowski, T.G. An environmental analysis of machining. In Proceedings of the ASME 2004 International Mechanical Engineering Congress and Exposition, Manufacturing Engineering and Materials Handling Engineering, Anaheim, CA, USA, 13–19 November 2004; pp. 643–652.
8. Xie, J.; Cai, W.; Du, Y.B.; Tang, Y.; Tuo, J.B. Modelling approach for energy efficiency of machining system based on torque model and angular velocity. *J. Clean. Prod.* **2021**, *293*, 126249. [CrossRef]
9. Jia, S.; Yuan, Q.; Lv, J.; Liu, Y.; Ren, D.; Zhang, Z. Therblig-embedded value stream mapping method for lean energy machining. *Energy* **2017**, *138*, 1081–1098. [CrossRef]
10. Mirjalili, S.; Mirjalili, S.M.; Lewis, A. Grey wolf optimizer. *Adv. Eng. Softw.* **2014**, *69*, 46–61. [CrossRef]
11. Saxena, A.; Soni, B.P.; Kumar, R.; Gupta, V. Intelligent grey wolf optimizer development and application for strategic bidding in uniform price spot energy market. *Appl. Soft Comput.* **2018**, *69*, 1–13. [CrossRef]
12. Miao, D.; Chen, W.; Zhao, W.; Demsas, T. Parameter estimation of PEM fuel cells employing the hybrid grey wolf optimization method. *Energy* **2020**, *193*, 571–582. [CrossRef]
13. Sattar, M.K.; Ahmad, A.; Fayyaz, S.; Ul Haq, S.S.; Saddique, M.S. Ramp rate handling strategies in Dynamic Economic Load Dispatch (DELD) problem using Grey Wolf Optimizer (GWO). *J. Chin. Inst. Eng.* **2019**, *43*, 200–213. [CrossRef]
14. Zhou, J.G.; Huo, X.J.; Xu, X.L.; Li, Y.S. Forecasting the carbon price using extreme-point symmetric mode decomposition and extreme learning machine optimized by the Grey Wolf Optimizer Algorithm. *Energies* **2019**, *12*, 950. [CrossRef]
15. Gokuldhev, M.; Singaravel, G.; Mohan, N.R. Ram. Multi-Objective local pollination-based Gray Wolf Optimizer for task scheduling heterogeneous cloud environment. *J. Circuits Syst. Comput.* **2020**, *29*, 2050100. [CrossRef]
16. Lu, C.; Gao, L.; Li, X.; Xiao, S. A hybrid multi-objective grey wolf optimizer for dynamic scheduling in a real-world welding industry. *Eng. Appl. Artif. Intell.* **2017**, *57*, 61–79. [CrossRef]
17. Qin, H.B.; Fan, P.F.; Tang, H.T.; Huang, P.; Fang, B.; Pan, S.F. An effective hybrid discrete grey wolf optimizer for the casting production scheduling problem with multi-objective and multi-constraint. *Comput. Ind. Eng.* **2019**, *128*, 458–476. [CrossRef]

18. Luo, S.; Zhang, L.X.; Fan, Y.S. Energy-efficient scheduling for multi-objective flexible job shops with variable processing speeds by grey wolf optimization. *J. Clean Prod.* **2019**, *234*, 1365–1384. [CrossRef]
19. Paul, M.; Ramanan, T.R.; Sridharan, R. Simulation modelling and analysis of dispatching rules in an assembly job shop production system with machine breakdowns. *Int. J. Adv. Manuf. Technol.* **2018**, *3*, 234–251. [CrossRef]
20. Hongbum, N.; Jinwoo, P. Multi-level job scheduling in a flexible job shop environment. *Int. J. Prod. Res.* **2014**, *52*, 3877–3887.
21. Li, Y.J.; Liu, J.J.; Chen, Q.X.; Mao, N. Lot splitting and scheduling algorithm of multi-level assembly job shops. *Compu. Integr. Manuf. Syst.* (accepted).
22. Lu, H.; He, L.; Huang, G.Q.; Wang, K. Development and comparison of multiple genetic algorithms and heuristics for assembly production planning. *IMA J. Manag. Math.* **2014**, *27*, 181–200. [CrossRef]
23. Wan, F.; Liu, J.H.; Ning, R.X.; Zhuang, C.B. Visual production scheduling technology for the complex product assembly process. *Compu. Inte. Manuf. Syst.* **2013**, *19*, 755–765.
24. Suharyanti, Y.; Ariyono, V. The effect of product structure complexity and process complexity on optimum lot size in multilevel product scheduling. In Proceedings of the Asia Pacific Industrial Engineering and Management Systems (APIEMS) Conference, Melaka, Malaysia, 7–10 December 2010.
25. Che, A.; Wu, X.Q.; Peng, J.; Yan, P.Y. Energy-efficient bi-objective single-machine scheduling with power-down mechanism. *Comput. Oper. Res.* **2017**, *85*, 172–183. [CrossRef]
26. Twomey, J.; Yildirim, M.B.; Whitman, L.; Liao, H.; Ahmad, J. *Energy Profiles of Manufacturing Equipment for Reducing ConSumption in a Production Setting*; Working Paper; Wichita State University: Wichita, KS, USA, 2008.
27. Wang, X.L.; Luo, W.; Zhang, H.; Dan, B.B.; Feng, L. Energy consumption model and its simulation for manufacturing and remanufacturing systems. *Int. J. Adv. Manuf. Technol.* **2016**, *87*, 1557–1569. [CrossRef]
28. Luan, X.N.; Zhang, S.; Chen, J.; Li, G. Energy modelling and energy saving strategy analysis of a machine tool during non-cutting status. *Int. J. Prod. Res.* **2019**, *57*, 4451–4467. [CrossRef]
29. Liu, Y.; Dong, H.B.; Niels, L.; Sanja, P.; Nabil, G. An investigation into minimising total energy consumption and total weighted tardiness in job shops. *J. Clean. Prod.* **2014**, *65*, 87–96. [CrossRef]
30. Peng, C.; Peng, T.; Zhang, Y.; Tang, R.; Hu, L. Minimising non-processing energy consumption and tardiness fines in a mixed-flow shop. *Energies* **2018**, *11*, 3382. [CrossRef]
31. Wu, X.L.; Cui, Q. Multi-objective flexible flow shop scheduling problem with renewable energy. *Compu. Inte. Manuf. Syst.* **2018**, *24*, 144–159. [CrossRef]
32. Gilles, M.; Mehmet, B.Y.; Twomey, J. Operational methods for minimization of energy consumption of manufacturing equipment. *Int. J. Prod. Res.* **2007**, *45*, 4247–4271.
33. Wang, H.; Jiang, Z.G.; Wang, Y.; Zhang, H.; Wang, Y.H. A two-stage optimization method for energy-saving flexible job-shop scheduling based on energy dynamic characterization. *J. Clean. Prod.* **2018**, *188*, 575–588. [CrossRef]
34. Zhang, Q.; Manier, H.; Manier, M.A. A genetic algorithm with tabu search procedure for flexible job shop scheduling with transportation constraints and bounded processing times. *Comput. Oper. Res.* **2012**, *39*, 1713–1723. [CrossRef]
35. Zhao, N.; Li, K.D.; Tian, Q.; Du, Y.H. Fast optimization approach of flexible job shop scheduling with transport time consideration. *Compu. Inte. Manuf. Syst.* **2015**, *21*, 724–732.
36. Rahman, H.F.; Nielsen, I. Scheduling automated transport vehicles for material distribution systems. *Appl. Soft. Comput.* **2019**, *82*, 105552. [CrossRef]
37. Benton, W.C.; Srivastava, R. Product structure complexity and inventory storage capacity on the performance of a multi-level manufacturing system. *Int. J. Prod. Res.* **1993**, *31*, 2531–2545. [CrossRef]
38. Mirjalili, S.; Saremi, S.; Mirjalili, S.M.; Coelho, L.S. Multi-objective grey wolf optimizer: A novel algorithm for multi-criterion optimization. *Expert Syst. Appl.* **2016**, *47*, 106–119. [CrossRef]
39. Deb, K.; Pratap, A.; Agarwal, S.; Meyarivan, T. A fast and elitist multi-objective genetic algorithm: NSGA-II. *IEEE Trans. Evol. Comput.* **2002**, *6*, 182–197. [CrossRef]
40. Zhao, M.; Song, X.Y.; Chang, C.G. Improved artificial bee colony algorithm and its application on optimization of emergency scheduling. *Appl. Res. Comput.* **2016**, *33*, 3596–3601.
41. Bosman, P.; Thierens, D. The balance between proximity and diversity in multi-objective evolutionary algorithms. *IEEE Trans. Evol. Comput.* **2003**, *7*, 174–188. [CrossRef]

Article

A Novel Health Prognosis Method for a Power System Based on a High-Order Hidden Semi-Markov Model

Qinming Liu [1,*]**, Daigao Li** [1]**, Wenyi Liu** [1]**, Tangbin Xia** [2] **and Jiaxiang Li** [1]

[1] Department of Industrial Engineering, Business School, University of Shanghai for Science and Technology, 516 Jungong Road, Shanghai 200093, China; 203781388@st.usst.edu.cn (D.L.); 193860959@st.usst.edu.cn (W.L.); 1913520624@st.usst.edu.cn (J.L.)

[2] SJTU-Fraunhofer Center, State Key Laboratory of Mechanical System and Vibration, School of Mechanical Engineering, Shanghai Jiao Tong University, Shanghai 200240, China; xtbxtb@sjtu.edu.cn

* Correspondence: qmliu@usst.edu.cn

Abstract: Power system health prognosis is a key process of condition-based maintenance. For the problem of large error in the residual lifetime prognosis of a power system, a novel residual lifetime prognosis model based on a high-order hidden semi-Markov model (HOHSMM) is proposed. First, HOHSMM is developed based on the hidden semi-Markov model (HSMM). An order reduction method and a composite node mechanism of HOHSMM based on permutation are proposed. The health state transition matrix and observation matrix are improved accordingly. The high-order model is transformed into the corresponding first-order model, and more node dependency information is stored in the parameter group to be estimated. Secondly, in order to estimate the parameters and optimize the structure of the proposed model, an intelligent optimization algorithm group is used instead of the expectation–maximization (EM) algorithm. Thus, the simplification of the topology of the high-order model by the intelligent optimization algorithm can be realized. Then, the state duration variables in the high-order model are defined and deduced. The prognosis method based on polynomial fitting is used to predict the residual lifetime of the power system when the prior distribution is unknown. Finally, the intelligent optimization algorithm is used to solve the proposed model, and experiments are performed based on a set of power system data sets to evaluate the performance of the proposed model. Compared with HSMM, the proposed model has better performance on the power system health prognosis problem and can get a relatively good solution in a short computation time.

Keywords: high-order hidden semi-Markov model; composite node; model reduction; state duration; polynomial fitting; residual life prognosis

---

Citation: Liu, Q.; Li, D.; Liu, W.; Xia, T.; Li, J. A Novel Health Prognosis Method for a Power System Based on a High-Order Hidden Semi-Markov Model. *Energies* **2021**, *14*, 8208. https://doi.org/10.3390/en14248208

Academic Editor: Andrea Mariscotti

Received: 28 October 2021
Accepted: 3 December 2021
Published: 7 December 2021

Publisher's Note: MDPI stays neutral with regard to jurisdictional claims in published maps and institutional affiliations.

1. Introduction

With the development of technology, the operation of power systems has an increasing demand for higher reliability, lower environmental risks, and higher human safety. Power system failures usually come at the cost of high maintenance costs and uncertain downtime. However, it is difficult to obtain accurate health status and predict failures of a power system in time. Thus, power system health prognosis is an important topic in reliability and maintenance engineering, which determines how to properly integrate positioning degradation data into power system fault detection and failure prevention [1]. Health prognosis involves evaluating the current state, classifying the current state of several failure modes, and predicting the residual lifetime of the power system. Residual lifetime refers to the remaining life from the current health status to the functional failure of the system.

As a probabilistic statistical method, the hidden Markov model (HMM) has a good randomness representation ability and potential structural relationship description ability. It is widely used in the field of complex system modeling. Carey first migrated HMM

---

from the field of speech recognition to the field of power system fault diagnosis, creating a new research direction of fault diagnosis [2]. Researchers proposed an improved hidden Semi-Markov model (HSMM) to get rid of the constraints of HMM characteristics, which further divided the macro-states into micro-states, and made a priori assumptions about the time distribution of each macro-state to successfully realize the application of HMM in the field of residual lifetime prediction. Liu et al. proposed an HSMM of information fusion, completed the diagnosis and life prediction of equipment, and achieved good results [3].

System health prognosis is a complex process, and the real-time operation of the power system is particularly difficult. Most of the methods in the field of reliability involve intensive calculation and processing of a large number of historical data [4,5]. The methods are usually divided into three categories: model-driven models, physical models, and data-driven models.

The design of the physical model considers the operating conditions to model the operating process of the system. The models integrate the physical features and numeric features of the system under monitoring and develop the mappings between parameters and prognostics features, such as rotating machinery [6,7] and lithium-ion batteries [8]. However, it is difficult to guarantee its accuracy. There is no perfect physical model to adapt to the system.

The data-driven model is suitable for systems with little prior knowledge or a complex structure because of the characteristics of using existing data to predict the health of the system. Moreover, they rely solely on the past observed trajectories [9], including neural networks [10–12], support vector machines [13,14], and Gaussian process regression [15]. Finding the effective part of the original data is an arduous task. The main disadvantages of this model are slow convergence and the ease in which it falls into local minimums; these shortcomings limit the application of this model.

Model-driven models focus on the prediction level fusion of information. For the first time, Dong et al. used HSMM to conduct research on various sensor diagnoses [16]. Then, several variants of Markovian-based models were applied to the diagnosis and prognostics of equipment. Yan et al. used HSMM-based equipment health estimation and proposed several new methods to effectively predict equipment RUL [17]. Li et al. proposed an improved HMM that is improved through performance degradation and successfully applied it to the reliability evaluation of wind turbine bearings [18]. Huang et al. proposed an improved HMM based on a predictive neural network and applied it to a motor drive system to evaluate the proposed model [19]. These studies performed well in different domains, whereas the modification was mostly done in the modeling part, and little consideration was given to signal preprocessing and parameters estimation' furthermore, the recognizing and training processes of model-driven models are usually time-consuming, so they are applied to offline health prognosis.

In order to better refer to the historical statistical information to improve the model recognition rate, a high-order hidden Markov model (HOHMM) was proposed by researchers. For the problem of parameter explosion and a more complex derivation of the high-order model, researchers have established an order reduction algorithm, ORED, to simplify HOHMM. Basically, the fast incremental algorithm was proposed to train HOHMM and the three problems were discussed, respectively [20]. Compared with the successive order reduction of the ORED algorithm, Hadar proposed a more complex algorithm based on the idea of equivalent transformation, which transformed any high-order model into the corresponding first-order model [21]. The above researchers provide a general research method for the research of the high-order model. However, in the process of reducing the order of HOHMM to HMM, it is necessary to re-deduce each variable of the model and re-evaluate the parameters based on the Baum–Welch algorithm. With the increase of the order of the model, the workload will increase explosively. Dong et al. established the HO-HMMAR model, deduced variables, and better solved the optimal portfolio problem [22]. Heng et al. established the daily average temperature evolution model and assumed that the asymmetric component was a high-order hidden Markov process. The results showed that the model can effectively capture the characteristics of temperature data under various conditions Polynomial fitting was a simple tool for nonlinear fitting, which has efficient data processing ability [23]. Ritesh et al. used the coefficients of polynomial fitting to generate the feature vector of iris recognition and verified the effectiveness of the polynomial method through the benchmark IITD and casia-v4 interval iris data [24]. With the popularization of neural networks and various machine learning methods, the flashpoint of polynomial fitting became dimmer and dimmer in the field of fault diagnosis. However, the polynomial fitting is one of the few methods that can directly write the analytical formula of a nonlinear relationship. It cannot be done by the neural network.

The problem of parameter estimation has always been the most difficult link of the probability model. The EM algorithm with a heavy workload depends on the initial value and can easily fall into local optimization; thus, more and more researchers are turning their attention to various intelligent optimization algorithms. The algorithm of Yu [13] and other genetic algorithms were combined within the grey model. The numerical results showed that the new method combined with a genetic algorithm (GA) can greatly improve the prediction accuracy. Krause [14] et al. used the function approximator based on a GA to estimate the parameters of the induction motor, which achieved better results than other methods. Zhang [15] and others used the artificial immune algorithm to optimize HMM and obtained the initial observation matrix with the highest identification degree. Zhang [16] and others used an adaptive genetic particle swarm optimization algorithm to optimize HSMM and obtained more accurate results than traditional HSMM. How an intelligent simulation algorithm can be used to simplify the research process of HMM is a point that should be paid attention to in future research.

In order to obtain more accurate and practical predictions under the premise of unknown distribution, the contribution of this paper is to propose an improved HOHSMM for the health prognosis of the power system. First, the framework of HOHSMM based on HSMM is established. By considering the ORED algorithm and Hadar's equivalent transformation, a model reduction method based on permutation is proposed, which uses the definition of the high-order hidden Markov group model to transform it into the corresponding first-order model, and the solution of three problems of the low-order model can be used in the high-order complex model. Moreover, the transition probability matrix and observation probability matrix are deformed so that the interdependence information of nodes in the high-order complex model is naturally integrated into the model parameters, and the effect of simplifying the model is achieved. Then, the auxiliary resident variables are defined and deduced, and the parameter group carrying more dependency information is estimated by using the intelligent optimization algorithm group. The maximizing occurrence probability of observation is described as objective, the decomposition dependency of the high-order model is represented by the parameter group, and the complexity is transferred from the model itself to the parameter group. The polynomial fitting method is used to fit each resident variable sequence, and the residual lifetime of the power system is predicted when the distribution is unknown. Finally, the proposed model is verified and evaluated with one power system data set. The results show that the method proposed in this paper is feasible and effective.

This article aims to develop a more superior power system health prognostics method. The paper is organized as follows: Section 2 introduces first-order and high-order HMM, and Section 3 develops an improved high-order hidden semi-Markov model. Then, the residual lifetime prognosis method of this paper is proposed in Section 4. Section 5 analyzes and discusses the case study of this article, and finally the article is concluded in Section 6.

2. Hidden Markov Model

2.1. First-Order HMM

The first-order HMM can be described as $\lambda = (\pi, A, B)$. It is composed of observable nodes and hidden state nodes. The health state of the node at time t is represented by

$State_t$, then the transition between the hidden state of the hidden state node conforms to the Markov property.

$$\text{Prob}(State_t|State_1 \cdots State_{t-1}) = \text{Prob}(State_t|State_{t-1})$$

Note that the total number of health states is N and the total number of observations is M. π is the initial state distribution vector in the model, and it is the probability value of each state of power system when $t = 1$, $\pi = [\overline{\pi}]_{1 \times N}$. A is the transition probability matrix, and $A = [a_{ij}]_{N \times N}$. a_{ij} is the probability value from state i to state j, and $a_{ij} = \text{Prob}(State_{t+1} = j|State_t = i)$, $\sum_{i=1}^{N} a_{ij} = 1$, $\sum_{j=1}^{N} a_{ij} = 1$. B is the observation probability matrix, and $B = [b_i(k)]_{N \times M}$; it is the probability value of the observation generated by the node under state i and $\text{Prob}(o_t = k|State_t = i)$. o_t represents the observation value at time t, and meets $\sum_{i=1}^{N} b_i(k)=1$ and $\sum_{k=1}^{M} b_i(k)=1$.

In order to solve the evaluation problem of HMM, that is, $\text{Prob}(O|\lambda)$, forward–backward variables based on HMM are usually developed, and the evaluation problem is decomposed into recursive expressions of forward–backward variables. In order to solve the learning problem, the Baum–Welch algorithm is usually used to solve the parameter set $-\lambda$, which may optimize the model and produce the current observations. Its idea is to establish the Lagrange multiplier equation between the current parameter group λ and the maximum parameter group λ, and calculate the partial derivative under the constraints of each parameter to optimize the current parameters. In order to solve the prognosis problem, dynamic programming is applied to HMM to produce the Viterbi algorithm to find the most likely health state path. Due to its own characteristics, the conventional HMM has the defects that it must obey the exponential distribution and cannot describe the deterioration of the power system.

2.2. The High-Order HMM

The high-order hidden Markov model (HOHMM) is a generalization of the first-order Markov model, and it retains more historical statistical information. It is assumed that the current health state of the research object is related to the previous health states. Taking an n-order hidden Markov model as an example, its health state at time $t(t > n + 1)$ is related to the health state at the previous n times, and it is expressed as

$$\text{Prob}(State_t|State_1 \cdots State_{t-1}) = \text{Prob}(State_t|State_{t-N} \cdots State_{t-1})$$

HOHMM is different from the conventional HMM, and it is described as $\lambda = (\pi, A, B, \breve{A}, \breve{B})$, where $\breve{A}$ is the state transition matrix only applicable to the previous n times and $\breve{B}$ is the observation probability matrix only applicable to the previous n times. HOHMM also has three problems from HMM. However, due to relatively complex node dependencies, the dependencies become more complex with the increase of orders, and the model parameters will increase exponentially. Thus, the research on HOHMM generally focuses on model order reduction. HOHMM improves the model recognition rate on the basis of retaining more historical statistical information, but it still fails to overcome the shortcomings of HMM.

3. Improved High-Order Hidden Semi-Markov Model

In this paper, by considering the shortcomings of HMM and the advantages of high-order modeling, an improved high-order hidden semi-Markov model (HOHSMM) is proposed based on HSMM.

Taking the second-order HSMM as an example, the model can be described as $\lambda = (\pi, A, B, \breve{A}, \breve{B})$. It is usually assumed that each sub-state conforms to the same time distribution, and the topology of a second-order HSMM is described in Figure 1.

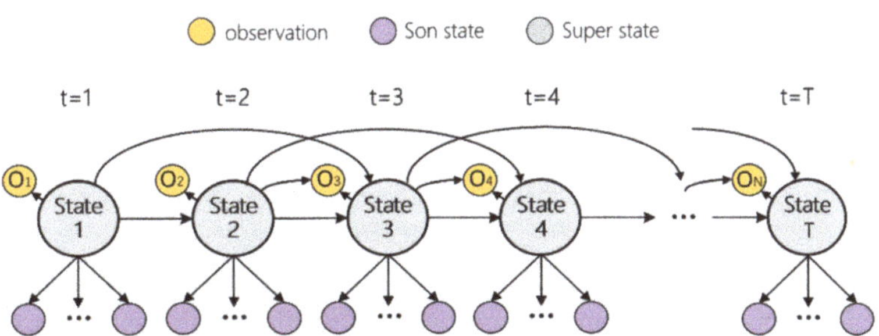

Figure 1. The illustration of two-order HSMM.

3.1. The Model Order Reduction Based on Permutation Mapping

Similarly, taking the second-order HOHSMM as an example, due to the structural changes of the second-order model, the model parameters and related algorithms can be changed accordingly. The respective algorithms for solving the three problems of the low-order model are not applicable to the high-order model. Thus, this paper proposes a model reduction method based on combinatorial mapping, and it is essential to merge the hidden state nodes corresponding to two adjacent time points in the second-order model into one node; then, the merged nodes can be modeled by the Markov process, as shown in Figure 2.

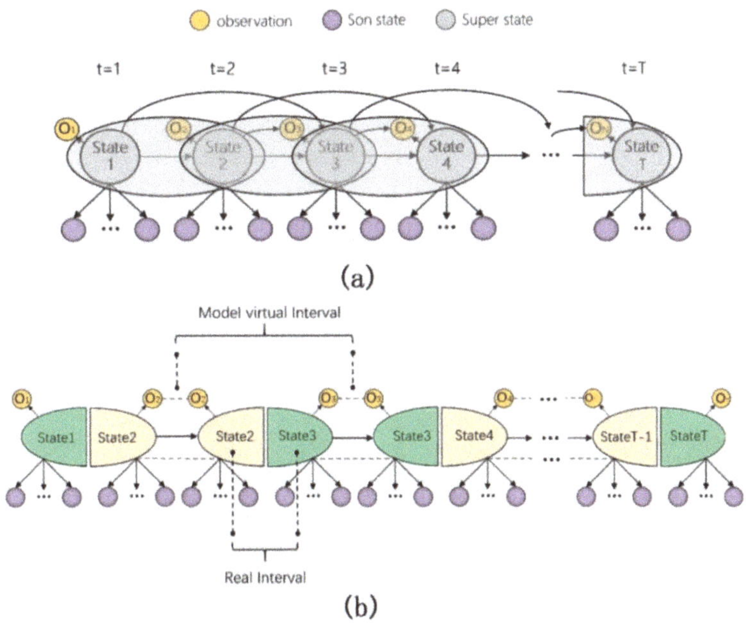

Figure 2. The illustration of the model reduction based on permutation. (**a**): Division of compound nodes; (**b**): Markov chain model with reduced order after compound nodes.

In this paper, a health state in HSMM generates a segment of observations, as opposed to a single observation in the HMM. Thus, the states in a segment semi-Markov model are called super states. Each super state consists of several single states, which are called son states, which can be seen in Figure 2. Figure 2a is a division of the model on the second-order HOHSMM, and the adjacent state nodes can be combined into a new, larger node. The relationship between the two health states in the new node has no impact on the Markov property of the whole model. The Markov property of the new model after order reduction can be described as Equation (1).

$$\text{Prob}(State_t, State_{t-1}|State_{t-1}\cdots State_1) = \text{Prob}(State_t, State_{t-1}|State_{t-1}, State_{t-2})(t \geq 3) \tag{1}$$

The description of time in the new model also changes t to $(t, t+1)$. If the first element of the time combination $(t, t+1)$ is the unique index of the time of the new model, the time can also be expressed as $\breve{t}$. $\breve{t}$ is different from t in the mathematical sense, but in the physical sense, $\breve{t} = t$. When the implicit state node of the new model combination time $\breve{t}$ is expressed as $(State_{\breve{t}}, State_{t+1})$, it generates two groups of observations, $o_{\breve{t}}$ and $o_{\breve{t}+1}$, where, $o_{\breve{t}+1}$ completely depends on $(State_{\breve{t}}, State_{t+1})$ and $o_{\breve{t}}$ completely depends on $(State_{t\check{}-1}, State_t)$. Moreover, at time $\breve{t} - 1$, the observations connected by dotted lines are the same observation in Figure 2b, and the sub-state set connected by dotted lines is also the same sub-state set. Thus, each group of observations can be obtained from the unique combined hidden state node, while the unique determined observations corresponding to the combined hidden state can be only retained (except the initial time) in its topology.

In the reduced order model, taking the combination of hidden states as the modeling object, different states are essentially an arrangement problem. Based on the above partial definitions of HMM parameters, if a second-order HOHSMM has N different super states, and considering the power system has performance degradation and the state is irreversible, then there are $C_N^2 + N$ states appearing in the original model. In this paper, a simple mapping between the arrangement scheme and natural numbers is introduced, as shown in Equation (2), where i, j are states, respectively.

$$Index = \text{Mapping}(i, j) = 4i + j - \sum_{z=0}^{i}(i, j \in (0, 1 \cdots N - 1), j \geq i) \tag{2}$$

In the parameters of the second-order HOHSMM original model, the transition probability matrix becomes the transition probability cube of $N \times N \times N$ and the observation probability matrix becomes the observation probability cube of $N \times N \times M$. The initial probability distribution remains unchanged and the initial transition probability matrix and the initial observation probability matrix are the same as the first-order model. The new model parameters after order reduction are composed of a transition probability matrix $(C_N^2 + N) \times (C_N^2 + N)$, observation probability matrix $(C_N^2 + N) \times M$, initial probability distribution, initial transition probability matrix, and initial observation probability matrix. Letting the reduced-order transition probability matrix be $\hat{A}$ with the element $\hat{a}_{ij}(i, j \in (0, \cdots N - 1))$, the reduced-order observation probability matrix is $\hat{B}$, and the element is $\hat{b}_{ij}(i, j \in (0, \cdots N - 1))$. For $\lambda = (\pi, A, B, \breve{A}, \breve{B})$, the element of π is $\pi_i(i \in (0, \cdots N - 1))$, the element of $\breve{A}$ is $\breve{a}_{ij}(i, j \in (0, \cdots N - 1))$, and the element of $\breve{B}$ is $\breve{b}_i(j)(i, j \in (0, \cdots N - 1))$.

Based on the order reduction method in Section 3.1, the transition probability matrix $\hat{A}$ $((C_N^2 + N) \times (C_N^2 + N))$ after dimension reduction can be obtained. The transition probability matrix $\hat{A}$ is sparse, and the actual amount of effective data in the matrix that is not 0 is

$$\sum_{j=0}^{N-1}\left(-j^2 + jN + N - j\right) \tag{3}$$

where j is the second state in the state arrangement (i, j). Taking the second-order HOHSMM with $n = 4$ as an example, the effective data of the reduced probability transfer matrix of order reduction are 20. In Table 1, the positions represented by 1 and 1* are the effective data, the composite state represented by the column header at the position of 1* is the main state, and the composite state represented by the column header at the position of 1 is defined as the transition state, while the location represented by 0 is invalid data.

Table 1. Sparse representation of two-order HOHSMM reduced-order transition probability matrix with 4 states.

From/To	Perm	(0,0)	(0,1)	(0,2)	(0,3)	(1,1)	(1,2)	(1,3)	(2,2)	(2,3)	(3,3)
Perm	index/index	0	1	2	3	4	5	6	7	8	9
(0,0)	0	1 *	1	1	1	0	0	0	0	0	0
(0,1)	1	0	0	0	0	1 *	1	1	0	0	0
(0,2)	2	0	0	0	0	0	0	0	1 *	1	0
(0,3)	3	0	0	0	0	0	0	0	0	0	1
(1,1)	4	0	0	0	0	1 *	1	1	0	0	0
(1,2)	5	0	0	0	0	0	0	0	1 *	1	0
(1,3)	6	0	0	0	0	0	0	0	0	0	1
(2,2)	7	0	0	0	0	0	0	0	1 *	1	0
(2,3)	8	0	0	0	0	0	0	0	0	0	1
(3,3)	9	0	0	0	0	0	0	0	0	0	1

*: Main state; Shadow: Valid data.

3.2. The Model Reasoning

In view of the idea of a forward–backward algorithm, a linger time (LT) mechanism is introduced, and an auxiliary variable $\xi_t(i)$ is established, as is described in Equation (4).

$$\xi_t(i,d) = \text{Prob}(O_{[1:t]}, LT((i,i)) = d \mid \lambda) \ (t \geq d) \tag{4}$$

The probability that the observation sequence $O_{[1:t]}$ is generated at the cut-off time t and has stayed in the current state for d time under a given model parameter group λ can be obtained by Equation (4).

In this section, the sparse representation of the transition probability matrix of the general second-order HOHSMM is given, and the significance of the main state and the transition state can be described, respectively. The transition among the different main health states of the conventional second-order model is a gradual process.

Main state → (Transition state sequence) → Next main state

This transformation process needs at most $j - i + 2$ time points to be fully described. Taking the transition from the health state (0,0) to (3,3) as an example, the transition process needs five time points to be described, as shown in Figure 3.

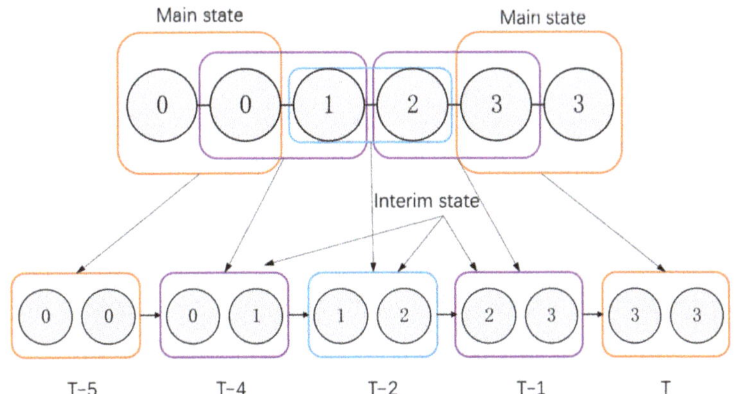

Figure 3. Schematic diagram of main state transition process.

In order to facilitate model reasoning, a transition variable $\vartheta_t(j,i,Road)$ is defined, representing the process intermediate value from the main state (j,j) to the time main state (i,i) of t time. The main states, by being transferred out, can be defined as acceptance states, and the main states being transferred in can be defined as return states. $\vec{t}$ is defined as the time point when the acceptance states appear.

(1) For $i = 0$ and $d = t$, $\xi_t(i)$ can be described as

$$\xi_t(0, d) = \mathrm{Prob}(O_{[1:t]}, LT(0) = d \,\big|\, \lambda) \tag{5}$$

The recursive initial value is

$$\xi_t(0, 1) = \pi_0 \check{b}_0(o_1) \check{b}_0(o_2) \check{a}_{00} \tag{6}$$

The inner transfer recursion is

$$\xi_t(0, d) = \xi_{t-1}(0, d-1) \hat{a}_{(0,0)(0,0)}(t-1) \, \hat{b}_{(0,0)}(o_t) \tag{7}$$

(2) For $i = 1$, $\xi_t(i)$ can be described as

$$\xi_t(1, d) = \mathrm{Prob}(O_{[1:t]}, LT(1) = d \,\big|\, \lambda) \tag{8}$$

The acceptance state can only be (0,0) and the transition state can only be (0,1). Then, the acceptance value is $\xi_{t-d}(0, t-d)$, $\vartheta_t(0, 1, Road) = \hat{a}_{(0,0)(0,1)}(t - d + 1)\hat{b}_{(0,1)}(o_{t-d+1})$. Thus, the recursive median is described as $\xi_t(1,0) = \xi_{t-d}(0, t-d)\hat{a}_{(0,0)(0,1)}(t - d + 1)\hat{b}_{(0,1)}$ $(o_{t-d+1})\hat{a}_{(0,1)(1,1)}(t - d + 2)\hat{b}_{(1,1)}(o_{t-d+2})$, and the recursive equation of internal transfer is

$$\xi_t(1, d) = \xi_{t-\Delta t}(1, d-1) \hat{a}_{(1,1)(1,1)}(t - \Delta t)\hat{b}_{(1,1)}(o_t) \tag{9}$$

(3) For $I = 2$, the acceptance states can be (0,0) and (1,1) and $\xi_t(i)$ can be described as

$$\xi_t(2, d) = \mathrm{Prob}\left(O_{[1:t]}, LT(2) = d \,\big|\, \lambda\right) \tag{10}$$

The inner transfer recursion is

$$\xi_t(2, d) = \xi_{t-1}(2, d-1) \hat{a}_{(2,2)(2,2)}(t-1) \, \hat{b}_{(2,2)}(o_t) \tag{11}$$

Sit1: When the acceptance state is (0,0), it needs four time points to describe from (0,0) to (2,2) and the acceptance value is

$$\xi_{\vec{t}}\left(0, \vec{t}\right) \; \vec{t} \in \{t - d - 3, t - d - 2\} \tag{12}$$

The corresponding $\vartheta_t(0, 2, Road)$ can be described as

$$\begin{cases} \hat{a}_{(0,0)(0,1)}\left(\vec{t}\right)\hat{b}_{(0,1)}\left(o_{\vec{t}+1}\right)\hat{a}_{(0,1)(1,2)}\left(\vec{t}+1\right)\hat{b}_{(1,2)}\left(o_{\vec{t}+2}\right)\hat{a}_{(1,2)(2,2)}\left(\vec{t}+2\right)\hat{b}_{(2,2)}\left(o_{\vec{t}+3}\right) \; \vec{t} = t - d - 3 \\ \hat{a}_{(0,0)(0,2)}\left(\vec{t}\right)\hat{b}_{(0,2)}(o_{t-d-1})\hat{a}_{(0,2)(2,2)}\left(\vec{t}+1\right)\hat{b}_{(2,2)}\left(o_{\vec{t}+2}\right) \; \vec{t} = t - d - 2 \end{cases} \tag{13}$$

The corresponding return value is given by Equation (14).

$$\begin{cases} \xi_{\vec{t}}\left(0, \vec{t}\right)\hat{a}_{(0,0)(0,1)}\left(\vec{t}\right)\hat{b}_{(0,1)}\left(o_{\vec{t}+1}\right)\hat{a}_{(0,1)(1,2)}\left(\vec{t}+1\right)\hat{b}_{(1,2)}\left(o_{\vec{t}+2}\right)\hat{a}_{(1,2)(2,2)}\left(\vec{t}+2\right)\hat{b}_{(2,2)}\left(o_{\vec{t}+3}\right) \; \vec{t} = t - d - 3 \\ \xi_{\vec{t}}\left(0, \vec{t}\right)\hat{a}_{(0,0)(0,2)}\left(\vec{t}\right)\hat{b}_{(0,2)}(o_{t-d-1})\hat{a}_{(0,2)(2,2)}\left(\vec{t}+1\right)\hat{b}_{(2,2)}\left(o_{\vec{t}+2}\right) \; \vec{t} = t - d - 2 \end{cases} \tag{14}$$

Sit2: When the acceptance state is (1,1), it needs three time points to describe from (1,1) to (2,2) and the corresponding $\vec{t} = t - d - 2$ acceptance value is

$$\sum_{du=1}^{\vec{t}-1} \xi_{\vec{t}}(1, d_u)$$

$\vartheta_t(1, 2, Road)$ is

$$\hat{a}_{(1,1)(1,2)}\left(\overrightarrow{t}\right)\hat{b}_{(0,1)}\left(o_{\overrightarrow{t}+1}\right)\hat{a}_{(1,2)(2,2)}\left(\overrightarrow{t}+1\right)\hat{b}_{(2,2)}\left(o_{\overrightarrow{t}+2}\right) \tag{15}$$

The return value is

$$\sum_{du=1}^{\overrightarrow{t}-1}\xi_{\overrightarrow{t}}(1, d_u)\hat{a}_{(1,1)(1,2)}\left(\overrightarrow{t}\right)\hat{b}_{(0,1)}\left(o_{\overrightarrow{t}+1}\right)\hat{a}_{(1,2)(2,2)}\left(\overrightarrow{t}+1\right)\hat{b}_{(2,2)}\left(o_{\overrightarrow{t}+2}\right) \tag{16}$$

(4)   For $I = 3$, the acceptance states can be (0,0), (1,1), and (2,2) and $\xi_t(i)$ can be described as

$$\xi_t(3, d) = \mathrm{Prob}\left(O_{[1:t]}, LT(3) = d\middle|\lambda\right) \tag{17}$$

Inner transfer recursion is

$$\xi_t(3, d) = \xi_{t-1}(3, d-1)\hat{a}_{(3,3)(3,3)}(t-1)\,\hat{b}_{(3,3)}(o_t) \tag{18}$$

Sit1: When the acceptance state is (0,0), it needs five time points to describe from (0,0) to (3,3), and the corresponding acceptance value is

$$\xi_{\overrightarrow{t}}\left(0, \overrightarrow{t}\right)\,\overrightarrow{t} \in \{t-d-4, t-d-3, t-d-2\} \tag{19}$$

$\vartheta_t(0, 3, Road)$ can be described as

$$\begin{cases} \hat{a}_{(0,0)(0,1)}\left(\overrightarrow{t}\right)\hat{b}_{(0,1)}\left(o_{\overrightarrow{t}+1}\right)\hat{a}_{(0,1)(1,2)}\left(\overrightarrow{t}+1\right)\hat{b}_{(1,2)}\left(o_{\overrightarrow{t}+2}\right)\hat{a}_{(1,2)(2,3)}\left(\overrightarrow{t}+2\right)\hat{b}_{(2,3)}\left(o_{\overrightarrow{t}+3}\right)\hat{a}_{(3,3)(3,3)}\left(\overrightarrow{t}+2\right)\hat{b}_{(3,3)}\left(o_{\overrightarrow{t}+3}\right) & \overrightarrow{t}=t-d-4 \\ \sum_{i=1}^{2}\hat{a}_{(0,0)(0,i)}\left(\overrightarrow{t}\right)\hat{b}_{(0,i)}\left(o_{\overrightarrow{t}+1}\right)\hat{a}_{(0,i)(i,3)}\left(\overrightarrow{t}+1\right)\hat{b}_{(i,3)}\left(o_{\overrightarrow{t}+2}\right) & \overrightarrow{t}=t-d-3 \\ \hat{a}_{(0,0)(0,3)}\left(\overrightarrow{t}\right)\hat{b}_{(0,i)}\left(o_{t-d-1}\right)\hat{a}_{(0,3)(3,3)}\left(\overrightarrow{t}+1\right)\hat{b}_{(3,3)}\left(o_{\overrightarrow{t}+2}\right) & \overrightarrow{t}=t-d-2 \end{cases} \tag{20}$$

The return value is the product of the corresponding acceptance value and the intermediate value.

Sit2: When the acceptance state is (1,1), it needs four time points to describe from (1,1) to (3,3), and the corresponding acceptance value is

$$\xi_{\overrightarrow{t}}\left(1, \overrightarrow{t}\right)\,\overrightarrow{t} \in \{t-d-3, t-d-2\} \tag{21}$$

The corresponding intermediate value $\vartheta_t(0, 3, Road)$ is

$$\begin{cases} \hat{a}_{(1,1)(1,2)}\left(\overrightarrow{t}\right)\hat{b}_{(1,2)}\left(o_{\overrightarrow{t}+1}\right)\hat{a}_{(1,2)(2,3)}\left(\overrightarrow{t}+1\right)\hat{b}_{(2,3)}\left(o_{\overrightarrow{t}+2}\right)\hat{a}_{(2,3)(3,3)}\left(\overrightarrow{t}+2\right)\hat{b}_{(3,3)}\left(o_{\overrightarrow{t}+3}\right) & \overrightarrow{t}=t-d-3 \\ \hat{a}_{(1,1)(1,3)}\left(\overrightarrow{t}\right)\hat{b}_{(1,3)}\left(o_{\overrightarrow{t}+1}\right)\hat{a}_{(1,3)(3,3)}\left(\overrightarrow{t}+1\right)\hat{b}_{(3,3)}\left(o_{\overrightarrow{t}+2}\right) & \overrightarrow{t}=t-d-2 \end{cases} \tag{22}$$

The recursive median is

$$\begin{cases} \xi_{\overrightarrow{t}}\left(1, \overrightarrow{t}\right)\hat{a}_{(1,1)(1,2)}\left(\overrightarrow{t}\right)\hat{b}_{(1,2)}\left(o_{\overrightarrow{t}+1}\right)\hat{a}_{(1,2)(2,3)}\left(\overrightarrow{t}+1\right)\hat{b}_{(2,3)}\left(o_{\overrightarrow{t}+2}\right)\hat{a}_{(2,3)(3,3)}\left(\overrightarrow{t}+2\right)\hat{b}_{(3,3)}\left(o_{\overrightarrow{t}+3}\right) & \overrightarrow{t}=t-d-3 \\ \xi_{\overrightarrow{t}}\left(1, \overrightarrow{t}\right)\hat{a}_{(1,1)(1,3)}\left(\overrightarrow{t}\right)\hat{b}_{(1,3)}\left(o_{\overrightarrow{t}+1}\right)\hat{a}_{(1,3)(3,3)}\left(\overrightarrow{t}+1\right)\hat{b}_{(3,3)}\left(o_{\overrightarrow{t}+2}\right) & \overrightarrow{t}=t-d-2 \end{cases} \tag{23}$$

Sit3: When the acceptance state is (2,2), it needs three time points to describe from (2,2) to (3,3), and the corresponding acceptance value is

$$\xi_{\overrightarrow{t}}\left(2, \overrightarrow{t}\right)\,\overrightarrow{t} \in \{t-d-2\} \tag{24}$$

The corresponding intermediate value $\vartheta_t(2, 3, Road)$ is

$$\hat{a}_{(2,2)(2,3)}\left(\vec{t}\right)\hat{b}_{(1,3)}\left(o_{\vec{t}+1}\right)\hat{a}_{(2,3)(3,3)}\left(\vec{t}+1\right)\hat{b}_{(3,3)}\left(o_{\vec{t}+2}\right) \tag{25}$$

The recursive median is

$$\zeta_{\vec{t}}\left(2,\vec{t}\right)\hat{a}_{(2,2)(2,3)}\left(\vec{t}\right)\hat{b}_{(1,3)}\left(o_{\vec{t}+1}\right)\hat{a}_{(2,3)(3,3)}\left(\vec{t}+1\right)\hat{b}_{(3,3)}\left(o_{\vec{t}+2}\right) \tag{26}$$

The general recursive equation of $\xi_t(i,d)$ can be expressed by Equation (27):

$$\zeta_t(i,d) = \sum_{j=0}^{i}\left[\left(\prod_{k=t-d}^{t}\hat{a}_{(i,i)(i,i)}\hat{b}_{(i,i)}(o_k)\right)\sum_{d_i=1}^{ct}\sum_{Road\in R}\xi_{ct}(j,d_i)\vartheta_t(j,i,Road)\right] \tag{27}$$

where $ct = t - d - lr + 1$, lr is the length of *Road*.

Based on the above classification and derivation of recursion, the auxiliary variables ξ_t in the higher state are decomposed into the expression of lower auxiliary variables. In the recursive process, all the transition process state paths meeting the main state transition can be generated, and $\xi_t(i,d)$ can calculated by traversing all possible transition state paths. The transition state paths included by each main state are given in Table 2, where x is not included in the path length.

Table 2. Path generator for each main state transition process.

From	To	Road	From	To	Road
(0,0)	(1,1)	[0,0,1,1]	(0,0)	(3,3)	[x,x,0,0,3,3]
(0,0)	(2,2)	[x,0,0,2,2]	(1,1)	(2,2)	[1,1,2,2]
(0,0)	(2,2)	[0,0–2,2]	(1,1)	(3,3)	[1,1–3,3]
(0,0)	(3,3)	[0,0–3,3]	(1,1)	(3,3)	[x,1,1,3,3]
(0,0)	(3,3)	[x,0,0,1,3,3]	(2,2)	(3,3)	[2,2,3,3]
(0,0)	(3,3)	[x,0,0,2,3,3]	–	–	–

By the given model parameters, the probability of generating observations O is

$$\text{Prob}(O|\lambda) = \sum_{i=0}^{N-1}\sum_{d=1}^{D}\zeta_T(i,d)$$

Auxiliary variable $\tau_t(index)$ is the probability of being in $\text{Mapping}^{-1}(index)$ at time t under the premise of the given model parameters and observations.

$$\tau_t(index) = \text{Prob}\left(State_t = \text{Mapping}^{-1}(index)\middle|\lambda, O\right)$$

Where $\tau_0(index)$ can be obtained by calculating $\pi, \breve{A}, \breve{B}$, and the $\tau_t(index)$ recursive equation can be shown as follows.

$$\tau_{t+1}(index) = \frac{\text{prob}\left(o_{\vec{t}+1}, State_t = \text{Mapping}^{-1}(index)\middle|\lambda\right)}{\sum_{indexi=0}^{9}\text{prob}\left(o_{\vec{t}+1}, State_t = \text{Mapping}^{-1}(indexi)\middle|\lambda\right)} = \frac{\hat{b}_{(index)}\left(o_{\vec{t}+1}\right)\sum_{i=0}^{9}\tau_t(i)\hat{a}_{(i)(index)}}{\sum_{indexi=0}^{9}\hat{b}_{(index)}\left(o_{\vec{t}+1}\right)\sum_{i=0}^{9}\tau_t(i)\hat{a}_{(i)(indexi)}} \tag{28}$$

For the corresponding low-order model, the probability in a single state $i(i \in (1,2,3,4))$ at time t can be described by the high-order model.

$$\text{Prob}(S_t = i) = \sum_{indexi\in I}\tau_{t-1}(indexi)$$

I is the index set corresponding to the second sub state i under different composite states, and the index correspondence is given in Table 3.

Table 3. Index correspondence.

Son State	Index Gather (*I*)	Perm Gather
0	{0}	{(0,0)}
1	{1,4}	{(0,1),(1,1)}
2	{2,5,7}	{(0,2),(1,2),(2,2)}
3	{3,6,8,9}	{(0,3),(1,3),(2,3),(3,3)}

For the parameter estimation problem, this paper selects the intelligent simulation algorithm group to replace the Baum–Welch algorithm and carries out a two-stage estimation. First, a one-stage likelihood function $L(\lambda, \breve{A}, \breve{B})$ is established, and it is described as

$$L(\lambda, \breve{A}, \breve{B}) = \text{Prob}(O_{[1:2]} | \lambda, \breve{A}, \breve{B})$$

For the situation of the current $\lambda, \breve{A}, \breve{B}$, if the probability of the first two observations is generated, then the optimal $\lambda, \breve{A}, \breve{B}$ in one stage is

$$\lambda, \breve{A}, \breve{B} = \underset{\lambda, \breve{A}, \breve{B}}{argmax} \text{Prob}(O_{[1:2]} | \lambda, \breve{A}, \breve{B})$$

The two-stage likelihood function is

$$L(A, B) = \text{Prob}(O_{[3:]} | \lambda, \breve{A}, \breve{B}, A, B)$$

In the situation of the current $\lambda, \breve{A}, \breve{B}, A, B$, the probability from the third to the final observation is generated, and the corresponding optimal parameters A, B in the second stage are

$$A, B = \underset{A,B}{argmax} \text{Prob}(O_{[3:]} | \lambda, \breve{A}, \breve{B}, A, B)$$

Different intelligent simulation algorithms are used to optimize and compare the two-stage likelihood function, and finally to select the best result.

4. Residual Life Prognosis under Uncertain Distribution

For the recursive reasoning on the duration of each state in Section 3.2, the system duration of each health state and the joint probability value of the observation can be obtained. In fact, there is a certain correlation between the observations generated by the system and the system duration in a single state, and it can be described by the observation and state transition matrix through a specific mode. In order to obtain the edge probability of the system duration in each health state, it is advisable to assume that system observation and dwell are independent of each other. For the above obtained $\xi_t(i, d)$, by calculating the generation probability of the corresponding observation, the conditional probability equation can be obtained.

$$\text{Prob}(LT(i) = d | \lambda) = \frac{\xi_t(i, d)}{\text{Prob}(O | \lambda)}$$

The time represented by each time point is used to calculate the probability that a certain state produces the duration, and a group of duration probability sequences are obtained. Generally, when the discrete data points are known, an a priori distribution assumption is conducive to study the continuity characteristics of the data. However, if the data distribution is incorrectly assumed, there will be great errors in the fitting and the original properties of the data will be lost. Thus, this paper proposes a method to fit the data under an unknown distribution based on polynomial regression. The polynomial regression can be described as

$$f^n(x) = w_0 + w_1 x + w_2 x^2 + w_3 x^3 \cdots + w_m x^m$$

Letting the m-order polynomial function fitted by duration probability sequences generated by health i be described as $Lds_m^i(t)$, the residual lifetime of system at time t can be obtained.

$$RUL_1(t) = \sum_{i=0}^{N-1} \left[\int_t^D \frac{Lds_n^i(x)R_i}{\int_0^D Lds_n^i(y)dy}(x-t)dx + \sum_{j=i+1}^{N-1} \int_t^D xLds_m^j(x)R_j dx \right] \mathrm{Prob}(s_t = i) \qquad (29)$$

where, D is the maximum duration of each state and $\mathrm{Prob}(s_t = i)$ is the probability in health state i at time t. R_i, R_j are the integral scaling coefficients after continuous discrete data.

5. Case Study

This paper verifies and evaluates the proposed model and method through the example of the system health diagnosis and life prediction of one power station. A LW42A-40.5 high voltage circuit breaker system was selected as the experimental platform; it was equipped with a CT14 spring operating mechanism, three-phase linkage during opening and closing, and SF6 was used as the arc extinguishing medium. An LW42A-40.5 high voltage circuit breaker system is mainly used in 35 kV power transmission and distribution systems for protection and control. The vibration signals of the system in the laboratory were collected by a hydraulic accelerometer installed parallel to the rotating shaft of the power system. In the application example, the power system was filled with 20, 40, 60, and 80 mg of micro dust, respectively, and a fixed length time window every 10 min to collect a vibration signal (P6) of about 1 min was used, as shown in Figure 4. Then, the vibration signal was divided into five layers by using 10 dB wavelets and the array of high-frequency and low-frequency wavelet coefficients were obtained. The dimensionally reduced wavelet coefficients were used as the input feature sequence vector of DGHMM. During the whole experiment, the states of the power system could be divided into four types: Baseline, Cont1, Cont2, and Cont3. Cont3 is the complete failure state of system. The whole experimental analysis platform was Python 3 and the platform running environment was Windows 10.

The power system state monitoring information was used to predict the state of health. In order to obtain a reliable health prognosis, the features to be monitored should be sensitive to vibration trends. We used wavelet transform to remove noise from the original signal and perform feature extraction in this paper [25]. It can generate a suitable framework to study the multi-scale transient representation of the signal. It is also good at time-frequency analysis and processing non-stationary signals. More than one characteristic should usually be used for a healthy prognosis. Here, a wavelet amplitude model based on the overall monitoring signal was used to demonstrate how to use features for health prognosis.

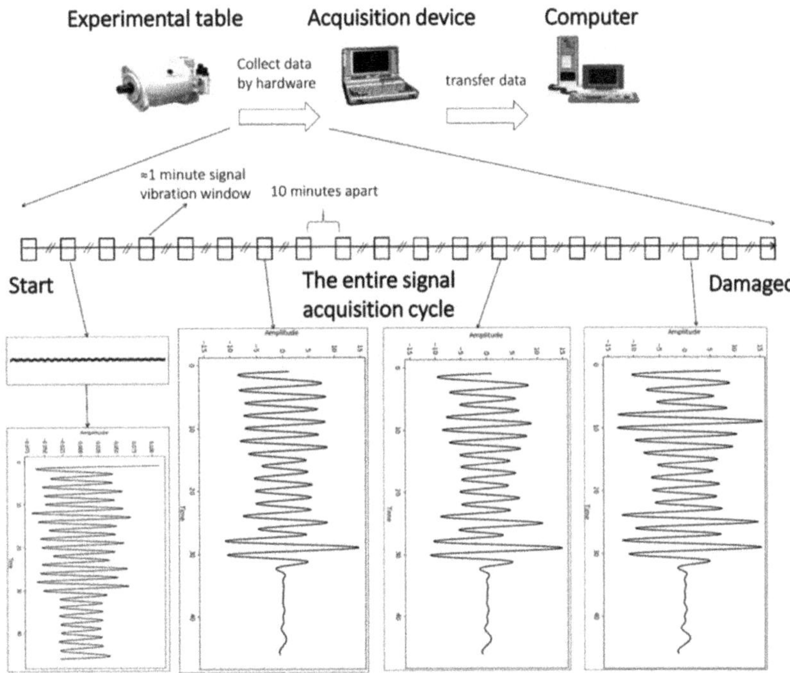

Figure 4. Experimental bench and signal acquisition.

5.1. Partial Data Preview

Using the vibration signal monitoring data from the hydraulic accelerometer, the health status diagnosis and residual life prognosis of the power system were carried out. Part of the vibration data (P6) after wavelet transform is shown in Table 4.

Table 4. Partial wavelet transform data of P6.

Time Spot	Sen 1	Sen 7	Sen 15	Sen 23	Sen 32
1	2.62	19.53	5.66	0.06	0.96
3	2.65	20.24	6.10	0.06	1.04
5	2.39	17.45	4.92	0.06	0.87
7	2.07	16.15	3.98	0.06	0.75
9	2.25	18.53	4.51	0.06	0.81
11	2.24	21.44	4.38	0.06	0.87
13	2.41	5.37	6.84	0.08	1.66
15	2.58	6.16	7.96	0.09	1.84
17	2.53	6.07	7.63	0.09	1.84
19	2.44	6.24	7.71	0.10	1.84
21	20.66	9.68	8.42	0.11	2.17
23	5.06	8.75	7.20	0.10	1.89
25	7.53	8.82	8.15	0.10	2.00
27	5.74	8.37	7.19	0.10	1.89
29	3.46	8.25	7.16	0.10	1.92
31	41.78	7.98	7.12	0.15	1.86
33	31.08	7.90	6.98	0.14	1.84
35	60.15	7.99	7.03	0.15	1.85
37	77.27	7.80	7.12	0.17	1.84

5.2. Parameter Estimation of the Intelligent Optimization Algorithm Group

Considering that the maximum expectation algorithm has a large dependence on the initial value, we used a genetic algorithm (GA), particle swarm algorithm (PSO), and artificial fish swarm algorithm (AFSA) to estimate the parameters of the proposed model under the same set of full observations, respectively.

The maximum number of iterations was 300, and the number of population (the number of particles in a particle swarm) was 30. The GA adopted a large parameter adaptive value encoding and decoding strategy. The PSO adopted the fixed inertia weight strategy, with an inertia weight of 0.5 and a learning factor of 2. The AFSA perception field was 1 and the crowding factor was 0.6. The objective optimization model was to maximize the occurrence probability of observation. The iterative process and results are shown in Figure 5. Under the premise of the same population number and iteration times, GA had good results with a continuous evolution and maximum likelihood. The likelihood values of the parameter groups found by the three optimization algorithms are shown in Table 5, and the reduced probability transition probability matrix is given in Table 6.

Table 5. The maximum likelihood of each algorithm.

Algorithm	Maximum Likelihood
GA	9.8665×10^{-8}
PSO	2.44412×10^{-13}
AFSA	7.22939×10^{-13}

Table 6. The optimal probability transfer matrix is represented sparsely.

From/To	Perm	(0,0)	(0,1)	(0,2)	(0,3)	(1,1)	(1,2)	(1,3)	(2,2)	(2,3)	(3,3)
Perm	index/ index	0	1	2	3	4	5	6	7	8	9
(0,0)	0	7.78×10^{-1}	4.45×10^{-2}	4.50×10^{-1}	2.67×10^{-2}	0.00	0.00	0.00	0.00	0.00	0.00
(0,1)	1	0.00	0.00	0.00	0.00	8.53×10^{-1}	1.40×10^{-1}	7.00×10^{-3}	0.00	0.00	0.00
(0,2)	2	0.00	0.00	0.00	0.00	0.00	0.00	0.00	7.44×10^{-1}	2.56×10^{-1}	0.00
(0,3)	3	0.00	0.00	0.00	0.00	0.00	0.00	0.00	0.00	0.00	1.00
(1,1)	4	0.00	0.00	0.00	0.00	8.66×10^{-1}	1.32×10^{-1}	2.45×10^{-3}	0.00	0.00	0.00
(1,2)	5	0.00	0.00	0.00	0.00	0.00	0.00	0.00	9.97×10^{-1}	3.44×10^{-3}	0.00
(1,3)	6	0.00	0.00	0.00	0.00	0.00	0.00	0.00	0.00	0.00	1.00
(2,2)	7	0.00	0.00	0.00	0.00	0.00	0.00	0.00	8.87×10^{-1}	1.13×10^{-1}	0.00
(2,3)	8	0.00	0.00	0.00	0.00	0.00	0.00	0.00	0.00	0.00	1.00
(3,3)	9	0.00	0.00	0.00	0.00	0.00	0.00	0.00	0.00	0.00	1.00

5.3. Residual Lifetime Prognosis

Based on the model reasoning, the duration of each state was analyzed, and the full probability equation was used for equivalent replacement in the calculation process. The different duration values of each state at different time points were separated according to the state to obtain the duration probability sequence of each state, and the polynomial regression was used to fit the sequence to obtain their respective analytical formulas. In the fitting process, the order m with less of a resonance effect on the prediction of the residual life of the subsequent state was preferentially selected (priority selection principle). The polynomial fitting of four different states at the final time point is given as an example in

Figure 6. The optimal values of fitting in the four states were 20, 15, 19, and 17 respectively, and the detailed polynomial regression coefficients are given in Table 7.

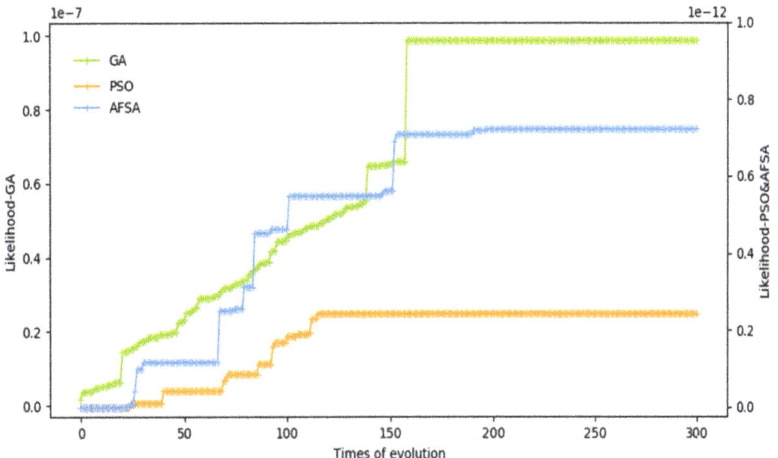

Figure 5. Swarm intelligence algorithm iteration process.

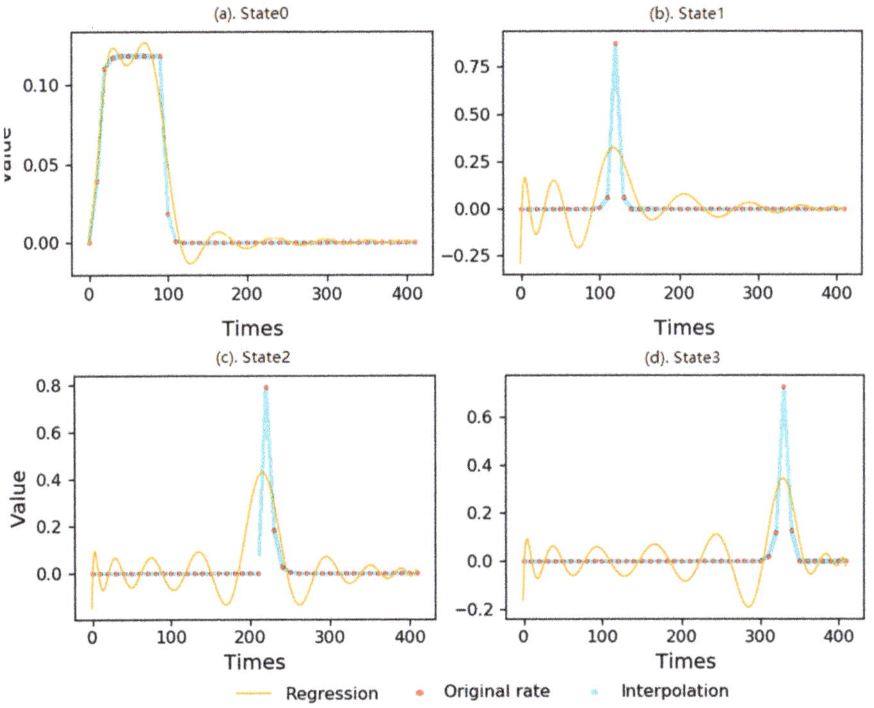

Figure 6. The polynomial fitting of each state at the final time.

Table 7. The polynomial fitting coefficients of each state at the final time.

Coeff. Index	State Type			
	State0(20)	State1(15)	State2(19)	State3(17)
0	-2.38819×10^{-45}	1.89034×10^{-32}	-3.01553×10^{-41}	-4.21246×10^{-37}
1	7.47909×10^{-42}	-6.14991×10^{-29}	1.02846×10^{-37}	1.23709×10^{-33}
2	-9.90244×10^{-39}	9.03133×10^{-26}	-1.54556×10^{-34}	-1.56243×10^{-30}
3	6.85507×10^{-36}	-7.90982×10^{-23}	1.31921×10^{-31}	1.0738×10^{-27}
4	-2.21019×10^{-33}	4.59824×10^{-20}	-6.68384×10^{-29}	-3.92341×10^{-25}
5	-1.73946×10^{-31}	-1.86838×10^{-17}	1.65986×10^{-26}	2.81649×10^{-23}
6	4.20376×10^{-28}	5.44157×10^{-15}	2.2419×10^{-24}	4.69387×10^{-20}
7	-1.12487×10^{-25}	-1.14607×10^{-12}	-3.79886×10^{-21}	-2.7376×10^{-17}
8	-3.1407×10^{-23}	1.73894×10^{-10}	1.72559×10^{-18}	8.37058×10^{-15}
9	3.36199×10^{-20}	-1.8709×10^{-8}	-4.83436×10^{-16}	-1.67201×10^{-12}
10	-1.2786×10^{-17}	1.3854×10^{-6}	9.40639×10^{-14}	2.29806×10^{-10}
11	2.98005×10^{-15}	-6.72669×10^{-5}	-1.31442×10^{-11}	-2.18798×10^{-8}
12	-4.69272×10^{-13}	1.982402×10^{-3}	1.32469×10^{-9}	1.41766×10^{-6}
13	5.09843×10^{-11}	$-3.1177213 \times 10^{-2}$	-9.50595×10^{-8}	-6.00431×10^{-5}
14	-3.77167×10^{-9}	$2.04641444 \times 10^{-1}$	4.72263×10^{-6}	1.5492×10^{-3}
15	1.82137×10^{-7}	$-2.9072277 \times 10^{-1}$	1.55018×10^{-4}	2.1526379×10^{-2}
16	-5.28119×10^{-6}	——	3.118172×10^{-3}	$1.26386828 \times 10^{-1}$
17	7.74612×10^{-5}	——	3.3825859×10^{-2}	$1.61790099 \times 10^{-1}$
18	4.00078×10^{-4}	——	$1.54167196 \times 10^{-1}$	——
19	4.512709×10^{-3}	——	$1.47377792 \times 10^{-1}$	——
20	6.50429×10^{-4}	——	——	——

In principle, negative values are not allowed for probability integration, but polynomial regression shows obvious fluctuation characteristics, positive and negative values have certain offset effect, and resonance also exists at the same time point in different states. It is not difficult to predict that the predicted residual life value may have certain fluctuation characteristics. In order to better fit the polynomial, the original data points were linearly interpolated combined with the characteristics of the original data. In Figure 7, the red point is the original data point, that is, the adaptive duration generation probability value, the light blue is the interpolation point, and the yellow line is the polynomial fitting line.

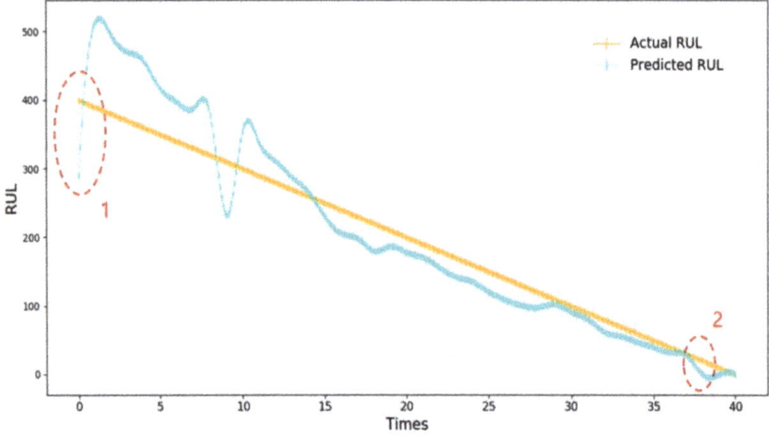

Figure 7. Residual lifetime prognosis of P6.

Finally, the residual life prognosis of P6 is shown in Figure 7, in which the discrete prediction points have been interpolated and smoothed. Mark 1 shows that there is a large deviation in the predicted value at the final time point. Corresponding to the phenomenon of the initial fluctuation range of the polynomial regression formula in each subgraph of Figure 7 being large but the fluctuation gradually decreasing with the increase of the time point, this is the result of the priority selection principle. The phenomenon of early damage appears at Mark 2, corresponding to the phenomenon of residence early reduction in Figure 6d.

Compared with the low-order HSMM results [3], the residual life prognosis by the proposed model in this paper is shown in Table 8. From the sampling time point, the overall effect of the residual life prognosis method based on HOHSMM and polynomial fitting is significantly better than that of the conventional HSMM. The case shows that the life prognosis method based on HOHSMM and polynomial regression is effective and feasible.

Table 8. Relative error analysis of the predicted RU.

Actual RUL	Model of This Paper		HSMM	
	Predicted RUL	The Relative Error (%)	Predicted RUL	The Relative Error (%)
300	357.010	19.00	302.558	0.85
260	266.729	2.59	299.643	15.25
220	180.068	18.15	297.954	35.43
170	142.383	16.25	194.981	14.69
150	120.147	19.90	192.081	28.05
120	98.159	18.20	188.666	57.22
110	102.573	6.75	102.471	6.84
90	79.956	11.16	100.291	11.43
50	39.600	20.80	97.675	95.35
Mean relative error /%	14.76		29.4592	

6. Conclusions

The diagnostic condition of mechanical systems is that the training data are sufficient and the samples of different categories in the training data have a balanced distribution. In the whole life cycle of a high-voltage circuit breaker system, it is in normal state most of the time, and the number of normal samples in the actually monitored signal will be more than the number of fault samples, resulting in the imbalance of training data and unknown distribution. Thus, for the problem of large error in the residual lifetime prognosis of power systems, a novel residual lifetime prognosis model based on HOHSMM and polynomial fitting was proposed. Based on HSMM, an order reduction method and composite node mechanism of HOHSMM based on permutation were proposed. The order reduction method of the permutation and combination model is simple and intuitive and uses the definition of the high-order hidden Markov group model. The high-order model can be transformed into the corresponding first-order model by changing the observation angle, and the solution of the three problems of the low-order model can be used in the high-order complex model. The intelligent optimization algorithm group can be used to replace the EM algorithm to estimate the parameters and optimize the structure of the proposed model, and the simplification of the topology of the high-order model by the intelligent optimization algorithm can be realized. The complex dependency information in the high-order model is transferred to the deformed parameter group. It effectively simplifies the model and provides a new idea for the study of this kind of model. Finally, a case was studied to verify the proposed model. From the experimental results, the comparison between the proposed model and HSMM showed several advantages of the proposed model, indicating that the remaining life prediction based on polynomial fitting has better performance for the health prognosis problem of the power system.

Energies **2021**, 14, 8208

Author Contributions: Conceptualization, Q.L., D.L. and W.L.; methodology, T.X.; investigation, Q.L. and D.L.; resources, J.L.; writing—original draft preparation, Q.L.; writing—review and editing, D.L.; visualization, W.L.; supervision, T.X.; funding acquisition, Q.L. All authors have read and agreed to the published version of the manuscript.

Funding: This research was funded by National Natural Science Foundation of China(Nos. 71632008, 71840003, and 51875359), the Natural Science Foundation of Shanghai (No. 19ZR1435600 and 20ZR1428600), the Humanity and Social Science Planning Foundation of the Ministry of Education of China (No. 20YJAZH068), and the Science and Technology Development Project of the University of Shanghai for Technology and Science (No. 2020KJFZ038).

Acknowledgments: The authors are indebted to the reviewers and the editors for their constructive comments which greatly improved the contents and exposition of this paper.

Conflicts of Interest: The authors declare no conflict of interest.

References

1. Duan:, C.; Makis, V.; Deng, C. Optimal Bayesian early fault detection for CNC equipment using hidden semi-Markov process. *Mech. Syst. Signal Process.* **2019**, *122*, 290–306. [CrossRef]
2. Carey, B.; Dan, M.; Tarik, A. Condition-based maintenance of machines using Hidden Markov models. *Mech. Syst. Signal Process.* **2000**, *14*, 597–612.
3. Liu, Q.; Dong, M.; Lv, W.; Geng, X.; Li, Y. A novel method using adaptive hidden semi-Markov model for multi-sensor monitoring equipment health prognosis. *Mech. Syst. Signal Process.* **2015**, *64–65*, 217–232. [CrossRef]
4. Huang, W.; Dietrich, D. An Alternative Degradation Reliability Modeling Approach Using Maximum Likelihood Estimation. *IEEE Trans. Reliab.* **2005**, *54*, 310–317. [CrossRef]
5. Liu, Q.; Dong, M.; Peng, Y. A novel method for online health prognosis of equipment based on hidden semi-Markov model using sequential Monte Carlo methods. *Mech. Syst. Signal Process.* **2012**, *32*, 331–348. [CrossRef]
6. Wu, B.; Li, W.; Qiu, M. Remaining useful life prediction of bearing with vibration signals based on a novel indicator. *Shock. Vib. Shock. Vib.* **2017**, *90*, 1–10. [CrossRef]
7. Kacprzynski, G.J.; Sarlashkar, A.; Roemer, M.J. Predicting remaining life by fusing the physics of failure modeling with diagnostics. *J. Miner. Met. Mater. Soc.* **2004**, *56*, 29–35. [CrossRef]
8. Guha, A.; Patra, A.; Vaisakh, K.V. Remaining useful life estimation of lithiumion batteries based on the internal resistance growth model. In Proceedings of the 2017 Indian Control Conference (ICC), Guwahati, India, 4–6 January 2017; pp. 33–38.
9. Yang, T.; Zheng, Z.; Qi, L. A method for degradation prediction based on Hidden semi-Markov models with mixture of Kernels. *Comput. Ind.* **2020**, *122*, 103295. [CrossRef]
10. Yang, Z.; Baraldi, P.; Zio, E. A multi-branch deep neural network model for failure prognostics based on multimodal data. *J. Manuf. Syst.* **2021**, *59*, 42–50. [CrossRef]
11. Behera, S.; Misra, R.; Sillitti, A. Multiscale deep bidirectional gated recurrent neural networks based prognostic method for complex non-linear degradation systems. *Inf. Sci.* **2021**, *554*, 120–144. [CrossRef]
12. Li, J.; Ping, Y.; Li, H.; Li, H.; Liu, Y.; Liu, B.; Wang, Y. Prognostic prediction of carcinoma by a differential-regulatory-network-embedded deep neural network. *Comput. Biol. Chem.* **2020**, *88*, 107317. [CrossRef] [PubMed]
13. Lv, X.; Wang, H.; Zhang, X.; Liu, Y.; Jiang, D.; Wei, B. An evolutional SVM method based on incremental algorithm and simulated indicator diagrams for fault diagnosis in sucker rod pumping systems. *J. Pet. Sci. Eng.* **2021**, *203*, 108806. [CrossRef]
14. Zeng, N.; Qiu, H.; Wang, Z.; Liu, W.; Zhang, H.; Li, Y. A new switching-delayed-PSO-based optimized SVM algorithm for diagnosis of Alzheimer's disease. *Neurocomputing* **2018**, *320*, 195–202. [CrossRef]
15. Saha, S.; Saha, B.; Saxena, A.; Goebel, K. Distributed prognostic health management with Gaussian process regression. In Proceedings of the 2010 IEEE Aerospace Conference, Big Sky, MT, USA, 6–13 March 2010; pp. 1–8.
16. Dong, M.; He, D. Hidden semi-Markov model-based methodology for multi-sensor equipment health diagnosis and prognosis. *Eur. J. Oper. Res.* **2007**, *178*, 858–878. [CrossRef]
17. Yan, Y.; Cai, J.; Li, T.; Zhang, W.; Sun, L. Fault prognosis of HVAC air handling unit and its components using hidden-semi Markov model and statistical process control. *Energy Build.* **2021**, *240*, 110875. [CrossRef]
18. Li, J.; Zhang, X.; Zhou, X.; Lu, L. Reliability assessment of wind turbine bearing based on the degradation-Hidden-Markov model. *Renew. Energy* **2018**, *132*, 1076–1087. [CrossRef]
19. Huang, D.; Ke, L.; Chu, X.; Zhao, L.; Mi, B. Fault diagnosis for the motor drive system of urban transit based on improved Hidden Markov Model. *Microelectron. Reliab.* **2018**, *82*, 179–189.
20. Du Preze, J.A. Efficient training of high-order hidden Markov models using first-order representations. *Comput. Speech Lang.* **1998**, *12*, 23–39. [CrossRef]
21. Hadar, U.; Messer, H. High-order hidden Markov models–estimation and implemen-tation. In Proceedings of the IEEE/SP 15th Workshop on Statistical Signal Processing (SSP 2009), Cardiff, UK, 31 August–3 September 2009; pp. 249–252.

22. Zhu, D.-M.; Lu, J.; Ching, W.-K.; Siu, T.-K. Discrete-time optimal asset allocation under Higher-Order Hidden Markov Model. *Econ. Model.* **2017**, *66*, 223–232. [CrossRef]
23. Heng, X.; Mamon, R. A self-updating model driven by a higher-order hidden Markov chain for temperature dynamics. *J. Comput. Sci.* **2016**, *17*, 47–61.
24. Vyas, R.; Kanumuri, T.; Sheoran, G.; Dubey, P. Efficient iris recognition through curvelet transform and polynomial fitting. *Optik* **2019**, *185*, 859–867. [CrossRef]
25. Dong, M.; He, D. A segmental hidden semi-Markov model (HSMM)-based diagnostics and prognostics framework and methodology. *Mech. Syst. Signal Process.* **2007**, *21*, 2248–2266. [CrossRef]

MDPI

Integrated Structural Dependence and Stochastic Dependence for Opportunistic Maintenance of Wind Turbines by Considering Carbon Emissions

Qinming Liu [1,*], Zhinan Li [1], Tangbin Xia [2], Minchih Hsieh [1] and Jiaxiang Li [1]

[1] Department of Industrial Engineering, Business School, University of Shanghai for Science and Technology, 516 Jungong Road, Shanghai 200093, China; 203791422@st.usst.edu.cn (Z.L.); 07586@usst.edu.cn (M.H.); 1913520624@st.usst.edu.cn (J.L.)

[2] State Key Laboratory of Mechanical System and Vibration, School of Mechanical Engineering, Shanghai Jiao Tong University, Shanghai 200240, China; xtbxtb@sjtu.edu.cn

* Correspondence: qmliu@usst.edu.cn

Citation: Liu, Q.; Li, Z.; Xia, T.; Hsieh, M.; Li, J. Integrated Structural Dependence and Stochastic Dependence for Opportunistic Maintenance of Wind Turbines by Considering Carbon Emissions. *Energies* **2022**, *15*, 625. https://doi.org/10.3390/en15020625

Academic Editor: Andrzej Teodorczyk

Received: 16 December 2021
Accepted: 11 January 2022
Published: 17 January 2022

Publisher's Note: MDPI stays neutral with regard to jurisdictional claims in published maps and institutional affiliations.

Abstract: Wind turbines have a wide range of applications as the main equipment for wind-power generation because of the rapid development of technology. It is very important to select a reasonable maintenance strategy to reduce the operation and maintenance costs of wind turbines. Traditional maintenance does not consider the environmental benefits. Thus, for the maintenance problems of wind turbines, an opportunistic maintenance strategy that considers structural correlations, random correlations, and carbon emissions is proposed. First, a Weibull distribution is used to describe the deterioration trend of wind turbine subsystems. The failure rates and reliability of wind turbines are described by the random correlations among all subsystems. Meanwhile, two improvement factors are introduced into the failure rate and carbon emission model to describe imperfect maintenance, including the working-age fallback factor and the failure rate increasing factor. Then, the total expected maintenance cost can be described as the objective function for the proposed opportunistic maintenance model, including the maintenance preparation cost, maintenance adjustment cost, shutdown loss cost, and operation cost. The maintenance preparation cost is related to the economic correlation, and the maintenance adjustment cost is described by using the maintenance probabilities under different maintenance activities. The shutdown loss cost is obtained by considering the structural correlation, and the operation cost is related to the energy consumption of wind turbines. Finally, a case study is provided to analyze the performance of the proposed model. The obtained optimal opportunistic maintenance duration can be used to interpret the structural correlation coefficient, random correlation coefficient, and sensitivity of carbon emissions. Compared with preventive maintenance, the proposed model provides better performance for the maintenance problems of wind turbines and can obtain relatively good solutions in a short computation time.

Keywords: wind turbines; opportunistic maintenance; structural correlation; random correlation; carbon emissions

1. Introduction

With the rapid development of wind-power generation technology, wind turbines have a more comprehensive range of applications as the leading equipment for wind-power generation. However, wind turbines operate in harsh environments, such as in high temperatures, extreme cold, and high altitudes. Compared with other repairable equipment, their operation and maintenance costs are higher, accounting for more than 15% of the life-cycle cost. Thus, it is essential to select a reasonable maintenance strategy to formulate a scientific maintenance plan to reduce the operation and maintenance costs of wind turbines.

Currently, an opportunistic maintenance strategy can be applied to the maintenance of large and complex equipment, such as general aircraft, rail transit trains, and port handling systems [1–3]. For opportunistic maintenance (OM), when a system subsystem is shut down or preventive maintenance needs to be adopted, if other subsystems meet the predetermined maintenance conditions, the opportunity to execute preventive maintenance in advance is obtained. It can dynamically combine the subsystems for preventive maintenance based on the deterioration trend of subsystems to reduce the frequent shutdown of the system and the scheduling of maintenance resources. A wind turbine is also a large complex system, and there are three different correlation relationships among its subsystems, including economic correlations, random failure correlations, and structural correlations. Thomas provides the definition of correlation [4]. Economic correlation refers to the replacement or maintenance costs of several components being less than the sum of their individual replacement or maintenance costs. Random failure correlation means that the failure rate or operation state of other components will be affected to a certain extent when some components fail. Structural correlation implies that some components must be replaced or at least removed in order to replace or repair failed components. Thus, the correlation of subsystems is considered in comparison with traditional maintenance strategies for maintaining a single component. The opportunistic maintenance strategy can better meet the maintenance demands of wind turbines [5].

For the opportunistic maintenance strategy for wind turbines, to study the economic correlation of subsystems [6–10], some references assume that the subsystems are independent. In the actual operation of wind turbines, there is a coupling relationship between subsystems, and failure does not occur independently. Thus, the references further study the random failure correlation among subsystems. Li et al. used the copula function to describe a joint risk model to establish a double-layer optimization model for wind turbine preventive maintenance. The lower-level optimization model that considers the degree of risk of wind turbines related to failure effectively reduces the risk degree in the whole life cycle of wind turbines [11]. Yeter et al. considered the operation state of components in wind turbines. The proportional failure model was used to describe the deterioration process of components and established the state opportunistic maintenance model by setting the state maintenance threshold and opportunistic maintenance threshold [12]. Zhong et al. considered the failure correlation of wind turbines and used the affected degree to describe the impact on one subsystem generated by the failure of other subsystems. The affected degree was also used as the decision-making factor of the maintenance strategy. The results showed that considering the fault correlation can solve over-repair and under-repair problems to a certain extent [13]. The above research considers one or two kinds of wind-turbine subsystem correlations and fails to consider the structural correlation for an opportunistic maintenance strategy for wind turbines. Moreover, research on structural correlation is about mechanical engineering. The literature established the multi-component maintenance model that considers structural correlation from the perspectives of the disassembly sequence, the renewal method of the maintenance plan, and the importance of decision variables [14–16]. However, the research on the opportunistic maintenance strategies that consider structural correlation for wind-turbine maintenance is still in its infancy.

The extension of wind turbines' operation time can reduce the preventive maintenance cost per unit, and the performance can also degrade. A poor operation is accompanied by increased energy consumption, intensified environmental damage, and increased enterprise operation cost. Thus, appropriate maintenance decisions can reduce operation costs and carbon emissions. Minne and Crittenden evaluated the impact of residential floor maintenance activities on the environment [17]. Giustozzi et al. proposed a road maintenance strategy with the optimization objectives of environmental impact, cost, and pavement performance [18]. Noland and Hanson used the life-cycle assessment method to assess the greenhouse gas emissions from road maintenance. The study found that the emissions from maintenance activities accounted for 10% of total emissions [19]. Sikos and Klemes studied the relationship between system reliability, maintainability, and environmental impact, and showed that system reliability and maintainability played important roles in the realization of cleaner production [20]. Liu et al. proposed a remanufacturing maintenance model to minimize the global warming potential based on reliability and replacement theory [21]. Chiara et al. established a regular preventive maintenance model, integrated the circular economy concept, selected the most suitable spare parts for maintenance activities from the perspective of sustainability, and proved the necessity of introducing sustainability considerations into routine maintenance procedures through described cases [22]. Jasiulewicz-Kaczmarek and Gola pointed out that the manufacturing industry was undergoing relevant changes because of the challenges brought by the sustainable economic development model. The maintenance function was also changing its role to better support value creation and expand its attention to environmental and social factors [23]. Hennequin and Ramirez Restrepo proposed a joint production and maintenance strategy model to minimize the impacts on the environment, inventory, and maintenance costs [24]. Afrinaldi et al. proposed a mathematical model to determine the optimal plan for preventive replacement of components and to minimize the economic and environmental impacts of the components [25]. Huang et al. integrated maintenance and energy saving into the same model, and introduced the opportunistic maintenance window mechanism [26]. Xia et al. proposed an oriented energy maintenance framework based on the energy saving windows [27]. The relevant work can be seen in Table 1. Most of the literature only evaluates the impact of maintenance activities on the environment in a specific context, and does not consider the impact of maintenance activities on failure rate and energy consumption, or the impact of recovery on the environment and cost rate.

Table 1. The relevant work.

Articles	Wind Turbines	Correlations			Considerations
		Economic	Random	Structural	
Song et al. [6]	√	√			Farm layout design
Ren et al. [7]	√	√			Carbon emissions
Li et al. [8]	√	√			Multiple-component age
Li et al. [9]	√	√			Maintenance intervals
Zhu et al. [10]	√	√			Logistic delay & weather condition
Li et al. [11]	√		√	√	Failure mode
Zhong et al. [13]	√	√			Reliability & cost
Zhou et al. [14]			√	√	Stochastic failures & disassembly sequence
Iung et al. [15]				√	Multi-dependent components
Wu et al. [16]				√	Component importance
Current study	√	√	√	√	Carbon emissions

To devise more accurate and practical maintenance strategies for wind turbines, this paper's contribution is to propose an opportunistic maintenance (OM) model that considers economic correlations, random correlations, and structural correlations among wind-turbine subsystems and carbon emissions. Random correlation describes the failure rate and reliability of wind turbines among the subsystems. The improvement factor is introduced into the failure rate and carbon emission model to describe imperfect maintenance. Moreover, the operation energy consumption of wind turbines increases with performance degradation. This paper further considers the reduction effect of wind-turbine recovery on costs and emissions. The benefits of wind-turbine recovery and the emissions of maintenance activities can be introduced into the proposed model by adopting the dynamic failure rate function and carbon emission function. The total expected maintenance cost can then be described as the objective function for the proposed opportunistic maintenance model, including the maintenance preparation cost, maintenance adjustment cost, shutdown loss cost, and operation cost. The operation cost is related to the energy consumption of wind turbines. Finally, a case study is provided to analyze the performance of the proposed model. The obtained optimal opportunistic maintenance interval can be used to interpret the structural correlation coefficient, random correlation coefficient, and sensitivity of carbon emissions. Compared with preventive maintenance, the proposed model provides better performance with wind-turbine maintenance problems and can provide relatively good solutions in a short computation time.

2. The Notation and Problem Description

2.1. Notation

β, α: Weibull distribution parameters
$\chi_{i,j}$: Random correlation coefficient
χ: Random correlation coefficient matrix
λ: Structural correlation coefficient matrix
C_{mm}: Total minor maintenance cost
C_{om}: Total opportunistic maintenance cost
C_{pm}: Total preventive maintenance cost
$c_{i,mm}$: Minor maintenance cost of subsystem i
$c_{i,om}$: Opportunistic maintenance cost of subsystem i
$c_{i,pm}$: Preventive maintenance cost of subsystem i
c_l: Downtime loss per unit
T: Life cycle
EC: Total expected maintenance cost
EC_F: Total expected maintenance preparation cost
EC_L: Total expected downtime loss
EC_M: Total expected maintenance adjustment cost
$f_i(t)$: Probability density distribution function of subsystem i
$g_i(t_i^o, t)$: Opportunistic maintenance probability density function of subsystem i
$h_i(t)$: Probability density function of subsystem i
$p_i(m)$: Minor maintenance probability of subsystem i
$p_i(o)$: Opportunistic maintenance probability of subsystem i
$p_i(p)$: Preventive maintenance probability of subsystem i
$R_i(t)$: Reliability function
t_i^o: Opportunistic maintenance time threshold
t_i^p: Preventive maintenance time threshold
t^d: Downtime
t^m: Maintenance time
t^{set}: Maintenance preparation time
w: Opportunistic maintenance interval

2.2. Problem Description

The extension of wind turbine operation time will reduce the preventive maintenance cost per unit, but the performance of the wind turbine will also degrade. For the life cycle of the wind turbine, the probability of unexpected failure becomes lower with an increase of the amount of preventive maintenance, and the amount of minor maintenance decreases. Most preventive maintenance (PM) strategies are undertaken, and the cost will increase. The preventive maintenance cost is inversely proportional to the minor maintenance cost, as shown in Figure 1. The extension of wind turbine operation time will increase energy consumption, environmental damage, and operation cost, as shown in Figure 2.

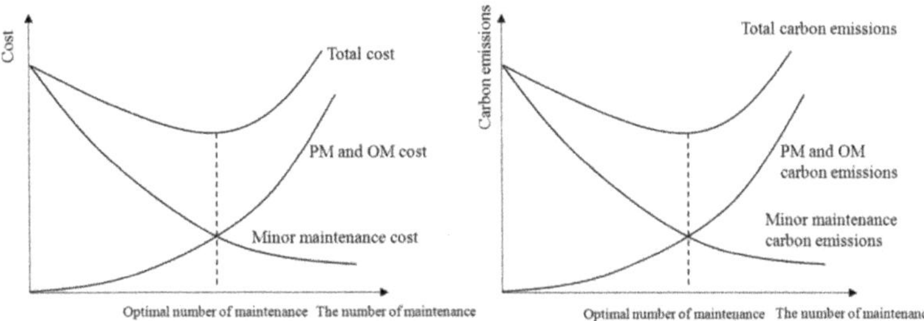

Figure 1. Cost and carbon emission change with the amount of maintenance.

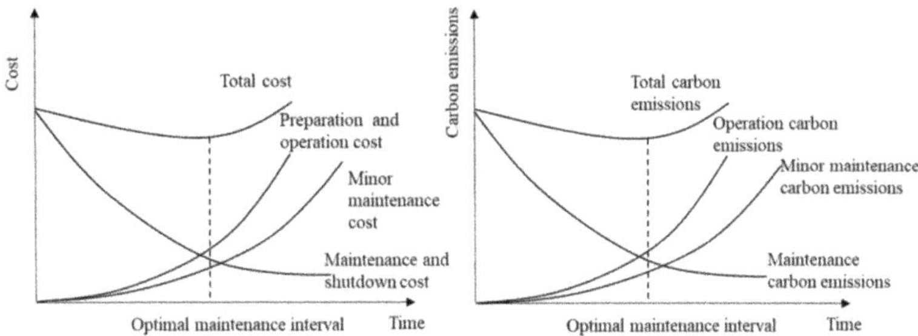

Figure 2. Cost and carbon emission change with the interval of maintenance.

The wind turbine is taken as the research object, and the optimization goal is to minimize the expected total cost by considering the carbon emissions per unit. When a wind turbine has a sudden failure, it is necessary to carry out minor maintenance to ensure regular operation. Minor maintenance can only restore the function of the wind turbine to the state before the failure, without changing the failure rate and emission rate. Actually, wind turbine maintenance cannot affect a permanent repair as good as new, and the improvement factor is introduced to describe the impact of the maintenance actions on the wind turbine. The wind turbine consists of subsystems, and when preventive maintenance is applied to one subsystem, opportunistic maintenance can be applied to other subsystems.

Let t_i^p denote the preventive maintenance time threshold of wind turbine subsystem i, and t_i^o the opportunistic maintenance time threshold. Thus, the opportunistic maintenance strategy of the wind turbine is shown in Figure 3.

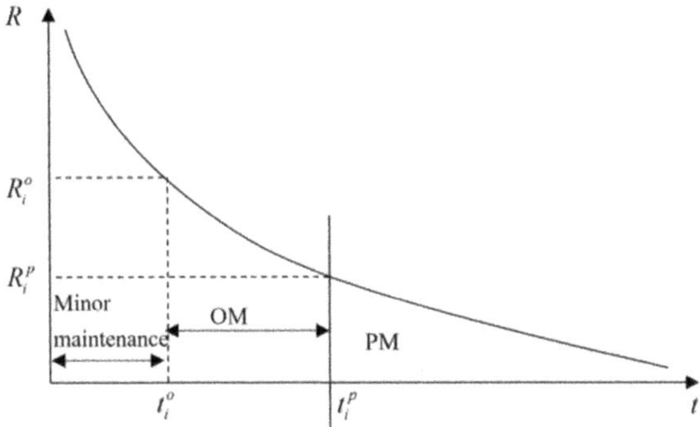

Figure 3. Opportunistic maintenance strategies of wind turbines.

Minor maintenance can be adopted for all failures of wind turbines before executing preventive maintenance and restore their operation state to that of before the fault. The instantaneous fault rate will not change. Preventive maintenance adopts the maintenance mode of replacement, which is regarded as a repair opportunity. If other subsystems meet the opportunistic maintenance conditions, they can be repaired simultaneously to the preventive maintenance of the subsystems when the wind turbine is shut down. Opportunistic maintenance can share maintenance resources during downtime and effectively reduce maintenance costs.

When the working age of subsystem i meets the condition $0 < t < t_i^p$, if the subsystem i fails, minor repairs can be undertaken. This kind of maintenance aims to restore the state of the subsystem without changing its instantaneous failure rate. When the working age of subsystem i meets the preventive maintenance condition $t \geq t_i^p$, replacement can be undertaken to maintain subsystem i, and it can restore it to its initial health state.

Additionally, during the wind-turbine shutdown, other subsystems obtain maintenance opportunities. When the working age meets the condition $t_i^o < t < t_i^p$, the conditions of opportunistic maintenance are met, and all parts required for opportunistic maintenance during one shutdown are recorded in the set $O, O = \{1, 2, \ldots, n\}$. Replacement can be undertaken to maintain subsystem i, and restore it to its initial health state.

3. The Model

3.1. Reliability Evolution

The traditional failure rate model usually takes the independence of component or system failure as the premise, and usually uses two parameters of a Weibull distribution to describe its degradation trend. The failure rate function and reliability function of subsystem i are respectively expressed as:

$$h_i^I(t) = \frac{\beta_i}{\alpha_i} \left(\frac{t}{\alpha_i} \right)^{\beta_i - 1} \tag{1}$$

$$R_i^I(t) = \exp \left[-\int_0^t h_i^I(t)dt \right] \tag{2}$$

where β_i and α_i are the shape parameter and scale parameter of each subsystem, respectively, and $\beta_i > 0, \alpha_i > 0$.

During actual operation, there is a coupling relationship among subsystems, and their faults are not independent of each other. Murthy et al. proposed three types of random failure correlations [28,29]. Type I random failure correlation indicates that if a component in the system fails, it will lead to the failure of other internal components according to a certain probability. Type II random failure correlation indicates that one component failure in the system will affect other internal components' failure rates to some extent. Type III is related to impact damage, specifically indicating that one component failure in the system will cause random damage to other internal components. The components will fail after the damage accumulates to a certain extent. In this paper, random correlation coefficient $\chi_{i,j}$ is introduced to describe the type II random failure correlation. Thus, the comprehensive instantaneous failure rate function of the subsystem can be expressed as:

$$h_i(t) = h_i^I(t) + \sum_j \chi_{i,j} h_j(t) \tag{3}$$

where, $i, j = 1, 2 \ldots n, i \neq j, \chi_{i,j} \in [0, 1]$. $\chi_{i,j} = 0$ indicates that there is no random failure correlation between two subsystems, and $h_i(t) = h_i^I(t)$. $\chi_{i,j} = 1$ indicates that the two subsystems are completely related, and the failure of one subsystem will inevitably lead to the failure of the other subsystem.

Based on the failure transfer relationship among subsystems in a wind turbine, the failure transfer relationship diagram is shown as Figure 4. The matrix χ is used to record the quantized random correlation intensity coefficient $\chi_{i,j}$.

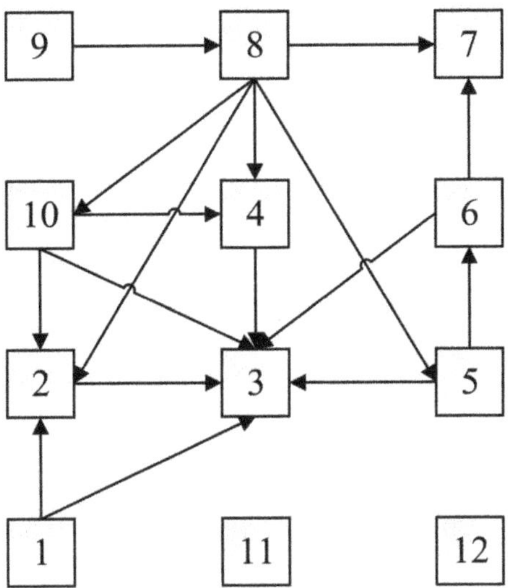

Figure 4. Fault transmission diagram of wind turbines.

$$\chi = \begin{bmatrix} 0 & 0 & 0 & 0 & 0 & 0 & 0 & 0 & 0 & 0 & 0 & 0 \\ \chi_{2,1} & 0 & 0 & 0 & 0 & 0 & 0 & \chi_{2,8} & 0 & \chi_{2,10} & 0 & 0 \\ \chi_{3,1} & \chi_{3,2} & 0 & \chi_{3,4} & \chi_{3,5} & \chi_{3,6} & 0 & 0 & 0 & \chi_{3,10} & 0 & 0 \\ 0 & 0 & 0 & 0 & 0 & 0 & 0 & \chi_{4,8} & 0 & \chi_{4,10} & 0 & 0 \\ 0 & 0 & 0 & 0 & 0 & 0 & 0 & \chi_{5,8} & 0 & 0 & 0 & 0 \\ 0 & 0 & 0 & 0 & \chi_{6,5} & 0 & 0 & 0 & 0 & 0 & 0 & 0 \\ 0 & 0 & 0 & 0 & 0 & \chi_{7,6} & 0 & \chi_{7,8} & 0 & 0 & 0 & 0 \\ 0 & 0 & 0 & 0 & 0 & 0 & 0 & 0 & \chi_{8,9} & 0 & 0 & 0 \\ 0 & 0 & 0 & 0 & 0 & 0 & 0 & 0 & 0 & 0 & 0 & 0 \\ 0 & 0 & 0 & 0 & 0 & 0 & 0 & \chi_{10,8} & 0 & 0 & 0 & 0 \\ 0 & 0 & 0 & 0 & 0 & 0 & 0 & 0 & 0 & 0 & 0 & 0 \\ 0 & 0 & 0 & 0 & 0 & 0 & 0 & 0 & 0 & 0 & 0 & 0 \end{bmatrix}$$

Based on Equations (2) and (3), the reliability of the subsystem under the influence of comprehensive failure can be calculated. Thus, the comprehensive reliability function of the subsystem i is obtained.

$$R_i(t) = \exp\left[-\int_0^t h_i(t)dt\right] = R_i^I(t) \prod_j [R_j(t)]^{\chi_{i,j}} \tag{4}$$

3.2. Improvement Factor

By adopting maintenance, the working age of a wind turbine will fall back to a previous stage. The fallback degree is related to maintenance cost and the amount of maintenance undertaken. The working age fallback factor is used to denote the effect of maintenance on the working age of wind turbines, and it can be obtained for subsystem i of a wind turbine.

$$\eta_l = \left(a\frac{c_{i,om}}{c_{i,pm}}\right)^{bl} \tag{5}$$

where, l represents l-th maintenance interval. a is the cost adjustment coefficient, and it is used to adjust the opportunistic maintenance cost rate, $0 \le a \le c_{i,pm}/c_{i,om}$. The opportunistic maintenance cost is higher, and the maintenance effect is better. $c_{i,om}$ is the opportunistic maintenance cost, $c_{i,pm}$ is the preventive replacement cost, and $b > 1$ represents the time adjustment function. $0 \leq \eta_l \leq 1$ and $\eta_l = 0$ represent minor maintenance and $\eta_l = 1$ shows preventive replacement. η_l is larger or smaller, and it indicates whether the working age fallback after OM is more or less. When other parameters are determined, η_l changes dynamically with the number of maintenance times. If the maintenance interval of a wind turbine is T, the effective working age of subsystem i after the first OM can be expressed as:

$$L_1^- = T_1 L_1^+ = t + (1 - \eta_1)T_1 \tag{6}$$

t is the time. In an l-th opportunistic maintenance interval, the value range of t is $t \in (0, T_l)$. By adopting OM, the working age falls back $\eta_1 T_1$, and the effective working age becomes $t + (1 - \eta_1)T_1$ after maintenance. It can then be deduced that the effective working age before and after the second OM can be expressed as:

$$L_2^- = L_1^+ + T_2 = t + (1 - \eta_1)T_1 + T_2 L_2^+ = t + (1 - \eta_1)T_1 + (1 - \eta_2)T_2 \tag{7}$$

The effective working age before and after $l - th$ $(l \geq 2)$ maintenance can be seen in Figure 5 and expressed as:

$$L_l^- = t + \sum_{j=1}^{l-1}(1 - \eta_j)T_j + T_l L_l^+ = t + \sum_{j=1}^{l}(1 - \eta_j)T_j \tag{8}$$

where, L_l^- represents the working age of subsystem i before $l - th$ maintenance and L_l^+ represents the working age of subsystem i after $l - th$ maintenance.

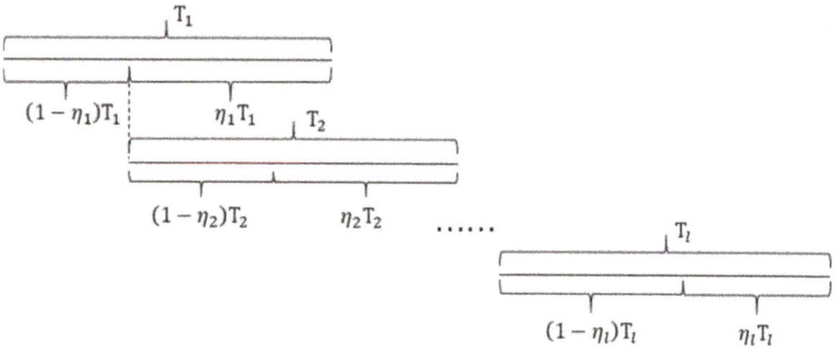

Figure 5. Changes of effective working age of subsystem i before and after maintenance.

The maintenance will increase the working age of subsystem i, and change its failure rate curve. The failure rate increasing factor is used to express the impact of maintenance on the failure rate curve for subsystem i. The relationship between the failure rate functions before and after maintenance for subsystem i is as follows:

$$h_{l+1}(t) = \gamma_l h_l(t) = \prod_{j=1}^{l} \gamma_j h_1\left(t + \sum_{j=1}^{l}(1 - \eta_j)T_j\right), t \in (0, T_{l+1}) \tag{9}$$

where, γ_l is the change rate of the failure rate of subsystem i after adopting l-th maintenance, $\gamma_l \geq 1$. The failure rate curve changes under the action of different improvement factors, as shown in Figure 6. The red line describes the failure rate curve that considers the failure rate increasing factor. The blue line describes the failure rate curve that considers the working age fallback factor. The green line describes the failure rate curve that considers the working age fallback factor and the failure rate increasing factor.

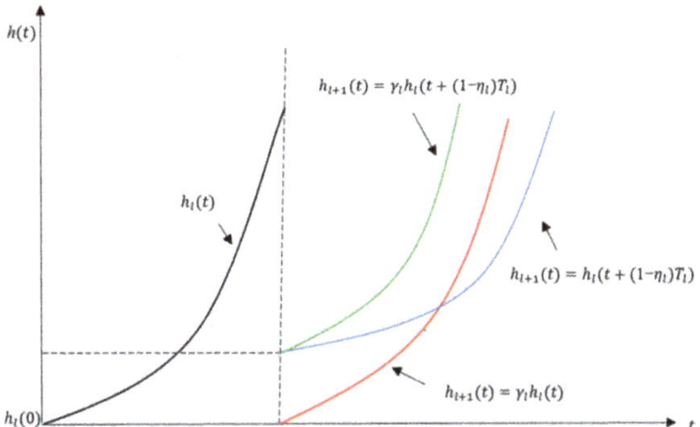

Figure 6. Change of failure rate curve before and after maintenance.

Considering the reliability limit R_{min} of subsystem i of a wind turbine, when the reliability is lower than R_{min}, maintenance is undertaken to restore the performance of subsystem i, and the reliability function can be obtained as follows:

$$\exp\left[-\int_0^{T_1} h_1(t)dt\right] = \exp\left[-\int_0^{T_2} h_2(t)dt\right] = \cdots = \exp\left[-\int_0^{T_N} h_N(t)dt\right] = R_{min} \quad (10)$$

Thus, the amount of minor maintenance for subsystem i in each maintenance interval is $-\ln R_{min}$, and each maintenance interval $[T_1, T_2, \cdots, T_N]$ can be obtained by solving Equation (10).

3.3. Carbon Emission Model

The impact on the environment during the life cycle of a wind turbine mainly consists of carbon emissions from the energy consumption of using and recycling, and the recovery is considered to reduce carbon emissions. Carbon emissions during the life cycles of wind turbines are mainly generated by the energy consumed by operation and maintenance, construction/installation, and the decommissioning of wind turbines. Thus, the carbon emissions during the wind turbine life cycle can be expressed as follows:

$$GT = G_{use}X + (N + N_c)G_m + \left(1 - \delta^{N-1}\right)G_p \quad (11)$$

where G_{use} is the carbon emission generated by consuming an unit energy, X is the total energy consumed during wind turbine operation, N is the maintenance cycle, N_c is the expected amount of minor maintenance undertaken, and G_m is the carbon emission generated by undertaking one maintenance activity. Assuming that the carbon emissions generated by minor maintenance, OM and PM, are the same, G_m is the same for each maintenance activity, δ^{N-1} is the recovery factor, and G_p is the carbon emission generated by manufacturing a piece of a wind turbine.

$$X = \int_0^{T_1} x_1(t)dt + \int_0^{T_2} x_2(t)dt + \cdots + \int_0^{T_N} x_N(t)dt = \sum_{l=1}^{N} \int_0^{T_l} x_l(t)dt \quad (12)$$

$$N_c = \int_0^{T_1} \lambda_1(t)dt + \int_0^{T_2} \lambda_2(t)dt + \cdots + \int_0^{T_N} \lambda_N(t)dt = \sum_{l=1}^{N} \int_0^{T_l} h_l(t)dt \quad (13)$$

where, $x_l(t)$ is the energy consumption function of wind turbine operation. It can be obtained as follows:

$$x_1(t) = yt + z \quad (14)$$

where, y and z are the parameters of the wind turbine energy consumption function. The maintenance activities on wind turbines can restore performance and also change carbon emissions. Thus, the improvement factor can be introduced into the energy consumption function, and this is more in line with the actual situation. Before and after l-th maintenance, the relationship between energy consumption functions can be obtained as follows:

$$x_{l+1}(t) = \prod_{j=1}^{l} \gamma_j x_1 \left(t + \sum_{j=1}^{l} (1 - \eta_j) T_j \right) = \prod_{j=1}^{l} \gamma_j \left(y \left(t + \sum_{j=1}^{l} (1 - \eta_j) T_j \right) + z \right) \tag{15}$$

The carbon emissions per unit during a wind turbine life cycle can be expressed as:

$$
\begin{aligned}
GWP = \frac{GT}{\sum_{l=1}^{N} T_l} &= \frac{G_{use} X + G_m (N + N_c) + G_P \left(1 - \delta^{N-1}\right)}{\sum_{l=1}^{N} T_l} \\
&= \frac{G_{use} \left(\sum_{l=1}^{N} \int_0^{T_l} x_l(t) dt \right) + G_m \left(N + \sum_{l=1}^{N} \int_0^{T_l} \lambda_l(t) dt \right) + G_P \left(1 - \delta^{N-1}\right)}{\sum_{l=1}^{N} T_l}
\end{aligned}
\tag{16}
$$

3.4. Expected Total Cost Model

Under the opportunistic maintenance strategy, the expected total maintenance cost of a wind turbine can be described as the objective function. For each maintenance cycle, the total maintenance cost is determined by the maintenance preparation cost C_F, maintenance operation cost TOC, maintenance adjustment cost C_M, and downtime cost C_L. The maintenance cost expectation is expressed as:

$$EC = EC_F + TOC + EC_M + EC_L \tag{17}$$

3.4.1. Maintenance Preparation Cost and Operation Cost

Before subsystem maintenance, maintenance personnel, vehicles, tower cranes, and other equipment are required to implement maintenance actions. Maintenance preparation costs mainly include labor services, vehicle transportation, and equipment rental fees. Assuming subsystem $i \in M$, M is the number of all the subsystems that require to be maintained in a maintenance activity. For the traditional maintenance strategy, the maintenance preparation cost depends on the total cost of the required maintenance subsystem, $EC_F = \sum_{i \in M} C_i^0$. Under the opportunistic maintenance strategy, the economic correlation is considered when the subsystems are maintained. When multiple subsystems are maintained together, they share maintenance resources and allocate maintenance preparation costs. The expression of the maintenance preparation cost can be shown as follows:

$$EC_F = C_i^0 \tag{18}$$

The wind-turbine performance degradation will increase its operation cost. The operation cost is related to its energy consumption. Let C_e denote the consumption cost, and the total operation cost can be shown as follows:

$$TOC = C_e X = C_e \sum_{l=1}^{N} \int_0^{T_l} x_l(t) dt \tag{19}$$

3.4.2. Maintenance Adjustment Cost

In order to restore the subsystem to the normal operation state, different maintenance activities will produce different maintenance costs by implementing the maintenance strategy. These costs are collectively referred to as the maintenance adjustment cost, which include minor maintenance costs, preventive replacement costs, and opportunistic replacement costs. The maintenance probability is used to calculate the maintenance adjustment cost.

$p(m)$ indicates the probability for minor maintenance when the subsystem fails before preventive maintenance. $p(p)$ indicates the probability that subsystem k, which possibly executed PM, has no failure in interval $t \in \left(t_k^o, t_k^p\right)$, and subsystem i $(i \neq k)$, which obtained OM, has good performance in this interval and does not need maintenance. $p(o)$ indicates the probability that subsystem k, which possibly executed PM in interval $t \in \left(t_k^o, t_k^p\right)$, is in good condition, but the subsystem r $(r \neq i \neq k)$ experiences failure in this interval and needs to be replaced. The opportunistic maintenance time t_i^o of subsystem i approximately obeys the exponential distribution, and its probability density function is:

$$g_i\left(t_i^o, t\right) = \frac{1}{\alpha_i} \exp\left[-\frac{t - t_i^o}{\alpha_i}\right] \tag{20}$$

The expectation function of the maintenance adjustment cost is obtained as follows:

$$EC_M = E\left(C_{mm} + C_{pm} + C_{om}\right) = \sum_{i=1}^{n}\left[c_{i,mm}p_i(m) + c_{i,om}p_i(o) + c_{i,pm}p_i(p)\right] \tag{21}$$

The maintenance probability $p_i(m)$, $p_i(p)$, $p_i(o)$ can be shown as follows:

$$p_i(m) = \begin{cases} \int_0^{t_k} f_i(t)dt, i = k \\ \int_{t_k}^{t_k^p} f_k(t) \prod_i \left[1 - \int_{t_i}^{t_k^p} g_i(t_i, u)du\right]dt, i \in O \end{cases}$$

$$p_i(o) = \int_{t_k}^{t_k^p}\left[\left(1 - \int_{t_i}^{t} f_k(u)du\right)g_r\left(t_r^o, t\right)\prod_{i \in O, i \neq r}\left(1 - \int_{t_i}^{t_k^p} g_i(t_i, u)du\right)\right]dt \tag{22}$$

$$p_i(p) = \left[1 - \int_{t_i}^{t_k^p} f_k(t)dt\right]\prod_{i \in O}\left[1 - \int_{t_i}^{t_k^p} g_i(t_i, t)dt\right]$$

3.4.3. Shutdown Loss Cost

The preventive replacement and opportunistic maintenance of a subsystem require the shutdown of a wind turbine to complete the maintenance activities. Thus, the wind turbine cannot generate power normally during this period, so the shutdown loss cost is calculated into the total maintenance cost. The downtime loss cost is equal to the loss cost per unit multiplied by the downtime, and can be obtained as follows:

$$C_L = c_l t^d = c_l\left(t^{set} + t^m\right) \tag{23}$$

For the maintenance process of wind turbines, the structural correlation is mainly demonstrated when multiple subsystems are maintained together, and one subsystem's disassembly and installation process will affect other subsystems to a certain degree. The lost time of subsystem maintenance equals the summation of preventive maintenance t_k^m and opportunistic maintenance t_i^m. The lost time of opportunistic maintenance is calculated by considering the structural correlation, and the subsystem that takes the longest time for opportunistic maintenance can be selected. The structure correlation matrix λ is used to represent the strong relationships among the structures of subsystems. Thus, t^m can be expressed as:

$$\lambda = \begin{bmatrix} \lambda_{1,1} & \lambda_{1,2} & \cdots & \lambda_{1,i} \\ \lambda_{2,1} & \lambda_{2,2} & \cdots & \lambda_{2,i} \\ \vdots & \vdots & \ddots & \vdots \\ \lambda_{i,1} & \lambda_{i,2} & \cdots & \lambda_{i,i} \end{bmatrix} \quad t^m = t_k^m + \lambda_{k,i}\max\{t_i^m\}, i \in O \tag{24}$$

where structural correlation coefficient $\lambda_{i,j} \in [0,1), \lambda_{i,j} = \lambda_{j,i}$. By combining with Equations (23) and (24), the expected shutdown loss cost can be obtained as follows:

$$EC_L = E\left(c_l t^d\right) = c_l\left[t^{set} + t_k^m p_k(p) + \lambda_{k,i}\max\{t_i^m\}p_i(o)\right] \tag{25}$$

Assuming that there are N maintenance cycles in the whole life cycle T, the expected total maintenance cost in the whole life cycle can be obtained as follows:

$$
\begin{aligned}
EC &= \sum_{l=1}^{N} EC^{(l)} = \sum_{l=1}^{N}\sum_{i=1}^{M}\left[C_i^0 + \left(c_{i,mm}p_i(m) + c_{i,om}p_i(o) + c_{i,pm}p_i(p)\right) + c_l t^d\right] + TOC \\
&= \sum_{l=1}^{N}\sum_{i=1}^{M}
\begin{bmatrix}
C_i^0 + \left(c_{k,mm}\int_0^{t_k} f_k(t)dt + c_{i,mm}\int_{t_k}^{t_k^p} f_k(t)\prod_i\left[1 - \int_{t_i}^{t_k^p} g_i(t_i,u)du\right]dt\right) \\
+c_{i,om}\int_{t_k}^{t_k^p}\left[\left(1 - \int_{t_i}^{t} f_k(u)du\right)g_r(t_r^o,t)\prod_{i\in O, i\neq r}\left(1 - \int_{t_i}^{t_k^p} g_i(t_i,u)du\right)\right]dt \\
+c_{i,pm}\left[1 - \int_{t_k}^{t_k^p}[f_k(t)]dt\right]\prod_{i\in O}\left[1 - \int_{t_i}^{t_k^p} g_i(t)dt\right] \\
+c_l\left[t^{set} + t_k^m p_k(p) + \lambda_{k,i}\max\{t_i^m\}p_i(o)\right]
\end{bmatrix}
+ TOC
\end{aligned}
\tag{26}
$$

4. The Procedure of Solving Model

In the opportunistic maintenance model, when the opportunistic maintenance threshold t_i^o is being changed, subsystem $i \in O$ meeting the opportunistic maintenance conditions will change, directly affecting the maintenance plan of the next cycle, and dynamically changing the whole maintenance process and the total maintenance cost. It is assumed that the length of the opportunistic maintenance interval of each subsystem is the same, and this lets w denote the length of the opportunistic maintenance interval, $t_i^o = t_i^o - w$. In $w \in \left[0, \min\left\{t_i^p\right\}\right]$, the total maintenance cost EC under the opportunistic maintenance threshold is minimized by traversing w. w^* is the optimal opportunistic maintenance duration and $t_i^{o*} = t_i^p - w^*$ is the optimal opportunistic maintenance threshold. Figure 7 is the solution flow chart, k describes the subsystem to possibly receive PM, and i $(j \neq k)$ describes the subsystem to possibly receive OM simultaneously. The specific calculation steps are as follows:

Step 1. Input the known condition, Weibull distribution parameters, failure correlation coefficient, and maintenance cost.

Step 2. Judge whether the running time t reaches the preventive maintenance threshold $t_k^p = \min\left\{t_i^p\right\}$, $i \in O$. If $t < t_k^p$, update the run time $t = t+1$ and cycle the step until the preventive maintenance conditions are met.

Step 3. For other subsystems $i(i \neq k)$ that have not experienced failure, calculate t_i^o and judge whether the subsystem needs maintenance based on the value of w. If $t_i^o < t < t_i^p$, subsystem i meets the opportunistic maintenance conditions, and group maintenance is performed on subsystems i and k. If the opportunistic maintenance conditions are not met by i, subsystem k shall receive preventive maintenance separately. Calculate maintenance probability $p(m), p(p), p(o)$ and maintenance cost expectation $EC^{(l)}$ of the l-th maintenance cycle. Update $T_l = t_k^p + t^d$.

Step 4. If $\sum_{l=1}^{N} T_l < T$, return to Step2 for the next cycle of maintenance. Until $\sum_{l=1}^{N} T_l \geq T$. Exit the cycle.

Step 5. Calculate the expected total maintenance cost EC in the whole life cycle, determine the optimal opportunistic maintenance interval w^*, and the optimal opportunistic maintenance threshold t_i^{o*}.

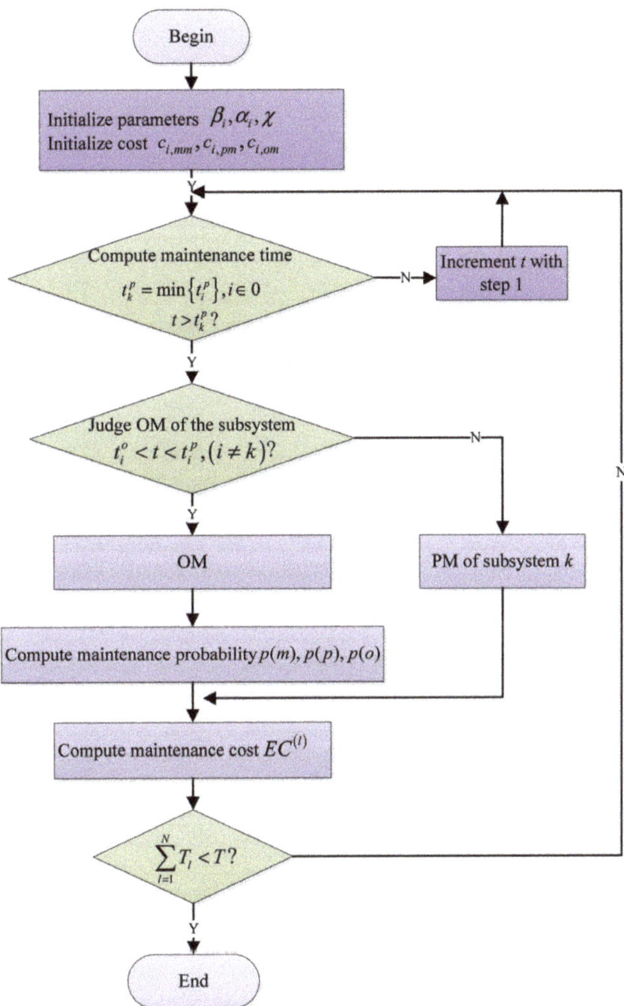

Figure 7. Solution flow chart.

5. Case Study

5.1. Data Preparation

This paper selects the same type of wind turbine in a wind farm for case analysis. The key components of the subsystem are used to analyze the impeller system, spindle system, gearbox, and generator, numbered 1–4. The specified service life of such wind turbines is 15 years. The Weibull distribution parameters, single maintenance cost, and maintenance time of the subsystem are shown in Table 2.

Table 2. Maintenance parameters.

No. i	Subsystem	Weibull Distribution Parameters		Maintenance Cost (Unit: 10,000 yuan)					Maintenance Time (Unit: h)	
		α_i	β_i	C_0	c_{mm}	c_{pm}	c_{om}	c_l	t^{set}	t_i^m
1	Impeller system	3000	3			63.2	63.2			240
2	Spindle system	3750	2			65.8	65.8			480
3	Gearbox	2400	3	3.5	0.65	180.0	180.0	0.24	2	360
4	Generator	3300	2			75.7	75.7			168

The joint risk degree among components is used to describe the correlation degree of random failure among components. The random correlation coefficient matrix χ is obtained by combining the failure correlation coefficient and the wind turbine failure transmission diagram (Figure 4).

$$\chi = \begin{bmatrix} 0 & 0 & 0 & 0 \\ 0.031 & 0 & 0 & 0 \\ 0.042 & 0.13 & 0 & 0.11 \\ 0 & 0 & 0 & 0 \end{bmatrix} \tag{27}$$

Based on the reliability requirements for failure-free operation of wind turbines and referring to the reliability image of subsystems (see Figure 8), in Figure 8, the unit of values in the abscissa is a day. The preventive maintenance time threshold of the subsystem is obtained under this requirement (see Table 3).

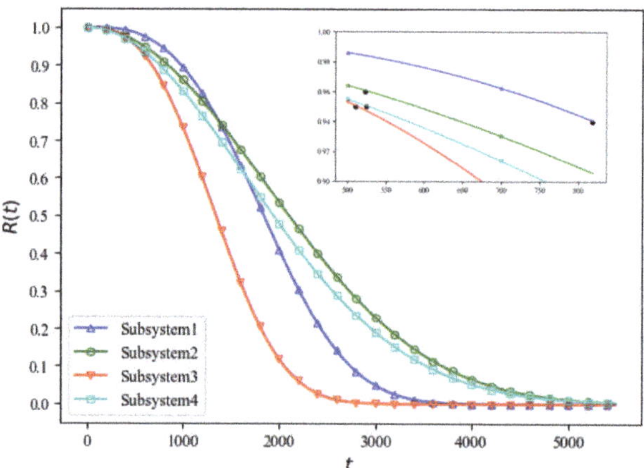

Figure 8. Reliability of four subsystems.

Table 3. Preventive maintenance reliability threshold and its corresponding time threshold.

Subsystem No. i	1	2	3	4
R_i^p	0.94	0.95	0.96	0.96
t_i^p (Unit: day)	823	536	619	529

5.2. Comparison of PM and OM

Compared with a preventive maintenance model based on the three-stage time delay theory proposed in the literature [30], opportunistic maintenance mainly considers the economic correlation among subsystems. The maintenance cost decreases with the reduction of downtime. Table 4 gives the maintenance results of wind turbines under two maintenance strategies: traditional preventive maintenance and opportunistic maintenance.

Under the preventive maintenance strategy, the total number of wind turbine shutdowns in the whole life cycle is thirty-four. In the opportunistic maintenance strategy, the total number of wind turbine downtimes is fifteen, and the number of wind turbine downtimes is reduced by nineteen, accounting for about 55.88%.

Table 4. Comparison of maintenance results between PM [30] and OM.

Cycle	PM [30]		OM(w = 220)		
	t (Unit: day)	Subsystem i	t (Unit: day)	PM Subsystem k	OM Subsystem i
1	529	4	529	4	2
2	536	2	619	3	1
3	619	3	1058	4	2, 3
4	823	1	1442	1	2, 3, 4
5	1058	4	1971	4	2
6	1072	2	2061	3	1
7	1238	3	2500	4	2, 3
8	1587	4	2884	1	2, 3, 4
⋮	⋮	⋮	—	—	—
15	2476	3	5384	4	2, 3
⋮	⋮	⋮	—	—	—
34	5360	2	—	—	—

Figure 9 shows the changing trend of total maintenance costs with increasing opportunistic maintenance interval w. In Figure 9, the unit of values in the abscissa is a day, and the unit of values in the ordinate is 10,000 yuan. For $w = 0$, opportunistic maintenance is not considered. Thus, frequent downtime leads to high maintenance preparation costs, and the expected value of the total maintenance cost is $EC = 6244.01$ (Unit: 10,000 yuan). After considering opportunistic maintenance, the probability of subsystems increases with the increase of w, and the cost of opportunistic maintenance increases continuously. Due to the reduction of downtime, the maintenance preparation cost and downtime loss cost are greatly reduced in opportunistic maintenance. For $w^* = 220$ (unit: day), the optimal value of the objective function is 2547.98 (unit: 10,000 yuan). Finally, compared with not considering opportunistic maintenance, the expected total maintenance cost can be decreased by 3696.03 (unit: 10,000 yuan), accounting for about 59.19%.

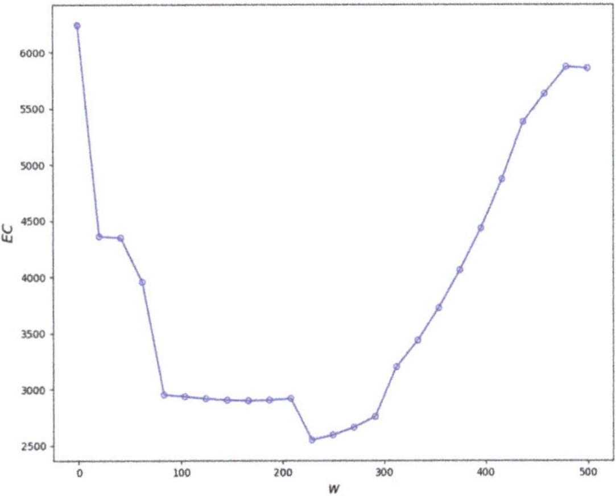

Figure 9. Expected cost under the opportunistic maintenance strategy.

5.3. OM Considering Structural Correlation and Random Correlation

For opportunistic maintenance, the impact of random correlation on wind turbine maintenance activities is considered. Type II random correlation accelerates the failure speed of subsystems to varying degrees. The preventive maintenance time threshold of a subsystem to meet the requirements of operation reliability will be affected by the random correlation coefficient and change. Table 5 shows the preventive maintenance time threshold calculated after considering the random correlation. Under the influence of subsystems 1, 2, and 4, subsystem 3 is 105 days ahead of that without considering the random correlation, becoming the subsystem that takes the lead in meeting the preventive maintenance reliability requirements, and the maintenance plan changes accordingly. Compared with the opportunistic maintenance strategy without considering random correlation, taking the maintenance plan when the length of opportunistic maintenance interval is 220 as an example (see Table 6), the expected total maintenance cost is 2888.51 (unit: 10,000 yuan). Due to the transformation of the maintenance action of subsystem 3 from OM to PM, the maintenance cost increases by 340.53.

Table 5. PM time threshold with the consideration of stochastic dependence.

Subsystem No. i	1	2	3	4
t_i^p (Unit: day)	823	530	514	529

Table 6. Maintenance results under OM considering stochastic dependence ($w = 220$).

Cycle	t (Unit: day)	PM Subsystem k	OM Subsystem i	Cycle	t (Unit: day)	PM Subsystem k	OM Subsystem i
1	514	3	2, 4	9	3229	1	3
2	823	1	3	10	3449	4	2
3	1043	4	2	11	3788	3	2, 4
4	1382	3	2, 4	12	4052	1	—
5	1664	1	—	13	4307	3	2, 4
6	1901	3	2, 4	14	4812	3	1, 2, 4
7	2406	3	2, 3, 4	15	5326	3	2, 4
8	2920	3	2, 4	—	—	—	—

Under OM by considering random correlation, when the opportunistic maintenance time interval is 240 (unit: day), the objective function obtains the optimal value 2576.13 (unit: 10,000 yuan) (see Figure 10). In Figure 10, the unit of values in the abscissa is a day, and the unit of values in the ordinate is 10,000 yuan. On this basis, the structural correlation among subsystems is further considered. For $w = 240$ (unit: day), There are two types of maintenance classification: $M = \{3, 2, 4\}$, PM for subsystem 3, and OM for subsystems 2 and 4, and $M = \{1, 2, 3, 4\}$, PM for subsystem 1, OM for subsystems 2, 3, and 4. Thus, when k executed PM equals 3 or 1, $\max\{t_i^m\} = t_2^m$. The structural correlation coefficients $\lambda_{3,2}$ and $\lambda_{1,2}$, and the impact on shutdown loss cost need to be considered.

$\lambda_{3,2}$ and $\lambda_{1,2}$ are taken in sequence in units of 0.1 in the interval $[0, 1]$ to calculate the shutdown loss cost. When the structural correlation coefficients are all 0, this means that only random correlation and economic correlation are considered. The shutdown loss cost is 993.24 (unit: 10,000 yuan). It can be seen from Table 7 that the shutdown loss cost decreases with the increase of the structural correlation coefficient. When $\lambda_{3,2}$ and $\lambda_{1,2}$ equal 0.9, the shutdown loss cost is 864.04 (unit: 10,000 yuan), which is reduced by 129.2 (unit: 10,000 yuan) compared with without considering the structural correlation. For $\lambda_{1,2} = 0.9, \lambda_{3,2} = 0$, the shutdown loss cost is reduced by 71.5 (unit: 10,000 yuan). For $\lambda_{1,2} = 0, \lambda_{3,2} = 0.9$, the shutdown loss cost is reduced by 57.7 (unit: 10,000 yuan). This shows that the structural correlation between subsystem 2 and subsystem 1 has a greater impact on the loss cost of a wind turbine.

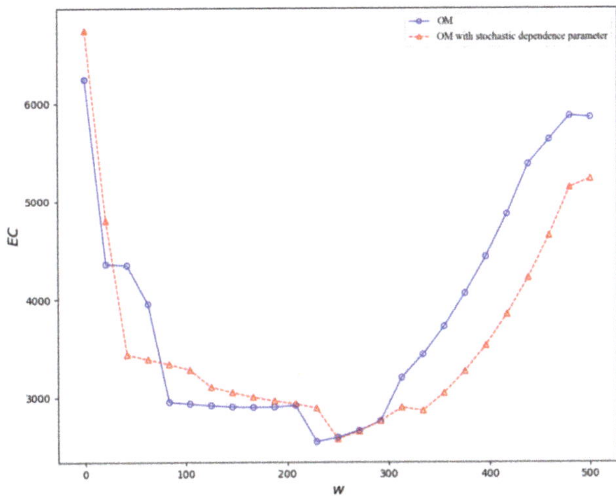

Figure 10. Expected cost under OM strategy before and after considering stochastic dependence.

Table 7. Downtime cost under different values of $\lambda_{3,2}$ and $\lambda_{1,2}$.

$\lambda_{3,2}$ \ $\lambda_{1,2}$	0	0.1	0.2	0.3	0.4	0.5	0.6	0.7	0.8	0.9
0	993.24	985.30	977.36	969.41	961.47	953.53	945.59	937.65	929.70	921.76
0.1	986.83	978.88	970.94	963.00	955.06	947.12	939.17	931.23	923.29	915.35
0.2	980.41	972.47	964.53	956.59	948.65	940.70	932.76	924.82	916.88	908.94
0.3	974.00	966.06	958.12	950.17	942.23	934.29	926.35	918.41	910.46	902.52
0.4	967.59	959.64	951.70	943.76	935.82	927.88	919.93	911.99	904.05	896.11
0.5	961.17	953.23	945.29	937.35	929.40	921.46	913.52	905.58	897.64	889.69
0.6	954.76	946.82	938.87	930.93	922.99	915.05	907.11	899.17	891.22	883.28
0.7	948.34	940.40	932.46	924.52	916.58	908.64	900.69	892.75	884.81	876.87
0.8	941.93	933.99	926.05	918.11	910.16	902.22	894.28	886.34	878.4	870.45
0.9	935.52	927.58	919.63	911.69	903.75	895.81	887.87	879.92	871.98	864.04

5.4. Carbon Emission Analysis

The optimization results of the wind turbine carbon emission model indicates that the number of PM reaches 11, and the objective function value of the carbon emission model is the smallest and is $GWP = 4015.7\text{g CO}_2-\text{eq}$. Economic benefit is an essential factor that enterprises must consider. Under the comprehensive consideration of environmental and economic benefits, the best amount of preventive maintenance is optimal, and the goal involving carbon emissions is relatively good.

5.5. Sensitivity Analysis

This paper mainly analyzes the sensitivity of three parameters: carbon emission G_{use} generated by wind turbine consumption per unit energy, carbon emission G_m generated by wind turbine maintenance activities, and carbon emission G_p generated by manufacturing one wind turbine.

(1) G_{use}. When G_{use} is being changed, the carbon emission per unit operation of a wind turbine will also change, as will the optimal solution of the wind turbine carbon emission model. The changing trend is shown in Figure 11.

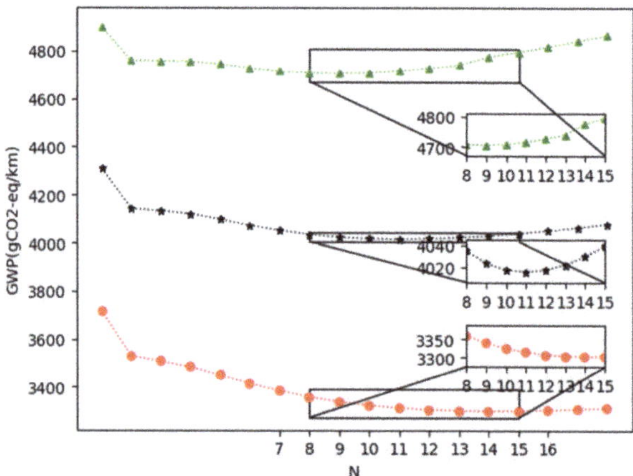

Figure 11. The change of GWP and N with the change of G_{use}.

In Figure 11, the abscissa represents maintenance cycle N (i.e., the total amount of wind turbine downtime). The top line (green line) is $G_{use} = 3195gCO_2-eq$, and the optimal $N^* = 10$. The middle line (black line) is $G_{use} = 2695gCO_2-eq$, and $N^* = 11$. The bottom line (red line) is $G_{use} = 1695gCO_2-eq$, and $N^* = 14$. When the value of G_{use} becomes larger (smaller), it means that the carbon emissions per unit of energy consumed by a wind turbine become larger (smaller), which will release more (less) greenhouse gas pollution into the environment during its operation. Thus, for a single carbon emission model, it is necessary to reduce (increase) the amount of preventive maintenance. Considering the cost rate comprehensively, when the value of G_{use} becomes smaller, the value of the optimal solution of preventive maintenance times is more sensitive to the change. When the value of G_{use} becomes larger, the value change of the optimal solution of preventive maintenance times is not sensitive.

(2) G_m. When G_m is being changed, the change of wind turbine carbon emissions per unit operation is shown in Figure 12.

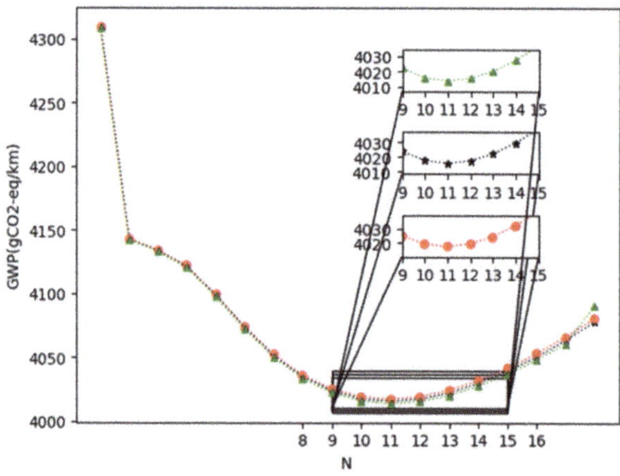

Figure 12. The change of GWP and N with the change of G_m.

In Figure 12, the top line (green line) is $G_m = 6450gCO_2-eq$, and $N^* = 11$. The middle line (black line) is $G_{use} = 11,450gCO_2-eq$, and $N^* = 11$. The bottom line (red line) is G_{use} = 16,450gCO$_2$−eq, and $N^* = 11$. It can be seen that when G_m is being changed, the change of the optimal solution of the carbon emission model and the optimization decision model is not sensitive, but the carbon emission per unit operation is changed.

(3) G_p. When G_p is being changed, the carbon emission per unit operation of a wind turbine will also change, and the optimal solution of the wind turbine carbon emission model will also change. The changing trend is shown in Figure 13.

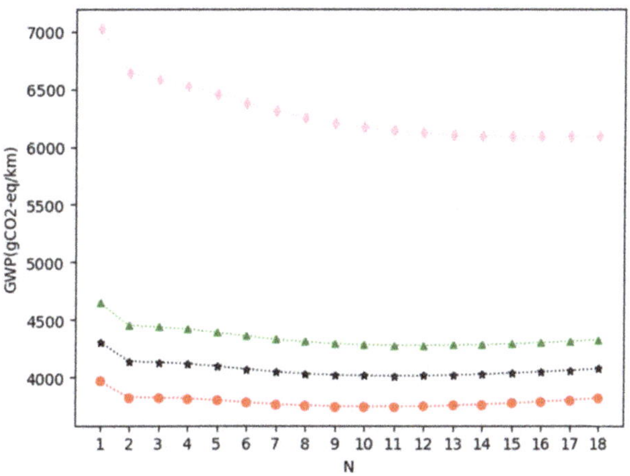

Figure 13. The change of *GWP* with the change of G_p.

From top to bottom, the first line (pink line) is G_p= 1.6×10^8gCO$_2$−eq, $N^* = 16$, the second line (green line) is $G_p = 9 \times 10^7$gCO$_2$−eq, $N^* = 12$, the third line (black line) is $G_p = 8 \times 10^7$gCO$_2$−eq, $N_g^* = 11$, and the fourth line (red line) is $G_p = 7 \times 10^7$gCO$_2$−eq, $N^* = 10$. When the value of G_p becomes larger (smaller), it means that the carbon emissions generated by manufacturing one wind turbine become larger (smaller). For a single carbon emission model, it is necessary to increase (reduce) the amount of preventive maintenance to reduce the carbon emission per unit operation in the whole life cycle of a wind turbine. Considering the cost rate comprehensively, when the value of G_p becomes larger, the value of the optimal solution of the preventive maintenance times is more sensitive. When the value of G_p becomes smaller, the value of the optimal preventive maintenance times is not sensitive.

6. Conclusions

This paper mainly studies the opportunistic maintenance strategy of wind turbines. The economic correlation, random correlation, and structural correlation among subsystems and carbon emissions can be considered in the proposed maintenance model. The stochastic correlation coefficient matrix is constructed by a failure chain to describe the reliability of the subsystems, and the structural correlation coefficient is used to describe the downtime loss cost in order to present the opportunistic maintenance model. Moreover, the operation energy consumption of wind turbines increases with their performance degradation. The environmental benefits are combined in the maintenance model of wind turbines. The working age fallback factor and failure rate increasing factor are introduced to establish the carbon emission model and the total expected cost model. This paper further considers the reduction effect of wind turbines recovery on cost and emission. The benefits of wind turbines can introduce recovery and emissions of maintenance activities into the proposed model by adopting the dynamic failure rate function and carbon emission function. The total expected maintenance cost could be described as the objective function for the proposed opportunistic maintenance model, including maintenance preparation cost, maintenance adjustment cost, shutdown loss cost, and operation cost. The operation cost is related to the energy consumption of wind turbines. Finally, a case study is provided to analyze the performance of the proposed model. Compared with preventive maintenance, the proposed model demonstrates better performance on wind turbines maintenance problems and can obtain a relatively good solution in a short computation time. The method proposed in this paper provides certain significance for guiding the selection of a wind turbine maintenance strategy.

The proposed model does not consider the complex external operation environment and external impacts. Thus, the joint optimization model between the carbon emission model and condition-based maintenance that considers the external operation environment and effect needs to be developed in the future.

Author Contributions: Conceptualization, Q.L. and Z.L.; methodology, T.X.; investigation, Q.L.; resources, J.L.; data curation, Z.L. and J.L.; writing—original draft preparation, Q.L.; writing—review and editing, Z.L.; supervision, T.X.; funding acquisition, M.H. All authors have read and agreed to the published version of the manuscript.

Funding: This research was funded by the National Natural Science Foundation of China (No. 71840003 and 51875359), the Natural Science Foundation of Shanghai (No. 19ZR1435600 and 20ZR1428600), the Humanity and Social Science Planning foundation of the Ministry of Education of China (No. 20YJAZH068), the science and technology development project of the University of Shanghai for Technology and Science (No. 2020KJFZ038) and the National Key R&D Program of China(2021YFF0900400).

Institutional Review Board Statement: Not applicable.

Informed Consent Statement: Not applicable.

Data Availability Statement: Not applicable.

Acknowledgments: The authors are indebted to the reviewers and the editors for their constructive comments, which greatly improved the contents and exposition of this paper.

Conflicts of Interest: The authors declare no conflict of interest.

References

1. Gorbunov, V.; Kuznetsov, S.; Savvina, A.; Poleshkina, I. Methodological aspects of avionics reliability at low temperatures during aircraft operation in the Far North and the Arctic. *Transp. Res. Procedia* **2021**, *57*, 220–229. [CrossRef]
2. Lin, B.; Wu, J.; Lin, R.; Wang, J.; Wang, H.; Zhang, X. Optimization of high-level preventive maintenance scheduling for high-speed trains. *Reliab. Eng. Syst. Saf.* **2019**, *183*, 261–275. [CrossRef]
3. Appoh, F.; Yunusa-Kaltungo, A.; Sinha, J.K. Hybrid adaptive model to optimise components replacement strategy, a case study of railway brake blocks failure analysis. *Eng. Fail. Anal.* **2021**, *127*, 105539. [CrossRef]
4. Thomas, L.C. A survey of maintenance and replacement models for maintainability and reliability of multi-item systems. *Reliab. Eng.* **1986**, *16*, 297–309. [CrossRef]
5. Allal, A.; Sahnoun, M.; Adjoudj, R.; Benslimane, S.; Mazar, M. Multi-agent based simulation-optimization of maintenance routing in offshore wind farms. *Comput. Ind. Eng.* **2021**, *157*, 107342. [CrossRef]
6. Song, S.; Li, Q.; Felder, F.A.; Wang, H.; Coit, D.W. Integrated optimization of offshore wind farm layout design and turbine opportunistic condition-based maintenance. *Comput. Ind. Eng.* **2018**, *120*, 288–297. [CrossRef]
7. Ren, Z.; Verma, A.S.; Li, Y.; Teuwen, J.J.; Jiang, Z. Offshore wind turbine operations and maintenance, a state-of-the-art review. *Renew. Sustain. Energy Rev.* **2021**, *144*, 110886. [CrossRef]
8. Li, M.; Jiang, X.; Negenborn, R.R. Opportunistic maintenance for offshore wind farms with multiple-component age-based preventive dispatch. *Ocean Eng.* **2021**, *231*, 109062. [CrossRef]
9. Li, M.X.; Wang, M.; Kang, J.C.; Sun, L.P.; Jin, P. An opportunistic maintenance strategy for offshore wind turbine system considering optimal maintenance intervals of subsystems. *Ocean Eng.* **2020**, *216*, 108067. [CrossRef]
10. Zhu, W.J.; Castanier, B.; Bettayeb, B. A dynamic programming-based maintenance model of offshore wind turbine considering logistic delay and weather condition. *Reliab. Eng. Syst. Saf.* **2019**, *190*, 106512. [CrossRef]
11. Li, H.; Diaz, H.; Soares, C.G. A developed failure mode and effect analysis for floating offshore wind turbine support structures. *Renew. Energy* **2021**, *164*, 133–145. [CrossRef]
12. Yeter, B.; Garbatov, Y.; Soares, C.G. Life-extension classification of offshore wind assets using unsupervised machine learning. *Reliab. Eng. Syst. Saf.* **2022**, *219*, 108229. [CrossRef]

13. Zhong, S.; Pantelous, A.A.; Goh, M.; Zhou, J. A reliability-and-cost-based fuzzy approach to optimize preventive maintenance scheduling for offshore wind farms. *Mech. Syst. Signal Processing* **2019**, *124*, 643–663. [CrossRef]
14. Zhou, X.; Huang, K.; Xi, L.; Lee, J. Preventive maintenance modeling for multi-component systems with considering stochastic failures and disassembly sequence. *Reliab. Eng. Syst. Saf.* **2015**, *142*, 231–237. [CrossRef]
15. Iung, B.; Do, P.; Levrat, E.; Voisin, A. Opportunistic maintenance based on multi-dependent components of manufacturing system. *CIRP Ann.* **2016**, *65*, 401–404. [CrossRef]
16. Wu, S.; Chen, Y.; Wu, Q.; Wang, Z. Linking component importance to optimisation of preventive maintenance policy. *Reliab. Eng. Syst. Saf.* **2015**, *146*, 26–32. [CrossRef]
17. Minne, E.; Crittenden, J.C. Impact of maintenance on life cycle impact and cost assessment for residential flooring options. *Int. J. Life Cycle Assess.* **2015**, *20*, 36–45. [CrossRef]
18. Giustozzi, F.; Crispino, M.; Flintsch, G. Multi-attribute life cycle assessment of preventive maintenance treatments on road pavements for achieving environmental sustainability. *Int. J. Life Cycle Assess.* **2012**, *17*, 409–419. [CrossRef]
19. Noland, R.B.; Hanson, C.S. Life-cycle greenhouse gas emissions associated with a highway reconstruction, a New Jersey case study. *J. Clean. Prod.* **2015**, *107*, 731–740. [CrossRef]
20. Sikos, L.; Klemeš, J. RAMS contribution to efficient waste minimisation and management. *J. Clean. Prod.* **2009**, *17*, 932–939. [CrossRef]
21. Liu, Z.-C.; Afrinaldi, F.; Zhang, H.-C.; Jiang, Q. Exploring optimal timing for remanufacturing based on replacement theory. *CIRP Ann.-Manuf. Technol.* **2016**, *65*, 447–450. [CrossRef]
22. Franciosi, C.; Lambiase, A.; Miranda, S. Sustainable maintenance, a periodic preventive maintenance model with sustainable spare parts management. *IFAC Pap.* **2017**, *50*, 13692–13697. [CrossRef]
23. Jasiulewicz-Kaczmarek, M.; Gola, A. Maintenance 4.0 technologies for sustainable manufacturing-an overview. *IFAC Pap.* **2019**, *52*, 91–96. [CrossRef]
24. Hennequin, S.; Restrepo, L.M.R. Fuzzy model of a joint maintenance and production control under sustainability constraints. *IFAC Pap.* **2016**, *49*, 1216–1221. [CrossRef]
25. Afrinaldi, F.; Taufik; Tasman, A.M.; Zhang, H.-C.; Hasan, A. Minimizing economic and environmental impacts through an optimal preventive replacement schedule, model and application. *J. Clean. Prod.* **2017**, *143*, 882–893. [CrossRef]
26. Huang, J.; Chang, Q.; Arinez, J.; Xiao, G.X. A maintenance and energy saving joint control scheme for sustainable manufacturing systems. *Procedia CIRP* **2019**, *80*, 263–268. [CrossRef]
27. Xia, T.; Xi, L.; Zhou, X.; Du, S. Modeling and optimizing maintenance schedule for energy systems subject to degradation. *Comput. Ind. Eng.* **2012**, *63*, 607–614. [CrossRef]
28. Murthy, D.N.P.; Nguyen, D.G. Study of two-component system with failure interactions. *Nav. Res. Logist. Q.* **1985**, *32*, 239–247. [CrossRef]
29. Nakagawa, T.; Murthy, D.N.P. Optimal replacement policies for a two-unit system with failure interactions. *Rairo Oper. Res.* **1993**, *27*, 427–438. [CrossRef]
30. Lv, X.; Liu, Q.; Li, Z.; Dong, Y.; Xia, T.; Chen, X. A New Maintenance Optimization Model Based on Three-Stage Time Delay for Series Intelligent System with Intermediate Buffer. *Shock. Vib.* **2021**, *2021*, 6694896. [CrossRef]

Article

Fault Diagnosis Technology for Ship Electrical Power System

Chaochun Yu [1], Liang Qi [1,*], Jie Sun [1], Chunhui Jiang [2], Jun Su [3] and Wentao Shu [2]

1   School of Electronic Information, Jiangsu University of Science and Technology, Zhenjiang 212003, China; just.stu@stu.just.edu.cn (C.Y.); 209030102@stu.just.edu.cn (J.S.)

2   Electrical Design Department, Zhenjiang Hongye Science & Technology Co., Ltd., Zhenjiang 212000, China; jch2003@163.com (C.J.); swt45617730@126.com (W.S.)

3   Mechanical Design Department, Zhenjiang Hongye Science & Technology Co., Ltd., Zhenjiang 212000, China; 13812451952@163.com

*   Correspondence: alfred_02030210@just.edu.cn; Tel.: +86-189-1280-1629

Abstract: This paper proposes a fault diagnosis method for ship electrical power systems on the basis of an improved convolutional neural network (CNN) to support normal ship operation. First, according to the mathematical model of the ship electrical power system, the simulation model of the ship electrical power system is built using the MATLAB/Simulink simulation software platform in order to understand the normal working state and fault state of the generator and load in the power system. Then, the model is simulated to generate the fault response curve, and the picture dataset of the network model is obtained. Second, a CNN fault diagnosis model is designed using TensorFlow, an open-source tool for deep learning. Finally, network model training is performed, and the optimal diagnosis results of the ship electrical power system are obtained to realize structural parameter optimization and diagnosis. The diagnosis results show that the established simulation model and improved CNN can provide support for fault diagnosis of the ship electrical power system, improve the operation stability and safety of the ship electrical power system, and ensure safety of the crew.

Keywords: fault diagnosis technology; improved convolutional neural network; ship electrical power system; Simulink; synchronous generator

Citation: Yu, C.; Qi, L.; Sun, J.; Jiang, C.; Su, J.; Shu, W. Fault Diagnosis Technology for Ship Electrical Power System. *Energies* **2022**, *15*, 1287. https://doi.org/10.3390/en15041287

Academic Editor: Xi Gu

Received: 26 December 2021
Accepted: 4 February 2022
Published: 10 February 2022

Publisher's Note: MDPI stays neutral with regard to jurisdictional claims in published maps and institutional affiliations.

1. Introduction

With the continuous development of modern technology, ship electrical power systems that can realize overall coordination of the energy of the entire ship are expected to constitute the development trend of ships in the future [1,2]. Ship electrical power systems are significantly different from land power systems [3]. In particular, they are strongly independent. Because ship electrical power systems have a smaller capacity than onshore power systems, bus voltage fluctuation may occur under the application or removal of a large load, which can easily cause serious faults. Any equipment fault in the system can affect the entire power grid. If potential safety hazards occur during operation, they will threaten the safety of the entire ship. Ship electrical power systems are regarded as the core of the entire ship. They are independent and have high requirements in terms of safe operation and fault diagnosis. They need faster and more accurate fault diagnosis than land power systems in case of system faults [4]. Therefore, fault diagnosis technologies are necessary to study ship electrical power systems [5].

Fault diagnosis of a ship electrical power system entails modeling and simulation of the system. Early modeling methods mainly involved physical modeling, i.e., a physical model was established using the similarity principle. At present, the mathematical modeling method is mainly used. This method can abstract the internal characteristics of the system into mathematical formulas, deduce the internal characteristics of the actual system, and diagnose faults through changes in the relationship between the independent variables and the dependent variables of the mathematical formulas. Research on modeling and simulation of ship electrical power systems is based on the power system model and involves a series of studies on how to maintain system stability. The ship electrical power system model is integrated with the modules of each basic unit. First, a mathematical model is established for each basic unit of the power system according to the structure and basic principles of the ship electrical power system. Then, a simulation model is built to form a complete ship electrical power system model [6]. The research process should include modeling and simulation technologies, automatic control, and other theoretical methods. In particular, the generator and its excitation are related to the voltage stability of the power system. The linear single variable control method, which was first used in excitation control, has been subsequently modified into the nonlinear multivariable control method. The development of the excitation control method has undergone several stages [7], from the earliest classical proportional integral and differential (PID) control method to the multivariable control method based on modern theory. At present, it is applied as an intelligent control method. In [8], the authors proposed a T-S fuzzy-weighting-based excitation switching control method for a tidal generator set, which can overcome the dynamic and static performance defects in the excitation control of the tidal generator set and improve its performance. In [9], feedback control was adopted for the field currents of the two-phase brushless exciter, and speed reference control was adopted for the excitation frequency and phase sequence; this method achieves a constant field current for the main generator. In [10], the authors presented a nonlinear coordinated excitation and static VAR compensator (SVC) control for regulating the output voltage and improving the transient stability of a synchronous generator infinite bus (SGIB) power system. In terms of the mathematical development of fault diagnosis methods for ship electrical power systems, the author in [11] developed a higher-order mathematical model of the generator to describe the generator state in greater detail. In [12], the authors proposed three different mathematical models for the mathematical modeling of a synchronous generator, used the models under different working conditions, and conducted a detailed comparative analysis of the models to improve the simulation accuracy.

In general, fault diagnosis methods are currently categorized into three main types [13], namely fault diagnosis based on analytical modeling, fault diagnosis based on signal processing, and fault diagnosis based on artificial intelligence. Analytical modeling includes state and parameter estimation as well as consistency testing. It has the characteristics of real-time diagnosis and the essence of deep human systems. However, it also has some defects, such as a large modeling error and significant noise interference. Signal processing includes spectrum analysis and wavelet transformation. It has the advantages of simple application and good real-time performance. However, it cannot deal with potential faults. Artificial intelligence [14] includes neural networks, fuzzy theory, genetic algorithms, rough sets, artificial immune systems and fuzzy cluster analysis algorithms, fault trees, and support vector machines, which have strong learning and reasoning abilities. To overcome key faults such as a broken rotor bar or electrical phase fault, a fault diagnosis method for the electric drive of an electric ship has been proposed [15]; however, the number of fault diagnosis objects is insufficient. In [4], the proposed load monitoring and fault detection method outlines a data-clustering-based approach to extract unique feature vectors from short-time Fourier transform analysis for any pulsed load; however, this method is not suitable for any general load curve integrated solution. In [16], the location and severity of a stator winding fault of a permanent magnet synchronous motor were modeled and detected, and a mathematical model that can describe both the health state and the fault state was established; however, the mathematical model is not suitable for other ship electrical power system equipment. In [17], a remote system was introduced for online condition monitoring and fault diagnosis of a gas turbine on an offshore oil well drilling platform on the basis of a kernelized information entropy model. In [18], a multi-class multi-core correlation vector machine fault diagnosis method based on manifold learning and swarm intelligence optimization was proposed to improve the predictive maintenance activities of diesel engines.

This paper proposes an improved network fault diagnosis model based on a convolutional neural network (CNN). This method can directly input the original image without feature decomposition and extraction. It has significant advantages, such as simple application, high operation speed, automatic parameter updates, and stable, convergent, and accurate results. These advantages enable the method to overcome existing drawbacks in the fault diagnosis of ship electrical power systems. First, based on the MATLAB/Simulink (The MathWorks Inc., Natick, MA, USA) simulation software platform, the ship electrical power system simulation model is established to understand the normal working state and fault state of the generator and load. Then, the fault response curve is generated and the picture dataset of the network model is obtained. Second, the CNN fault diagnosis model is designed using TensorFlow, an open source tool for deep learning. Finally, network model training is performed, and optimal diagnosis results are obtained to realize structural parameter optimization and diagnosis.

The remainder of this paper is organized as follows. Section 2 describes the model and simulation of the ship electrical power system. Section 3 discusses the development of the improved CNN. Section 4 presents and analyzes the experimental results. Finally, Section 5 concludes the paper.

2. Ship Electrical Power System

2.1. Model of Excitation Control System

At present, most ships use AC electric propulsion systems. Because the electrical equipment of the ship mainly comprises inductive loads, the load current will cause demagnetization of the synchronous generator, and the load change will alter the power grid voltage of the ship [19]. To ensure stable operation of the power system, the excitation system will adjust the excitation current supplied to the generator according to the load change, make the generator terminal voltage return to the given value, and realize stability of the generator terminal voltage.

Currently, the most common excitation mode of generators in ship electrical power systems is phase compound excitation of brushless excitation systems. The combination of brushless excitation and automatic voltage regulation (AVR) can effectively enhance the forced excitation capacity and reaction speed of the excitation system. The principle of the excitation system is that the rotating exciter rotates at the same speed as the magnetic field of the main generator, and the auxiliary exciter rotates with the rotating exciter.

AVR plays an important role in phase compound excitation systems [20]. The power supply unit in AVR converts the output voltage of the generator into the voltage required by the system through voltage transformation, rectification and other links, and then transmits the results to the PID control unit and phase control unit. The synchronous control unit controls and adjusts the phase change of the excitation current output through the exciter to maintain the same change as the phase change of the applied thyristor excitation voltage. The voltage difference detection unit monitors the error signal between the reference voltage and the actual voltage in the system. The PID control unit amplifies the voltage error signal. The phase control unit mainly amplifies the signal. The main thyristor rectifier unit mainly rectifies the armature current of the static exciter.

Based on the above-mentioned principles and by referring to the excitation system model recommended by IEEE [21], the mathematical model of the excitation system can be obtained as follows:

(1) Mathematical model of phase compound excitation device is represented by d and q components as follows:

$$U_r = \sqrt{(U_d - KI_d x)^2 + (U_q - KI_q x)^2} \tag{1}$$

where U_r is the output voltage of the phase compound excitation device, U_d is the armature terminal voltage of generator axis d, U_q is the armature terminal voltage of generator axis q, $K = 9\sqrt{2}/\pi$, and x is the moving reactance.

(2)   Mathematical model of voltage difference

The model of voltage difference is an adder model, which compares the effective voltage signal input of other elements and obtains a voltage difference signal as a control signal. The difference between the output terminal voltage and the set voltage satisfies Equation (3):

$$U_{tf} = \sqrt{U_d^2 + U_q^2} \cdot \frac{1}{1 + T_r s} \tag{2}$$

$$\Delta U = U_{ref} + \frac{U_{f0}}{K_e} - U_{tf} + U_{stab} - U_{ff} \tag{3}$$

where U_{tf} is phase compound excitation voltage signal, T_r is the time constant of the filter, U_{ref} is the reference voltage of the automatic voltage regulating device, ΔU is the voltage difference between the output terminal voltage and the set voltage, U_{f0} is the initial excitation voltage, k_e is the effective gain of the exciter, U_{stab} is the grounding voltage of power system (0), and U_{ff} is the feedback output voltage.

(3)   Amplifier mathematical model The mathematical equation that the amplifier satisfies is:

$$U_a = U_c \frac{K_a}{1 + T_a s} \tag{4}$$

where U_a is the output voltage of the amplifier, U_c is the voltage output by the compensator, K_a is the gain of the amplifier, and T_a is the time constant of the amplifier.

(4)   Mathematical model of lead–lag compensator The mathematical equation that the lead–lag compensator satisfies is:

$$U_c = \Delta U \frac{T_c S + 1}{1 + T_b s} \tag{5}$$

where T_c and T_b are the lead compensation time constant and lag compensation time constant of the compensator, respectively.

(5)   Mathematical model of proportional saturation element The mathematical equation of proportional saturation element is:

$$\text{In } 0 \leq E_f \leq E_{fmax}, E_f = E_d, E_{fmax} = \begin{cases} constant, \ K_p = 0 \\ K_p U_{tf}, \ K_p \geq 0 \end{cases} \text{ conditions}$$

$$\text{Then, } E_{fd} = U_a + U_r \tag{6}$$

where E_{fd} is the voltage output by the voltage regulating device, and E_f and E_{fmax} are the proportional saturation element output voltage value and maximum voltage value, respectively.

(6)   Simplified exciter mathematical model Exciter is usually represented by one-order inertia element, as shown in Equation (7):

$$U_f = E_{fd} \frac{1}{K_e + T_e} \tag{7}$$

where U_f is the output voltage of the exciter and T_e is the time constant of the exciter.

(7)   The mathematical equation that the feedback stabilization element satisfies is:

$$U_{ff} = U_f \frac{K_f s}{1 + T_f s} \tag{8}$$

where K_f and T_f are the gain and time constant of the feedback element.

2.2. Simulation Model of Diesel-Driven Synchronous Generator and Its Excitation System

According to the above-mentioned mathematical model, the simulation model of the excitation control system [22] is established using simulation software as follows:

As shown in Figure 1, the main regulator and lead–lag compensator, as the main controller and damping feedback link, constitute a closed-loop PID control loop. Input vref is the set value of the synchronous generator terminal voltage, vd and vq are the voltage values on the d-axis and q-axis of the generator, respectively, and vstab is the synchronous generator ground zero voltage. The output terminal Vf of the excitation control system is the excitation voltage. In addition to the corresponding parameters of each link shown in Figure 1, the initial excitation voltage Vf0 is set to 1, and the upper and lower limits of the proportional saturation link for the limiting amplitude are 5 and 0, respectively.

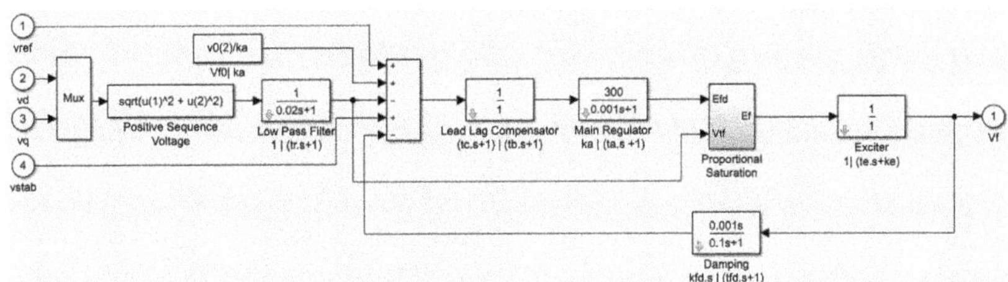

Figure 1. Excitation control system simulation model.

The power unit of the power system of the ship adopts diesel fuel combination for combined power. The main power unit is the synchronous generator driven by diesel, while the auxiliary power unit is the asynchronous generator driven by a gas turbine. The synchronous generator is the synchronous generator simulation model in the MAT-LAB/Simulink/SimPowerSystems module library, and its parameters are listed in Table 1. To ensure stability and correctness of the simulation graphics, it is also necessary to use the built-in function of the powergui module in Simulink in order to initialize the generator.

Table 1. Simulation parameters of synchronous generator.

Parameter	Value
Rated capacity, P_n	1560 kVA
Rated voltage, V_n	480 V
Power factor, $\cos\theta$	0.8
Rated frequency, f_n	60 Hz
Polar logarithm, p	2
Stator winding resistance, R_s	0.022 (pu)
Direct axis synchronous reactance, X_d	2.49 (pu)
Quadrature axis synchronous reactance, X_q	1.22 (pu)
Direct axis transient reactance, $X_d{'}$	0.15 (pu)
Direct axis subtransient reactance, $X_d{''}$	0.17 (pu)
Quadrature axis subtransient reactance, $X_q{''}$	0.19 (pu)
D − axis transient open circuit time constant, $T_{d0}{'}$	4.4754
D − axis subtransient open circuit time constant, $T_{d0}{''}$	0.0667
Q − axis subtransient open circuit time constant, $T_{q0}{''}$	0.1

2.3. Simulation Model of Gas-Turbine-Driven Asynchronous Generator

The auxiliary power unit gas turbine has a complex structure. To better simulate the operation, the turbine part of the of gas turbine is used to replace the entire gas turbine [23], as shown in Figure 2.

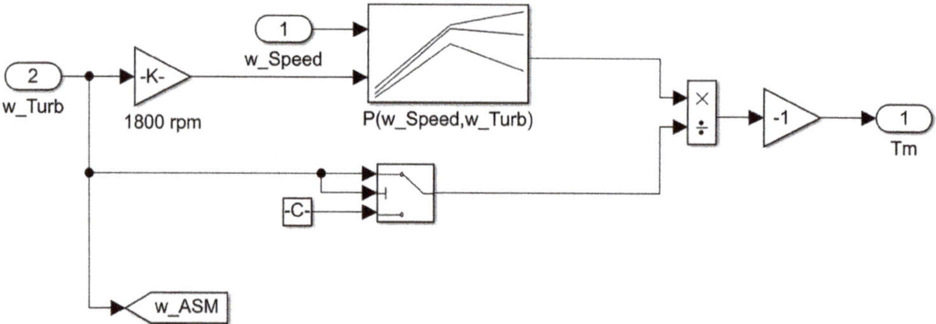

Figure 2. Turbine simulation model.

The turbine uses a two-dimensional look-up table to calculate the turbine torque output (TM) as a function of the gas speed (w speed) and turbine speed (w turbo) of the gas turbine.

The asynchronous generator is the asynchronous generator simulation model in the MATLAB/Simulink/SimPowerSystems module library. Its parameters are listed in Table 2.

Table 2. Simulation parameters of asynchronous generator.

Parameter	Value
Rated capacity, P_n	750 kVA
Rated voltage, V_n	480 V
Power factor, $\cos\theta$	0.8
Rated frequency, f_n	60 Hz
Polar logarithm, p	2
Stator resistance, R_s	0.022 (pu)
Stator inductance, L_s	0.11 (pu)
Rotor resistance, $R_r{}'$	0.021 (pu)
Rotor inductance, $L_r{}'$	0.11 (pu)
Mutual inductance, L_m	3.7 (pu)

2.4. Discrete Frequency Regulator

Figure 3 shows the simulation model of the discrete frequency regulator [24]. The frequency is controlled by the discrete frequency regulator module. The controller uses a standard three-phase phase locked loop (PLL) system to measure the system frequency. The measured frequency is compared with the reference frequency (60 Hz) to obtain the frequency error. The error is then integrated to obtain the phase error. Further, the proportional differential (PD) controller uses the phase error to generate an output signal representing the required secondary load power. The signal is converted into an 8 bit digital signal to control the switching of eight three-phase secondary loads. To minimize voltage interference, switching occurs when the voltage crosses zero.

2.5. Ship Electrical Power System Simulation Model

The ship electrical power system model consists of the diesel-driven main generator set module, gas-turbine-driven auxiliary generator set module, and power load. The main load is 300 kW [25]. The secondary load block consists of eight groups of three-phase resistors connected in series with the GTO thyristor switch. The nominal power of each unit follows a binary series; thus, the load can be varied from 0 to 510 kW in steps of 2 kW. The GTO is simulated by an ideal switch. In summary, the simulation model of the ship electrical power system can be obtained as shown in Figure 4. A, B and C are phase lines.

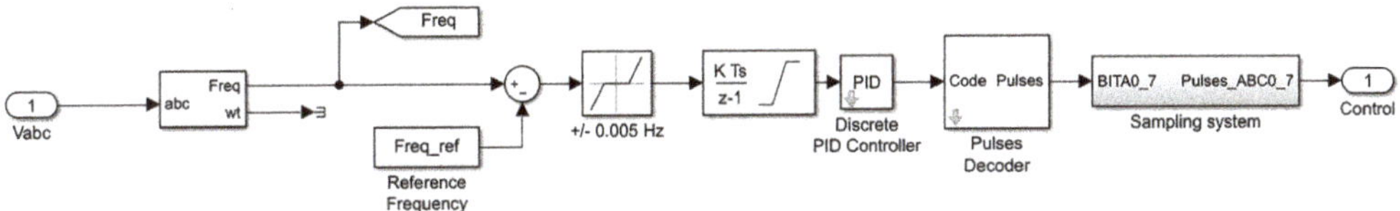

Figure 3. Discrete frequency regulator simulation model.

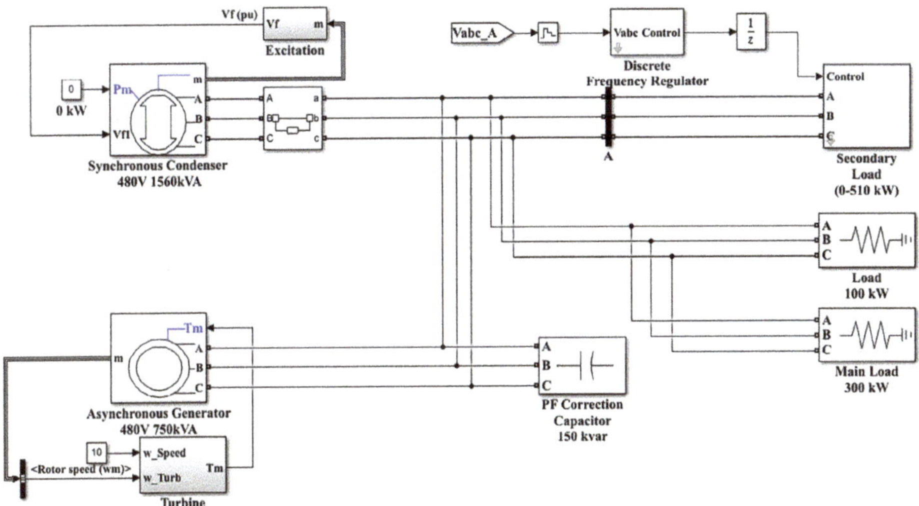

Figure 4. Ship electrical power system simulation model.

2.6. Typical Fault Simulation of Ship Electrical Power System

Ship electrical power systems are different from land power systems. In particular, ship electrical power systems are strongly independent. A component fault in the ship power grid may affect the entire power grid. Short-circuit faults and open-phase faults are typical faults that pose serious threats. Such faults affect not only individual equipment but also the power grid of the entire ship. In severe cases, they may lead to paralysis of the ship power grid. As shown in Table 3 short-circuit faults include single-phase short-circuit grounding, two-phase short circuit, two-phase short-circuit grounding, and three-phase short circuit, while open-phase faults include single-phase disconnection and two-phase disconnection. It is necessary to simulate and diagnose the state of short-circuit and open-phase faults as typical faults in ship electrical power systems.

There are many reasons for short-circuit faults [26], such as line aging, component damage, corrosion and ring breaking in the working environment of the ship, and component insulation. Short-circuit faults pose serious threats, as they may burn important equipment, interfere with nearby communication, endanger life, and even lead to power grid collapse and loss of power supply for the entire ship. For ship electrical power systems, any type of short-circuit fault can not only threaten the safety of the entire ship power grid but also produce various problems such as electromagnetic interference. Open-phase faults, similar to short-circuit failure, will also has devastating effects.

Based on the simulation model of the ship electrical power system, the "three-phase fault" fault module of Simulink is added at the synchronous generator, asynchronous generator, main load, and secondary load ends to set the time of fault occurrence and simulate typical single-phase short-circuit grounding faults, two-phase short-circuit faults, two-phase short-circuit grounding faults, and three-phase short-circuit faults [27]. A "three-phase breaker" is used to set the occurrence time of open-phase faults and simulate typical single-phase disconnection and two-phase disconnection.

Table 3. Schematic diagram of faults.

Faults Type	Schematic Diagram
Single-phase short-circuit grounding	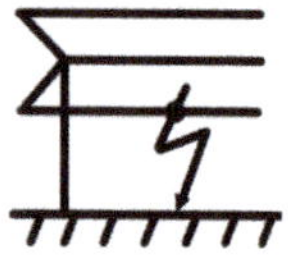
Two-phase short circuit	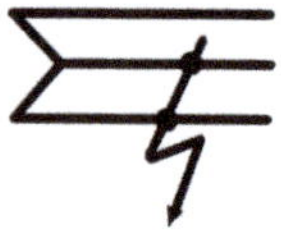
Two-phase short-circuit grounding	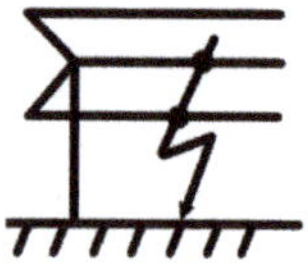
Three-phase short circuit	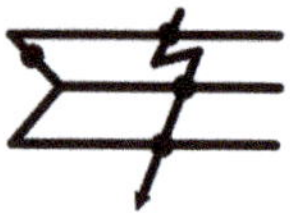
Single-phase disconnection	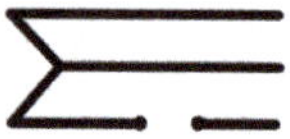
Two-phase disconnection	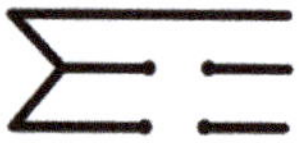

This section describes single-phase short-circuit grounding faults at the synchronous generator end. During normal operation, the three phases are symmetrical; hence, regardless of which phase of ABC is grounded, the changes in the short-circuit faults are basically the same. The single-phase short-circuit grounding fault simulation is performed by taking phase a as an example. The fault module is set to phase a grounding in 10 s, the fault module is cut off in 10.5 s, and the total simulation time is 25 s. Before 10 s, the entire power system was in stable operation. At 10 s, the fault module caused phase a short-circuit grounding fault, and the fault was removed after 0.5 s. The simulation results are shown as response curves in Figure 5. When different faults occur in different equipment of the power system of the ship, the amplitudes of the four output response curves in the fault time period are different from the normal waveform, and the differences are significant. Therefore, this is used as the basis for fault diagnosis.

The abscissa represents time, and its unit is second (s). Vf (pu) is the unit value of the excitation voltage, Frequency (Hz) is the system frequency, rotor speed wm (pu) is the unit value of the synchronous generator speed, and ASM speed (pu) is the unit value of the asynchronous generator speed.

It can be seen that in the initial stage of the ship electrical power system, the system tends to be in stable operation after starting, and the speed, excitation voltage and frequency of the two generators have not changed, and the entire power system has not fluctuated significantly. At 10 s, the fault module "Three-Phase Fault" causes the single-phase short-circuit grounding fault of Phase A, and the short-circuit current in phase A, generator speed, excitation voltage and frequency fluctuate greatly. The single-phase short-circuit grounding fault affects the stability of the power system and makes the ship unable to run normally. At 10.5 s, the fault module "Three-Phase Fault" is removed, and the single-phase short-circuit grounding fault disappears. After adjustment by the excitation control system, the ship electrical power system returns to normal; the speed, excitation voltage, and frequency of the two generators return to stability. Similarly, the simulation results of two-phase short-circuit, two-phase short-circuit grounding, three-phase short circuit, single-phase disconnection, and two-phase disconnection at the synchronous generator end are shown in Figure 6. The faults at the asynchronous generator, main load, and secondary load ends will not be repeated.

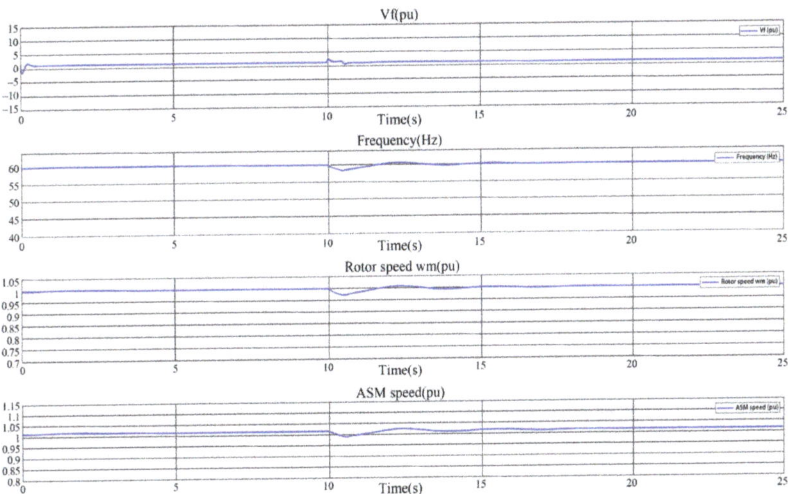

Figure 5. Response curve of single-phase short-circuit grounding fault at synchronous generator terminal.

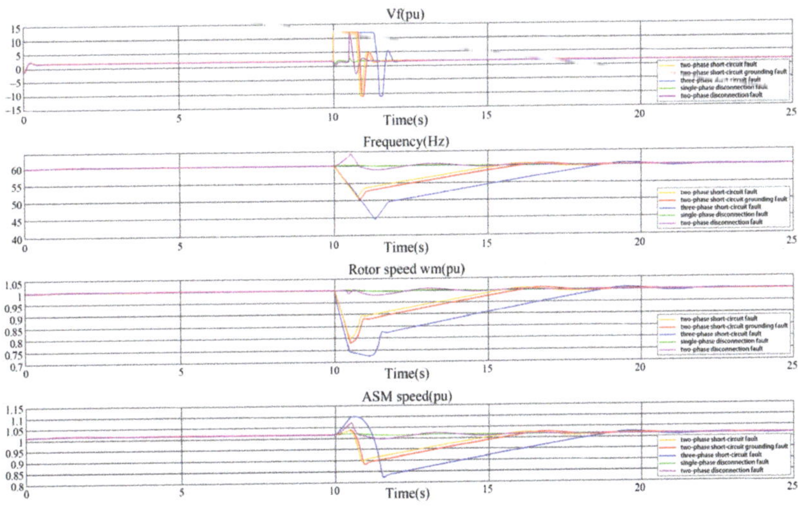

Figure 6. Response curve of five different faults at synchronous generator end.

3. Construction of Improved CNN

3.1. CNN

CNN has emerged as a research hotspot in many scientific fields, especially pattern classification [28]. CNN is composed of a series of layers, as well as data flows between the layers. The basic structure is as follows: input layer, convolution layer, activation function, pooling layer, and fully connected layer, i.e., INPUT-CONV-RELU -POOL-FC.

The convolution layer is a feature extraction layer. The input of each neuron is connected to the local receptive field of the previous layer and extracts local features. The convolution layer mainly convolutes the image according to the convolution kernel and reduces noise [29]. It also involves the principle of "weight sharing". The calculation formula of the convolution layer is as follows:

$$x_j^l = \sum_{i \in M_j} x_i^{l-1} \times k_{ij}^l + b_j^l \tag{9}$$

where l denotes the number of layers, M_j represents the feature graph set of the previous layer associated with the $j$the feature graph of the current layer, x_j^l is the jth characteristic diagram output by the lth layer, x_i^{l-1} is the ith characteristic diagram of the output of the $l-1$th layer, k_{ij}^l is the convolution kernel between the jth characteristic graph of the lth layer and the ith characteristic graph of the previous layer, and b_j^l is the offset of the jth characteristic graph of the lth layer.

The activation function is used to add nonlinear factors, because the convolution method is used to deal with linear operations, i.e., assign weights to each pixel. The expression of the linear model is not sufficient; hence, an activation function is required. Common activation functions include the sigmoid function, tanh function, ReLU function, and leaky ReLU function.

The pooling layer is a feature mapping layer. After adding bias, a new feature map is obtained in the pooling layer through a nonlinear function [30]. The functions of pooling are as follows: (i) reducing the size of the characteristic diagram and simplifying the computational complexity of the network; (ii) feature compression to extract the main features. The operation formula of the pooling layer is as follows:

$$x_j^l = \beta_j^l \text{subdown}\left(x_i^{l-1}\right) + b_j^l \tag{10}$$

where $\text{subdown}(\cdot)$ represents the pooled down-sampling function, β_j^l is the ratio column offset, and b_j^l is the additive bias.

The fully connected layer is used to connect all the features and send the output value to a classifier (such as a softmax classifier) for classification.

Finally, the test accuracy and error loss function value of the model are output. The structure and parameters of CNN are shown in Figure 7.

3.2. Improved CNN

The traditional model has a complex structure, massive parameters, and low running speed. Moreover, the convergence speed of the classification results is affected by the method of initializing the parameters and the updating of the network weights, and there are oscillation problems in the accuracy and loss rate curves. In summary, this study makes the following improvements and proposes a CNN model with better performance, which can avoid the above-mentioned issues.

(1) All local response normalization (LRN) layers are removed and the initial value program is changed. It is proven by practice that the normalization operation of batch normalization (BN) is used after simple parameter initialization. The use of BN is conducive to the convergence of the samples and the stability of the network.

(2) The number of nodes in the fully connected layer is adjusted; based on the reduction and updating of parameters and weights, the running speed is improved and the calculation time is shortened.

(3) Nesterov-accelerated adaptive moment estimation (NAdam) is used to update the weights of the neural networks iteratively on the basis of the training data, and the weights can be updated iteratively according to the output results.

(4) Kaiming initialization is used to initialize the normal_initializer. After testing, the results will be improved.

CNN model parameters and improved CNN model parameters are listed in Tables 4 and 5, respectively.

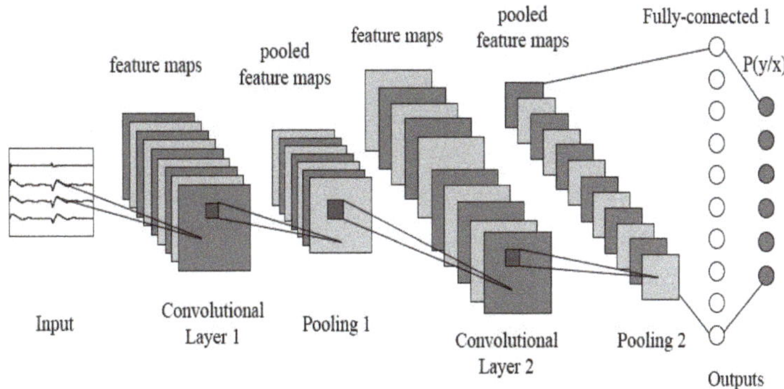

Figure 7. CNN structure.

Table 4. The parameters of the CNN.

Network Layer	Input	Filter	Output
Conv 0	$28 \times 28 \times 1$	$3 \times 3 \times 64$	$28 \times 28 \times 64$
Maxpooling 0	$28 \times 28 \times 64$	2×2	$14 \times 14 \times 64$
Conv 1	$14 \times 14 \times 64$	$3 \times 3 \times 32$	$14 \times 14 \times 32$
Maxpooling 1	$14 \times 14 \times 32$	2×2	$7 \times 7 \times 32$
Conv 2	$7 \times 7 \times 32$	$5 \times 5 \times 28$	$7 \times 7 \times 28$
Conv 3	$7 \times 7 \times 28$	$5 \times 5 \times 14$	$7 \times 7 \times 14$
Maxpooling 2	$7 \times 7 \times 14$	2×2	$3 \times 3 \times 14$
FC 1	126		118
FC 2	128		10

Table 5. The parameters of the improved CNN.

Network Layer	Input	Filter	Output
Conv 0	$28 \times 28 \times 1$	$5 \times 5 \times 128$	$28 \times 28 \times 128$
Maxpooling 0	$28 \times 28 \times 128$	2×2	$14 \times 14 \times 128$
Conv 1	$14 \times 14 \times 128$	$3 \times 3 \times 4$	$14 \times 14 \times 4$
Maxpooling 1	$14 \times 14 \times 4$	2×2	$7 \times 7 \times 4$
Conv 2	$7 \times 7 \times 4$	$5 \times 5 \times 64$	$7 \times 7 \times 64$
Conv 3	$7 \times 7 \times 64$	$5 \times 5 \times 128$	$7 \times 7 \times 128$
Maxpooling 2	$7 \times 7 \times 128$	2×2	$3 \times 3 \times 128$
FC 1	1152		512
FC 2	512		10

3.3. Flow of CNN Algorithm

The algorithm flow of the proposed CNN fault diagnosis model is shown in Figure 8. It is mainly divided into four stages:

(1) Sample image preprocessing: First, the sample dataset is constructed. Second, the size and color of the image are processed to facilitate network learning.
(2) Design network fault diagnosis model: Network programs are written and built in the Python (Python Software Foundation, Delaware, USA) compilation environment and TensorFlow (Google Brain, San Francisco, USA) learning framework.
(3) Training optimization network model: The weight and threshold are adjusted repeatedly according to the back propagation (BP) algorithm in order to minimize the error signal.
(4) The optimized model is tested on the sample image dataset to output the diagnosis results.

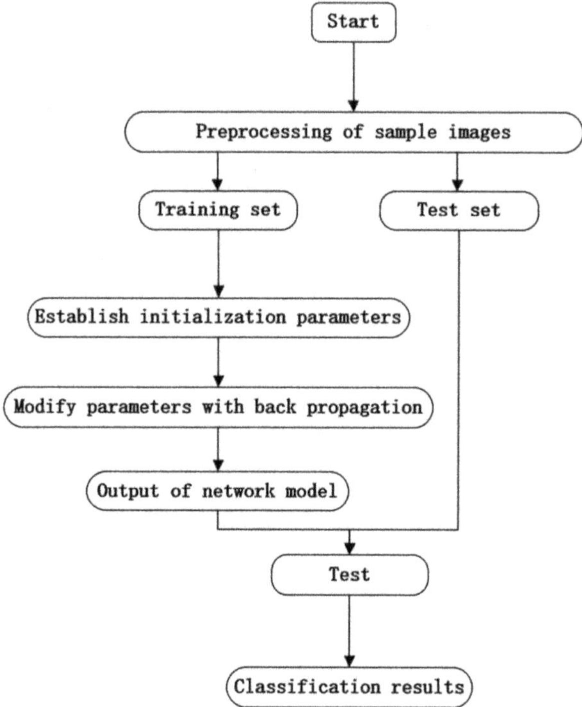

Figure 8. Algorithm flow.

3.4. Data Preprocessing

This study employs MATLAB/Simulink to build the ship electrical power system and takes the waveform of fault response curve as the input of the network fault model. Data preprocessing is divided into the following parts: data acquisition, image culling and data normalization.

(1) Data acquisition: In the simulation model, different faults in the "three-phase fault" fault module are set for different generators and loads to output the fault response curve. The file type is the JPG-format picture set recognized by CNN.
(2) Image culling: After converting the data into JPG-format pictures, some problems such as image overlap or feature blur will occur. These interfering images must be selected and eliminated to ensure the accuracy of the network training and test results.
(3) Data normalization: Owing to the difference between the orders of magnitude of the images, this difference will affect the results of the data analysis. To eliminate the influence between dimensions, data normalization is required. The min–max standardization method used in this study is the commonly used linear transformation

of data. The result value is mapped to the interval of (0,1). The transformation function is as follows:

$$x^* = \frac{x - min}{max - min} \tag{11}$$

where *max* and *min* are the maximum and minimum values of the sample data.

After preprocessing of the above-mentioned data, the image set with significant characteristics is used as the input of the CNN. The CNN model performs convolution, pooling, and other operations on the picture set to generate the output of the network model.

4. Experimental Results and Analysis

This study employed a Windows 10 system (Microsoft, Redmond, DC, USA) with the Python 3.7.11 (Python Software Foundation, Wilmington, DE, USA) compiling environment and TensorFlow learning framework to write the network programs. Because the image dataset used in this study was simple and regular, and the amount of data was small, the conventional method was used to adjust the parameters in order to optimize the CNN model.

4.1. Learning Rate

In the training network model, the learning rate is an important super-parameter that controls the speed of network weight adjustment. In general, the higher the learning rate, the faster is the learning of the network. However, if it is too high and reaches extreme values, the accuracy will be reduced; the loss value will stop falling and oscillate repeatedly at a certain position. The lower the learning rate, the slower the decrease in the loss gradient and the longer the convergence time. Therefore, it is crucial to choose an appropriate learning rate. By referring to numerous experiments as well as the literature, the learning rate of the network model was set to 0.0001.

4.2. Experimental Results

After several experiments, the optimal network fault diagnosis model is finally obtained. Compared with the original CNN model proposed, the average accuracy of the identification and classification of the ship electrical power system is up to 99%. This method makes the fault diagnosis of the ship electrical power system more convenient and reliable. The accuracy and loss variation diagrams were generated using PyCharm and TensorFlow frameworks, as shown in Figure 9a,b, respectively. The accuracy and loss variation diagrams of the original CNN are shown in Figure 9c,d, respectively. The accuracy of each fault category is listed in Table 6.

As can be seen from Figure 9a,b, the overall identification accuracy of the improved model for ship electrical power systems faults increases, and the loss rate decreases as the number of training epochs increases. After the improved model is trained once, the average accuracy of fault diagnosis reaches 97%, and the loss value is less than 0.1. After 4 times of model training, the average accuracy of fault diagnosis is 99%, and the loss value is less than 0.05. A comparison of Figure 9a–d shows that the recognition accuracy of the improved CNN after the first training epoch is higher than that of the original network after four training epochs. At the same time, the convergence speed of the loss value curve of the improved model is higher than that of the original model; the fluctuation range is smaller and is more stable after convergence. As can be seen from Table 6, the accuracy of the improved CNN for different faults at different locations is higher than that of the original network, indicating that the improved CNN provides good classification results for the fault identification of the ship electrical power systems; thus, it has considerable potential for the fault diagnosis of ship electrical power systems.

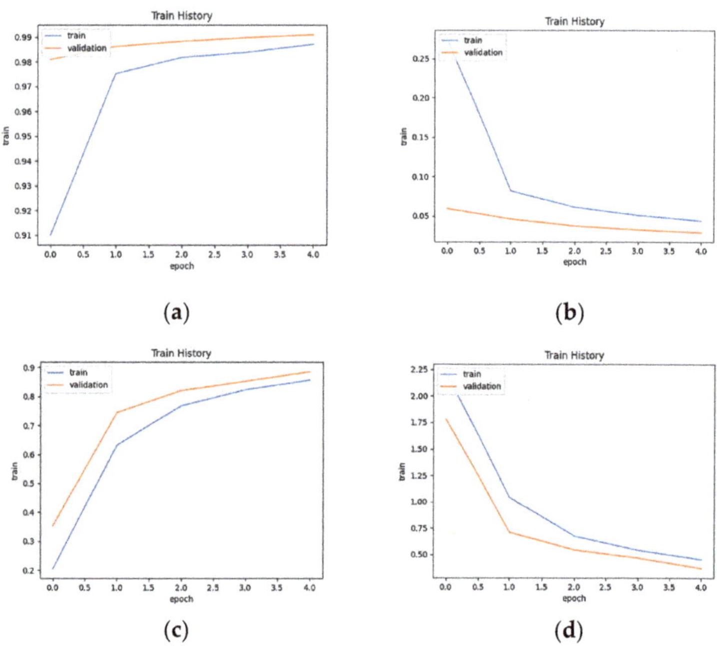

(a)

(b)

(c)

(d)

Figure 9. Experimental results of training network and original network: (**a**) accuracy of training network, (**b**) loss of training network, (**c**) accuracy of original network and (**d**) loss of original network.

Table 6. Accuracy of each fault category.

Fault Category	Fault Location	Improved CNN	CCN
Single-phase short-circuit grounding	Synchronous generator	99%	90%
	Asynchronous generator	99%	86%
	Main load	98%	91%
	Secondary load	99%	86%
Two-phase short- circuit	Synchronous generator	99%	87%
	Asynchronous generator	99%	90%
	Main load	99%	91%
	Secondary load	98%	88%
Two-phase short-circuit grounding	Synchronous generator	98%	90%
	Asynchronous generator	99%	90%
	Main load	99%	87%
	Secondary load	99%	91%
Three-phase short circuit	Synchronous generator	99%	86%
	Asynchronous generator	98%	87%
	Main load	99%	85%
	Secondary load	99%	88%
Single-phase disconnection	Synchronous generator	99%	86%
	Asynchronous generator	99%	89%
	Main load	98%	90%
	Secondary load	99%	88%
Two-phase disconnection	Synchronous generator	98%	90%
	Asynchronous generator	98%	87%
	Main load	99%	91%
	Secondary load	99%	87%

5. Conclusions

A fault diagnosis method for ship electrical power systems was proposed on the basis of an improved CNN to support the normal operation of ships. According to the results, the following conclusions can be drawn:

(1) To achieve multi-class fault diagnosis of different components of ship electrical power systems, this paper proposed an improved CNN fault diagnosis model, which can completely eliminate the subjectivity of manual feature extraction and expert experience, directly take the original fault data as the model input, automatically extract the fault features layer by layer in a nonlinear manner, and automatically output the fault classification results. Thus, "end-to-end" diagnosis from the original data to the fault category can be realized. The algorithm has a high fault recognition rate, and the evaluation accuracy is 99%.

(2) Through image culling, the accuracy of the network training results is improved significantly.

(3) The method used to realize fault diagnosis of the ship electrical power systems can also be used for fault diagnosis of other parts of more complex power systems. The simulation results showed that the improved model outperforms the original network. In particular, this method achieves high accuracy and reliability in ship electrical power systems fault diagnosis.

Nevertheless, there remains a scope for improvement in terms of the training time of the model. In the future, the simulation model will be improved such that it is more in line with actual ship electrical power systems. Based on the optimized CNN model, fault diagnosis accuracy can be improved further.

Author Contributions: Conceptualization, J.S. (Jie Sun); methodology, C.Y. and J.S. (Jie Sun); software, C.Y. and J.S (Jie Sun); validation, C.J., J.S (Jun Su) and W.S.; formal analysis, J.S. (Jie Sun) and J.S. (Jun Su); investigation, C.Y., J.S. (Jie Sun) and W.S.; resources, J.S. (Jie Sun) and C.J.; data curation, W.S.; writing—original draft, C.Y., L.Q. and J.S. (Jie Sun); writing—review & editing, C.Y. and L.Q.; visualization, J.S. (Jun Su); supervision, L.Q. and C.J.; project administration, L.Q. and C.J. All authors have read and agreed to the published version of the manuscript.

Funding: This research was funded by National Natural Science Foundation of China NO.51875270 and Industry University Research Collaboration of Jiangsu Province NO.BY2020031.

Conflicts of Interest: The authors declare no conflict of interest.

References

1. Al-Falahi, M.D.A.; Tarasiuk, T.; Jayasinghe, S.G.; Jin, Z.; Enshaei, H.; Guerrero, J.M. AC Ship Microgrids: Control and Power Management Optimization. *Energies* **2018**, *11*, 1458. [CrossRef]
2. Skjong, E.; Rødskar, E.; Molinas Cabrera, M.M.; Johansen, T.A.; Cunningham, J. The Marine Vessel's Electrical Power System: From its Birth to Present Day. *Proc. IEEE* **2015**, *103*, 2410–2424. [CrossRef]
3. Yu, C.; Huang, J.; Qi, L.; Gu, J. Health Status Evaluation of Radar Transmitter Based on Fuzzy Comprehensive Evaluation Method. In Proceedings of the 36th Youth Academic Annual Conference of Chinese Association of Automation (YAC), Nanchang, China, 28–30 May 2021; pp. 196–202. [CrossRef]
4. Maqsood, A.; Oslebo, D.; Corzine, K.; Parsa, L.; Ma, Y. STFT Cluster Analysis for DC Pulsed Load Monitoring and Fault Detection on Naval Shipboard Power Systems. *IEEE Trans. Transp. Electrif.* **2020**, *6*, 821–831. [CrossRef]
5. Ellefsen, A.L.; Æsøy, V.; Ushakov, S.; Zhang, H. A Comprehensive Survey of Prognostics and Health Management Based on Deep Learning for Autonomous Ships. *IEEE Trans. Reliab.* **2019**, *68*, 720–740. [CrossRef]
6. Sulligoi, G.; Vicenzutti, A.; Menis, R. All-Electric Ship Design: From Electrical Propulsion to Integrated Electrical and Electronic Power Systems. *IEEE Trans. Transp. Electrif.* **2016**, *2*, 507–511. [CrossRef]
7. Gunes, M.; Dogru, N. Fuzzy Control of Brushless Excitation System for Steam Turbogenerators. *IEEE Trans. Energy Convers.* **2010**, *25*, 844–852. [CrossRef]
8. Jiang, Y.; Luo, J. The Excitation Switching Control Method of Tidal Generator Based on T-S Fuzzy Weighting. *J. Coast. Res.* **2020**, *103*, 1010. [CrossRef]
9. Jiao, N.; Liu, W.; Meng, T.; Peng, J.; Mao, S. Detailed Excitation Control Methods for Two-Phase Brushless Exciter of the Wound-Rotor Synchronous Starter/Generator in the Starting Mode. *IEEE Trans. Ind. Appl.* **2017**, *53*, 115–123. [CrossRef]
10. Psillakis, H.E.; Alexandridis, A.T. Coordinated Excitation and Static Var Compensator Control with Delayed Feedback Measurements in SGIB Power Systems. *Energies* **2020**, *13*, 2181. [CrossRef]
11. Baek, S.M. Sensitivity Analysis Based Optimization for Linear and Nonlinear Parameters in AVR to Improve Transient Stability in Power System. *Int. J. Control. Autom.* **2015**, *8*, 69–80. [CrossRef]
12. Zhao, H.; Guo, C.; Wu, Z. Modeling and Simulation of marine electric propulsion system. *China Shipbuild.* **2006**, *47*, 51–56.

13. Jardine, A.; Lin, D.; Banjevic, D. A review on machinery diagnostics and prognostics implementing condition-based maintenance—ScienceDirect. *Mech. Syst. Signal Processing* **2006**, *20*, 1483–1510. [CrossRef]
14. Liu, R.; Yang, B.; Zio, E.; Chen, X. Artificial intelligence for fault diagnosis of rotating machinery: A review. *Mech. Syst. Signal Processing* **2018**, *108*, 33–47. [CrossRef]
15. Silva, A.A.; Gupta, S.; Bazzi, A.M.; Ulatowski, A. Wavelet-based information filtering for fault diagnosis of electric drive systems in electric ships. *ISA Trans.* **2017**, *78*, 105–115. [CrossRef] [PubMed]
16. Liu, W.; Liu, L.; Chung, I.Y.; Cartes, D.A.; Zhang, W. Modeling and detecting the stator winding fault of permanent magnet synchronous motors. *Simul. Model. Pract. Theory* **2012**, *27*, 1–16. [CrossRef]
17. Wang, W.; Xu, Z.; Tang, R.; Li, S.; Wu, W. Fault Detection and Diagnosis for Gas Turbines Based on a Kernelized Information Entropy Model. *Sci. World J.* **2014**, *2014*, 617162. [CrossRef]
18. Li, Z.; Jiang, Y.; Duan, Z.; Peng, Z. A new swarm intelligence optimized multiclass multi-kernel relevant vector machine: An experimental analysis in failure diagnostics of diesel engines. *Struct. Health Monit.* **2018**, *17*, 1503–1519. [CrossRef]
19. Cuculić, A.; Vučetić, D.; Prenc, R.; Ćelić, J. Analysis of Energy Storage Implementation on Dynamically Positioned Vessels. *Energies* **2019**, *12*, 444. [CrossRef]
20. Szczerba, Z. Automatic voltage regulator in power generation unit. *Przeglad Elektrotechniczny* **2009**, *85*, 55–57.
21. Ju, P. *Theory and Method of Power System Modeling*; Science Press: Beijing, China, 2010.
22. Gao, H.; Hu, X.; Wang, B.; Qi, H. Decentralized Fuzzy PID Excitation controller Combined with Turbine Regulating for Voltage Stability in Power Systems. *Prz. Elektrotechniczny* **2012**, *88*, 284–288.
23. Yee, S.K.; Milanovic, J.V.; Hughes, F.M. Overview and Comparative Analysis of Gas Turbine Models for System Stability Studies. *IEEE Trans. Power Syst.* **2008**, *23*, 108–118. [CrossRef]
24. Kim, H.; Degner, M.W.; Guerrero, J.M.; Briz, F.; Lorenz, R.D. Discrete-Time Current Regulator Design for AC Machine Drives. *IEEE Trans. Ind. Appl.* **2010**, *46*, 1425–1435.
25. Ji, H.-K.; Wang, G.; Kil, G.-S. Optimal Detection and Identification of DC Series Arc in Power Distribution System on Shipboards. *Energies* **2020**, *13*, 5973. [CrossRef]
26. Jeon, W.; Wang, Y.P.; Jeong, J.H.; Lyu, S.K.; Jung, S.Y. Power Characteristic Analysis of Assumed Short Circuit Instance of Electric Ship Propulsion System. *Korea Ocean. Eng. Soc.* **2008**, *32*, 323–329.
27. Martín, S.S.; Fernández, M.B. Model and performance simulation for overcurrent relay and fault-circuit-breaker using Simulink. *Int. J. Electr. Eng. Educ.* **2006**, *43*, 80–91. [CrossRef]
28. Huang, J.; Qi, L.; Gu, J.; Lu, Z.; Sun, J.; Yu, C. Servo Motor Fault Diagnosis Based on Data Fusion. In Proceedings of the 33rd Chinese Control and Decision Conference (CCDC), Kunming, China, 22–24 May 2021; pp. 104–110. [CrossRef]
29. Guo, Y.; Chen, Y.; Tan, M.; Jia, K.; Chen, J.; Wang, J. Content-aware convolutional neural networks. *Neural Netw.* **2021**, *143*, 657–668. [CrossRef]
30. Arulmozhi, P.; Abirami, S. DSHPoolF: Deep supervised hashing based on selective pool feature map for image retrieval. *Vis. Comput.* **2021**, *37*, 2391–2405. [CrossRef]

Article

An Improved Hidden Markov Model for Monitoring the Process with Autocorrelated Observations

Yaping Li [1,*], Haiyan Li [1], Zhen Chen [2] and Ying Zhu [2]

1   College of Economics and Management, Nanjing Forestry University, Nanjing 210037, China; lihaiyan117@njfu.edu.cn
2   Department of Industrial Engineering & Management, Shanghai Jiao Tong University, Shanghai 200240, China; chenzhendr@sjtu.edu.cn (Z.C.); zhuyingme@sjtu.edu.cn (Y.Z.)
*   Correspondence: yapingli@njfu.edu.cn

Abstract: With the development of intelligent manufacturing, automated data acquisition techniques are widely used. The autocorrelations between data that are collected from production processes have become more common. Residual charts are a good approach to monitoring the process with data autocorrelation. An improved hidden Markov model (IHMM) for the prediction of autocorrelated observations and a new expectation maximization (EM) algorithm is proposed. A residual chart based on IHMM is employed to monitor the autocorrelated process. The numerical experiment shows that, in general, IHMMs outperform both conventional hidden Markov models (HMMs) and autoregressive (AR) models in quality shift diagnosis, decreasing the cost of missing alarms. Moreover, the times taken by IHMMs for training and prediction are found to be much less than those of HMMs.

Keywords: hidden Markov model (HMM); autocorrelation; residual chart

Citation: Li, Y.; Li, H.; Chen, Z.; Zhu, Y. An Improved Hidden Markov Model for Monitoring the Process with Autocorrelated Observations. *Energies* **2022**, *15*, 1685. https://doi.org/10.3390/en15051685

Academic Editor: Abu-Siada Ahmed

Received: 17 January 2022
Accepted: 16 February 2022
Published: 24 February 2022

Publisher's Note: MDPI stays neutral with regard to jurisdictional claims in published maps and institutional affiliations.

1. Introduction

Process monitoring plays an essential role in intelligent manufacturing [1,2]. Statistical process control (SPC) is a quality control technique that uses statistical methods to monitor and control processes. The aim of SPC is to ensure that the process runs efficiently, producing more products that meet specifications, while reducing waste at the same time. Shewhart control charts are a key SPC tool used to determine whether a process is in control. If the observation value is within the upper and lower control limits, the process is in control; otherwise, the process is out of control. Shewhart charts often assume that the data collected from the process are independent. However, this assumption is not true in a variety of processes. For example, consecutive measurements on chemical and pharmaceutical processes or product characteristics are often highly autocorrelated, or the chronological measurements of every characteristic on every unit in automated test and inspection procedures are often autocorrelated. It has been shown in numerous studies that conventional charts do not work well, in the form of giving too many false alarms if the observations exhibit even a small amount of autocorrelation over time [3–9]. Clearly, better approaches are needed. In the following paragraphs, three common ways to solve the problems related to autocorrelation are introduced.

The first approach to solving an autocorrelation problem is simply to sample from the observation data stream at a lower frequency [10]. This seems to be an easy solution, although it has some shortcomings. Concretely speaking, it makes inefficient use of the available data. For example, if every tenth observation is sampled, approximately 90% of the information is discarded. In addition, since only every tenth datum is used, this approach may delay the discovery of a process shift.

The second approach is to re-estimate the real process variance, aiming to revise the upper and lower control limits. See, for example, [9,11–19].

The third approach is to use residual charts. The statistics in these charts are residuals; values are calculated by subtracting predicted values from observed values. In the implementation of residual charts, the key step is choosing a reasonable prediction model to obtain predicted values. Autoregressive integrated moving average (ARIMA) models are mostly used to model autocorrelated data. See, for example, [3,8,20–26]. In addition, multistage or multivariate autocorrelated processes are mostly studied by using Hotelling T^2 control charts, such as in [27–30]. Some exceptions include Pan et al., who proposed an overall run length (ORL) to replace T^2 charts [31], and S. Yang and C. Yang, who used a residual chart and cause-selecting control chart [32].

With the rapid development of artificial intelligence, machine-learning methods are becoming more and more popular in SPC in the case of observation autocorrelation [33–38]. Most of the related literature focuses on neural networks methods [39–42]. Some successful applications have also been introduced, for example, the failure diagnosis of wind turbine [43], and the prediction of the remaining useful life (RUL) of bearings [44]. Another machine-learning approach proposed in SPC is the hidden Markov model (HMM). Lee et al. proposed a modified HMM, combined with a Hotelling multivariate control chart to perform adaptive fault diagnosis [45]. The HMMs, whose training sets contain autocorrelated data, were employed to forecast observation values for residual charts in process monitoring [46]. However, although HMMs are supposed to deal with the autocorrelated processes, the essence of the model itself is inconsistent with the case of autocorrelations. This is because one key assumption in conventional HMMs is that the observations are independent of each other. Therefore, it is worth developing a modified HMM, by accounting for observation autocorrelations in models themselves, and observing whether it is better than a traditional HMM.

Therefore, to realize the goal of monitoring the autocorrelated process well, the residual chart based on an improved hidden Markov model (IHMM) with autocorrelated observations considered is developed. Due to the autocorrelations, the conventional expectation maximization (EM) algorithm for HMMs is not appropriate. A new EM algorithm is developed for the solutions. The Shewhart residual chart is employed in quality shift detection in conjunction with the IHMM. The residual is defined as the deviation of the predicted value by the IHMM and the current real observation value. Through the residual chart, we are able to see whether the process is in control. If the chart initiates an alarm, one running length is obtained. Thus, the average running lengths (ARLs) can be calculated with sufficient samples. The ARL is set as the comparison index for different models, including IHMM, HMM and AR.

The rest of this paper is organized as follows: Section 2 introduces the development of the IHMM model and its algorithm. In addition, the comparison of the prediction performances of different approaches is presented in Section 2. In Section 3, residual charts are introduced. In Section 4, numerical examples and interesting results are presented. The conclusions and possible areas for future research are given in Section 5.

2. Model Development

2.1. Hidden Markov Models

Denote a Markov chain with a finite state set $\{s_1, s_2, \cdots, s_N\}$ by $\{S_n, n = 1, 2, \cdots\}$. Let a_{ij} be the probabilities that the Markov chain enters state s_j from state s_i $(1 \leq i, j \leq N)$ and $\pi_i = P\{S_1 = s_i\}, i = 1, 2, \cdots, N$ be the initial state probabilities. Denote a finite set of signals by ζ, and suppose a signal from ζ is sent each time the Markov chain enters a state. Suppose that when the Markov chain enters state s_j independently of previous states and signals, the signal sent is o_r with probability $p(o_r|s_j)$ that meets $\sum_{o_r \in \zeta} p(o_r|s_j) = 1$. That is, if O_n represents the nth signal, then it can be written by

$$P\{O_1 = o_r|S_1 = s_j\} = p(o_r|s_j), \tag{1}$$

$$P\{O_n = o_r|S_1, O_1, \cdots, S_{n-1}, O_{n-1}, S_n = s_j\} = p(o_r|s_j). \tag{2}$$

Such a model, in which the signal sequence $O_1, O_2, \cdots$ can be observed while the underlying Markov chain state sequence $S_1, S_2, \cdots$ cannot be observed, is called a hidden Markov model [47].

Normally, an HMM contains the following elements [48]:

- A finite hidden state set $s = \{s_1, s_2, \cdots, s_N\}$, where N is the hidden state number.
- A set of possible observation values (signals) $q = \{q_1, \cdots, q_K\}$, where K is the number of possible observation values. Note that if the values of the observations are continuous, K should be infinite.
- An observation sequence, $o = (o_1, \cdots, o_T)$, where T is the number of observations, $o_t \in q, 1 \le t \le T$.
- A distribution of state transition probabilities, $A = \{a_{ij}\}$, where

$$a_{ij} = P\{S_{t+1} = s_j | S_t = s_i\}, 1 \le i,j \le N, \sum_j a_{ij} = 1. \tag{3}$$

- A distribution of initial state probabilities, $\pi = \{\pi_i\}$, where

$$\pi_i = P\{S_1 = s_i\}, i = 1, 2, \cdots, N, \sum_i \pi_i = 1. \tag{4}$$

- A conditional probability distribution of the observations, given $S_t = s_i$, $B = \{b_i(q_k)\}$, where

$$b_i(q_k) = P\{o_t = q_k | S_t = s_i\}, 1 \le i \le N, 1 \le t \le T, \sum_{q_k} b_i(q_k) = 1. \tag{5}$$

In general, an HMM contains three elements, denoted by $\lambda = \{A, B, \pi\}$.

2.2. The Improved HMM with Autocorrelated Observations

2.2.1. The Selection for the Order of Autocorrelation

Different from the traditional HMMs, in which current observations are assumed to be independent of previous states and observations, autocorrelated observations are considered. The observations are assumed to follow a Gaussian distribution, and the current observations are required to be dependent, not only on the current hidden state, but also on the previous observations.

The autocorrelated data from production may be multi-order, and it seems not cost-effective for judging this detailed order by using engineering experience and professional knowledge. Therefore, it is hoped to find a suitable order that will keep the implementation of the IHMM feasible, efficient and cost-effective.

The numerical experiment on the order selection for autocorrelated data in SPC application is designed as follows:

Step 1. Generate stationary autocorrelated data by a dth-order autoregressive model, $\text{AR}(d), d \ge 2$;

Step 2. Use $\text{AR}(p), p = d, d-1, \ldots, 1$ models to fit the data, respectively, and obtain the parameters that need to be estimated by least-squares estimation;

Step 3. Generate 2000 stationary data sequences with a same length of 2000 under the processes with a mean shift magnitude of δ;

Step 4. Predict each sequence by using $\text{AR}(p), p = d, d-1, \ldots, 1$ models, respectively, and calculate their residuals;

Step 5. Determine the central lines (CLs), upper control limits (UCLs) and lower control limits (LCLs) of p residual charts;

Step 6. Generate stationary autocorrelated data by a dth-order autoregressive model;

Step 7. Calculate the average running lengths (ARLs) of p residual charts.

In this study, twelve AR(2) and eight AR(3) models are used to generate data. The control limit coefficients of residual charts are 3. The shift magnitudes of the mean are set by 0, 1.5 and 3, respectively.

When data are generated by AR(2) models for an in-control process, the ARL of the residual charts is 371.3803 if the data are fitted by AR(2) models, and the ARL is 361.1688 if the data are fitted by AR(1) models. Thus, the probability of the type I error is increased by 2.83%. Similarly, when data are generated by AR(3) models, the ARL of the residual charts is 371.8250 if the data are fitted by AR(3) models, the ARL is 358.5401 if the data are fitted by AR(2) models, and the ARL is 345.8113 if the data are fitted by AR(1) models. Thus, the probabilities of the type I error are increased by 3.7% and 7.51%, respectively.

The ARLs of different situations are shown in Figure 1, in which the numbers on the x-axis represent the combinations of autocorrelation coefficients.

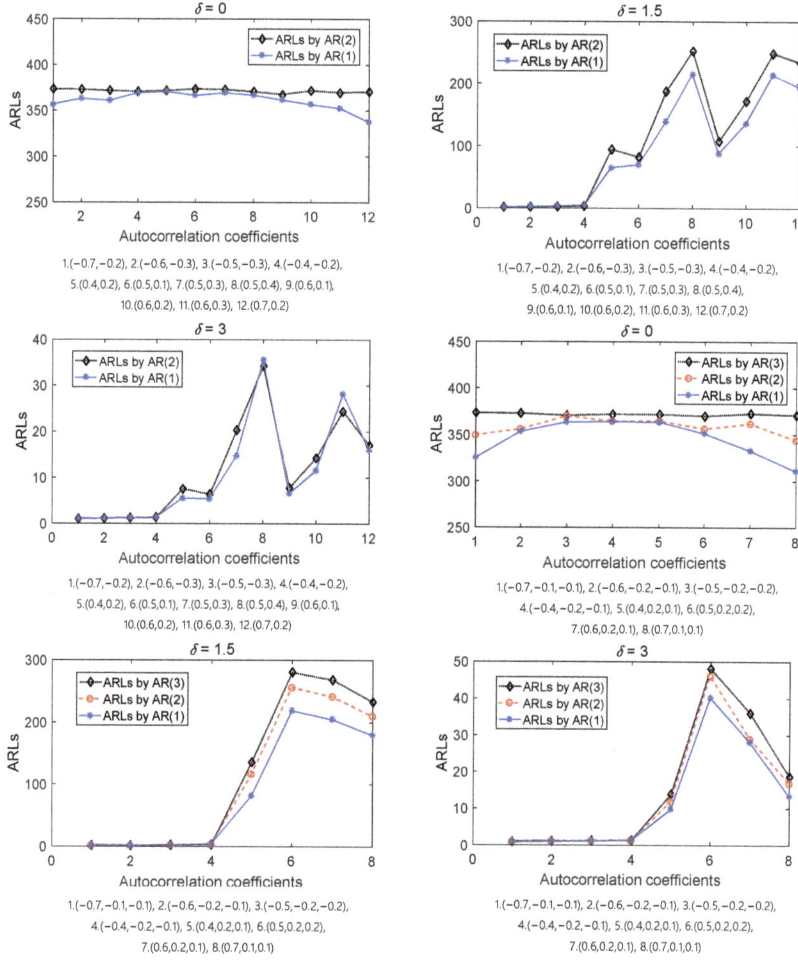

Figure 1. The ARL results of the order selection experiment.

From Figure 1, it is seen that generally, AR(1) models outperformed other AR(d), $d \geq 2$ models in shift detection regardless of the autocorrelation order. Although AR(1) models lead to a slight increase in type I error, it seems to be insignificant compared with their good performances in detecting quality shifts. Therefore, only the first-order autocorrelation is considered in this study. As a result, there is no need to judge what order the autocorrelation is so that the modeling cost can be saved.

2.2.2. The Development of the IHMM

According to the analysis in Section 2.2.1, the construction of the IHMM, in which the current observation is related to its previous observation, is shown in Figure 2.

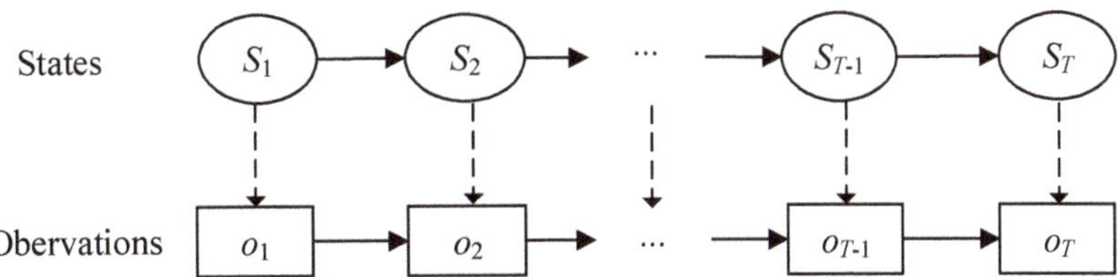

Figure 2. The construction of the IHMM.

In Figure 2, arrows pointing to the right indicate the correlation between neighboring states or observations. It is intuitive that the IHMM will become a traditional HMM if there is no correlation between neighboring observations.

Let $o_i(t)$ be the observation value at time t given state s_i; $o_i(t)$ can be fitted with the following function:

$$o_i(t) = \varsigma_i + c_i o_{t-1} + \varepsilon_i, 1 \le i \le N, 1 \le t \le T,\tag{6}$$

where c_i is the first-order autocorrelation coefficient, ς_i is the constant term and ε_i is white noise, following a normal distribution with mean zero and variance σ_i^2.

Let $\mu_i(t)$ be the mean of $o_i(t)$ given state s_i, $x_t = (1, o_{t-1})'$, then Equation (6) can be written as:

$$\mu_i(t) = C_i x_t,\tag{7}$$

where C_i is a 1×2 matrix consisting of $\{\varsigma_i\}$ and $\{c_i\}$.

Hence, we conclude that the states are a Markov process, and that the conditional observations given state s_i follows a Gaussian distribution with mean $C_i x_t$ and variance σ_i^2, i.e.,

$$P\{S_{t+1}|S_1, \cdots, S_t, o_1, \cdots, o_t\} = p(S_{t+1}|S_t),\tag{8}$$

$$b_i(o_l|o_{l-1}) - P(o_l|o_{l-1}, S_l - s_i) - \frac{1}{\sqrt{2\pi}\sigma_i} \exp\left(-\frac{(o_t - C_i x_t)^2}{2\sigma_i^2}\right).\tag{9}$$

If the current observation is not related to its previous observations, $\mu_i(t)$ is the constant ς_i. Thus, the IHMM becomes a traditional HMM.

2.3. Parameter Estimation

The aim of using an IHMM is to forecast observation values. Firstly, we estimate the parameters to obtain an optimal $\hat{\lambda}$ by maximizing the probability $P(o|\lambda)$.

The EM algorithm is a popular method to estimate the parameters for HMMs. However, since autocorrelation between observations is considered, the traditional algorithm could not be used directly. We redefine the parameters as $\lambda = \{A, C, \sigma^2, \pi\}$ where $C = \{C_i\}, \sigma^2 = \{\sigma_i^2\}$, and the definitions of C_i and σ_i^2 are provided in Section 2.2.2. The flowchart demonstrating how to estimate the parameters in IHMM with an improved EM algorithm is shown in Figure 3.

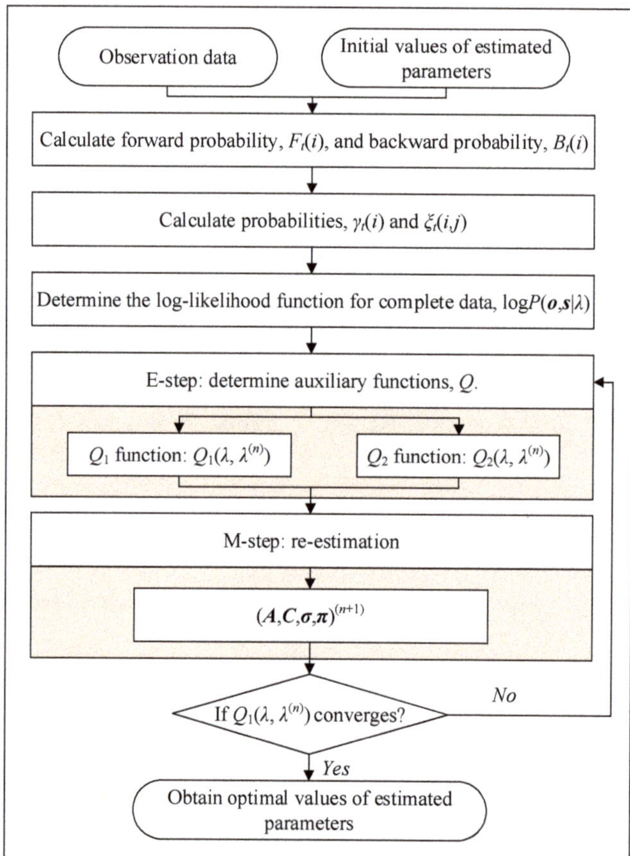

Figure 3. The flowchart of the improved EM algorithm.

Firstly, we introduce two types of probabilities: forward probability, $F_t(i)$, and backward probability, $B_t(i)$, defined as:

$$F_t(i) = P(o_1, \cdots, o_t, S_t = s_i | \lambda),\tag{10}$$

$$B_t(i) = P(o_{t+1}, \cdots, o_T | S_t = s_i, \lambda).\tag{11}$$

The calculation of the two probabilities here is similar to that of traditional HMMs. Slightly different from traditional HMMs, the initial values are defined by $F_{d+1}(i) = \pi_i b_i(o_{d+1} | o_1, \cdots, o_d)$ and $B_T(i) = 1$, respectively.

Based on the forward and backward probabilities, some intermediate probabilities are computed. Given λ and o, denote the probability that the process is in state s_i at time t by $\gamma_t(i)$ and the probability that the process is in state s_i at time t and in state s_j at time $t + 1$ by $\xi_t(i, j)$. $\gamma_t(i)$ and $\xi_t(i, j)$ can be written as Equations (12) and (13). Please refer to reference [48] for the details of the derivations.

$$\gamma_t(i) = \frac{F_t(i)B_t(i)}{\sum_j F_t(i)B_t(i)},\tag{12}$$

$$\xi_t(i, j) = \frac{F_t(i)a_{ij}b_j(o_{t+1})B_{t+1}(j)}{\sum_{i=1}^{N}\sum_{j=1}^{N} F_t(i)a_{ij}b_j(o_{t+1})B_{t+1}(j)}.\tag{13}$$

Next, we develop an improved EM algorithm for the IHMM, step by step.

Step 1. Determine the log-likelihood function for complete data.

The complete data are $(o, S) = (o_1, \cdots, o_T, s_1, \cdots, s_T)$, and its log-likelihood function is $\log P(o, s|\lambda)$.

Step 2. E-step: determine Q functions.

$$Q_1\left(\lambda, \lambda^{(n)}\right) = E_s\left[\log P(o, s|\lambda)\big| o, \lambda^{(n)}\right] = \sum_s \log P(o, s|\lambda) P\left(o, s\big|\lambda^{(n)}\right), \tag{14}$$

where λ is the parameter that will maximize the Q_1 function, while $\lambda^{(n)}$ is the current parameter value.

With $P(o, s|\lambda) = \pi_{s_1} b_{s_1}(o_1) a_{s_1 s_2} b_{s_2}(o_2) \cdots a_{s_{T-1} s_T} b_{s_T}(o_T)$, $Q_1\left(\lambda, \lambda^{(n)}\right)$ can be written by

$$\begin{aligned}
Q_1\left(\lambda, \lambda^{(n)}\right) &= \sum_s \pi_{s_1} \log P\left(o, s\big|\lambda^{(n)}\right) \\
&+ \sum_s \left(\sum_{t=1}^{T-1} \log a_{s_t s_{t+1}}\right) P\left(o, s\big|\lambda^{(n)}\right) \\
&+ \sum_s \left(\sum_{t=1}^{T} \log b_{s_t}(o_t)\right) P\left(o, s\big|\lambda^{(n)}\right).
\end{aligned} \tag{15}$$

For the re-estimation of $b_i(q_k)$, one more auxiliary function Q_2 is proposed by taking the conditional expectation of the log-likelihood of the observation sequence:

$$Q_2\left(\lambda, \lambda^{(n)}\right) = \sum_s \sum_{t=1}^{T} \gamma_t(i) \left(\ln\left(\frac{1}{\sqrt{2\pi}}\right) + \ln\frac{1}{\sigma_i} - \frac{(o_t - C_i x_t)^2}{2\sigma_i^2}\right). \tag{16}$$

Step 3. M-step: re-estimation.

Re-estimate λ that maximizes $Q_1\left(\lambda, \lambda^{(n)}\right)$, that is

$$\lambda^{(n+1)} = \underset{\lambda}{\arg\max} Q_1\left(\lambda, \lambda^{(n)}\right). \tag{17}$$

The state transition probabilities are derived as:

$$a_{ij}^{(n+1)} = \frac{\sum_{t=1}^{T-1} \xi_t(i, j)}{\sum_{t=1}^{T-1} \gamma_t(i)}. \tag{18}$$

The initial state probabilities are derived as:

$$\pi_i^{(n+1)} = \gamma_1(i). \tag{19}$$

By re-estimating λ that maximizes $Q_2\left(\lambda, \lambda^{(n)}\right)$, $c_{i,\tau}^{(n+1)}, \varsigma_i^{(n+1)}, \sigma_i^{2(n+1)}$ can be written as:

$$c_i^{(n+1)} = \frac{\sum_{t=1}^{T} \gamma_t(i) o_{t-1} \left(o_t - \frac{\sum_{t=1}^{T} \gamma_t(i) o_t}{\sum_{t=1}^{T} \gamma_t(i)}\right)}{\sum_{t=1}^{T} \gamma_t(i) o_{t-1} \left(o_{t-1} - \frac{\sum_{t=1}^{T} \gamma_t(i) o_{t-1}}{\sum_{t=1}^{T} \gamma_t(i)}\right)}, \tag{20}$$

$$\varsigma_i^{(n+1)} = \frac{\sum_{t=1}^{T} \gamma_t(i) o_t - \sum_{t=1}^{T} \gamma_t(i) o_{t-1} c_i^{(n+1)}}{\sum_{t=1}^{T} \gamma_t(i)}, \tag{21}$$

$$\sigma_i^{2(n+1)} = \frac{\sum_{t=1}^{T} \gamma_t(i) \left(o_t - c_i^{(n+1)} o_{t-1} - \varsigma_i^{(n+1)}\right)^2}{\sum_{t=1}^{T} \gamma_t(i)}. \tag{22}$$

Repeat Step 2 and Step 3 until the log-likelihood function converges.

2.4. Prediction

Once the parameter is determined by the improved EM algorithm, the model can be employed to forecast the expected value of the next observation.

The conditional probability distribution of the observation o_{T+1} can be derived as

$$
\begin{aligned}
P(o_{T+1}|o_T,\cdots,o_1,\lambda) &= \sum_{i=1}^{N} P(o_{T+1}|S_T = s_i, o_T,\cdots,o_1,\lambda) P(S_T = s_i|o_T,\cdots,o_1,\lambda) \\
&= \sum_{i=1}^{N}\sum_{j=1}^{N} P\Big(o_{T+1}\Big|S_{T+1} = s_j, o_T,\cdots,o_1,\lambda\Big) P\Big(S_{T+1} = s_j|S_T = s_i,\lambda\Big) \gamma_T(i) \\
&= \sum_{i=1}^{N}\sum_{j=1}^{N} \gamma_T(i) a_{ij} b_j(o_{T+1}|o_T,\cdots,o_{T+1-d}).
\end{aligned}
\tag{23}
$$

Therefore, the expectation of o_{T+1} is computed by

$$
\hat{o}_{T+1} = \int o_{T+1} P(o_{T+1}|o_T,\cdots,o_1,\lambda) do_{T+1}.
\tag{24}
$$

Given an observation sequence $o = (o_1,\cdots,o_T)$, $\hat{o}_t$ is predicted by

$$
\hat{o}_t = \int o_t P(o_t|o_{t-1},\lambda) do_t = \int o_t \sum_{i=1}^{N}\sum_{j=1}^{N} \gamma_{t-1}(i) b_j a_{ij}(o_t|o_{t-1}) do_t, t \geq d.
\tag{25}
$$

Thus, the predicted values of the observations can be denoted by $\hat{o} = (o_1,\hat{o}_2,\cdots,\hat{o}_T)$.

2.5. Performance Comparison

The mean squared error (MSE), absolute mean error (AME) and mean absolute percentage error (MAPE) are common tools for measuring, fitting and predicting accuracy [49]. Both MSE and AME values determine the average deviation between fitting values and original values, while MAPE provides a measurement for testing the relevant difference between them. In this study, we use MSE as the criterion to evaluate the models. The equation for MSE is:

$$
\mathrm{MSE} = \frac{1}{LT} \sum_{l=1}^{L} \sum_{t=1}^{T} (o_t - \hat{o}_t)^2,
\tag{26}
$$

where L is the number of predicted samples, and T is the length of each sample.

We assume that a variable from a production process follows a normal distribution with mean 100 and variance 25 when the process is under control, and that the observations are first-order autocorrelated with a correlation coefficient of 0.6. We use IHMM, HMM and AR(1) methods to predict observation values, respectively. The MSEs for the three models are 15.1276, 16.1867 and 15.7861, respectively. Since these MSEs are very close, we conclude that the predicted performances of the three approaches are similar. The predicted results of an observation sequence with a length of 50 from the in-control process are shown in Figure 4, from which we can see that the three models have close performances. However, the time taken for prediction using the three models are quite different. For the prediction of an observation sequence with a length of 50, IHMM is 14.6523 s, HMM is 93.6521 s, and AR(1) is almost instantaneous under the environment of win10 OS (Microsoft, Redmond, WA, USA) with CPU of Intel(R) Core(TM) i7-7500U (Santa Clara, CA, USA).

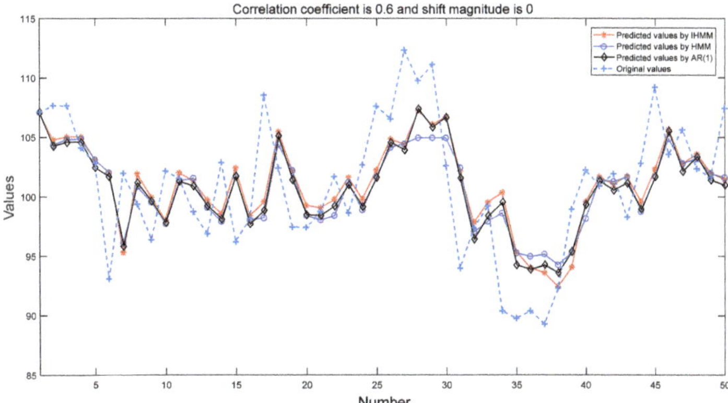

Figure 4. The predicted results of three models under an in-control process.

Then, we suppose the process has a shift magnitude of 3. We still use the three methods to predict observation values, respectively. The predicted results of an observation sequence with a length of 50 from the out-of-control process are shown in Figure 5, from which we can see distinctly different performances of the three models. By observing the distances between the lines with different colors, obviously, if the MSEs are calculated, the MSE from AR(1) is much less than that from IHMM, and the MSE from IHMM is much less than that from HMM. The IHMM only results in a medium-level performance in the prediction for the autocorrelated process. However, it is very interesting that the performances of corresponding residual charts have the best performances in detecting quality shifts. This seems to suggest that the residual charts integrating IHMM can achieve a surprising effect. This result is verified in Section 3.

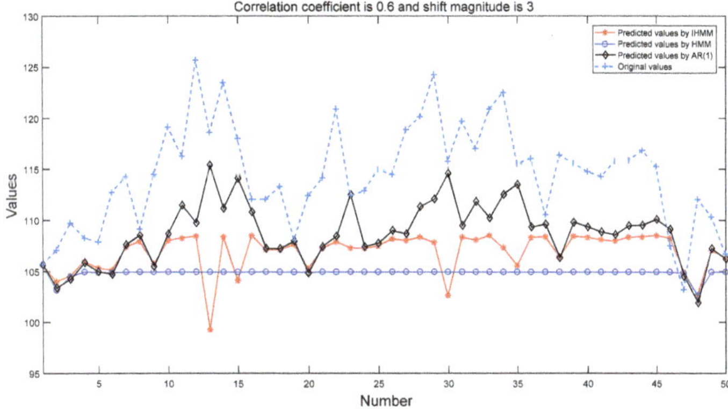

Figure 5. The predicted results of three models under an out-of-control process.

3. Statistical Process Control with Residual Chart

Residual control charts are an effective tool for online monitoring in the presence of autocorrelations. A residual chart called e chart is developed in our study.

Residuals are obtained by subtracting the predicted values of observations from the original values, that is $e = o - \hat{o} = (e_1, \cdots, e_T)$. The control limits of the e chart are given by

$$UCL = \mu_e + k\sigma_e, \tag{27}$$

$$LCL = \mu_e - k\sigma_e. \tag{28}$$

where UCL represents the upper limit, while LCL represents the lower limit, k is the number of σ_e, μ_e represents the mean of e, and σ_e represents the standard deviation of e. μ_e and σ_e can be obtained by simulations based on sufficient samples.

If the value of e_t drops within UCL and LCL, the process is judged to be in control; otherwise, it is judged to be out of control.

4. Numerical Examples

We consider that the variable from a production process followed a normal distribution with mean 100 and variance 25 when the process is in control and that the observations were first-order autocorrelated. The correlation coefficient varies between -0.6 and 0.6 with increments of 0.3. Two shift magnitudes of 1.5 and 3 are considered. According to the definition of residual charts, the ARLs of in-control processes for all predicted methods are 370, so we focus our discussion on the out-of-control processes. By conducting multiple experiments, we find that it is appropriate to make the state number with 5 for both IHMM and HMM. The experimental results are shown in Figures 6 and 7.

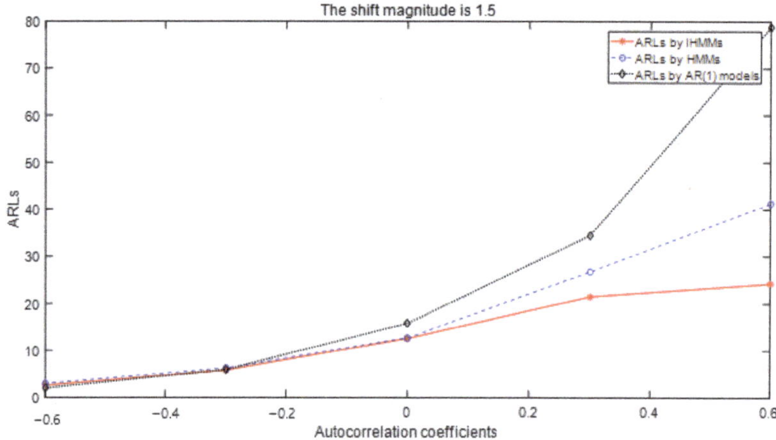

Figure 6. The ARLs of residual charts obtained by different models when the shift magnitude is 1.5.

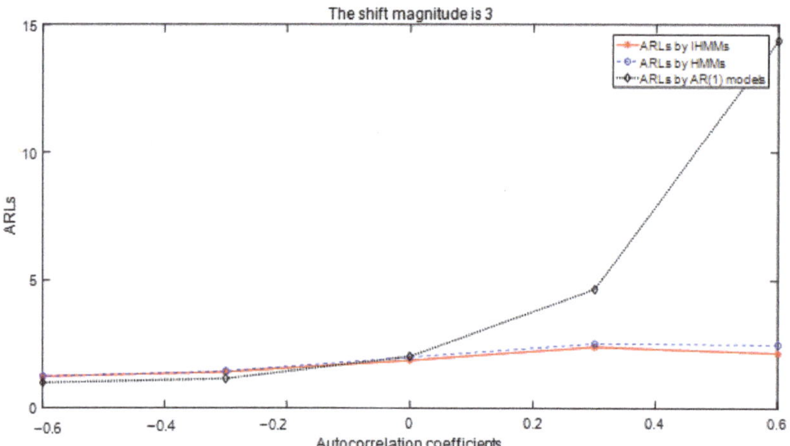

Figure 7. The ARLs of residual charts obtained by different models when the shift magnitude is 3.

As shown in Figures 6 and 7, when correlation coefficient changes from positive to negative, ARLs decrease dramatically, regardless of the approach used. Compared with positive correlations, the ARLs of negative correlations are relatively very small, and ARLs obtained by different models are very close to each other. Thus, the following discussions focus on positive correlations.

As pointed out in Section 3, although the predictions of both the IHMM and HMM are inferior to AR(1) models, the performances of residual charts from the former models are much better than the latter ones. As seen in Figures 4 and 5, when the coefficients are larger than zero, the ARLs by IHMMs are shorter than those by HMMs, and by HMMs shorter than by AR(1) models.

As correlation coefficients increase, the ARLs generally increase, regardless of the approach used, especially as the shift magnitude decreases.

Generally speaking, when detecting quality shifts, the performances of IHMM, HMM and AR(1) models are ranked with IHMMs first, HMMs second and AR(1) last. Moreover, as pointed out in Section 2, the times taken by IHMMs are much shorter than HMMs under the same running environments.

5. Conclusions

In this paper, an IHMM with autocorrelated observations and a new EM algorithm are proposed. Residual charts in conjunction with the IHMM are employed for detecting quality shifts. The results demonstrates that: (1) the IHMM outperforms the HMM and AR(1) method with positive correlations; (2) the IHMM has similar performances with the HMM and AR(1) methods with negative correlations; (3) compared with positive correlations, the ARLs of the IHMM under negative correlations are relatively very small, as well as those of the HMMs and AR(1) models; (4) the IHMMs take a much shorter time than HMMs, for both training and prediction, but still longer than the AR(1) models.

Future research might focus on further experimental validations for the IHMM and its algorithm. The strict Gaussian distribution of observations could be extended to other probability distributions. Since multistage systems are commonplace in the manufacturing industry, it is worth extending this approach in this research direction.

Author Contributions: Conceptualization, Y.L. and Z.C.; methodology, Y.L. and Z.C.; software, Y.L.; validation, Y.L. and H.L.; formal analysis, Y.L. and H.L.; investigation, Y.L. and Y.Z.; resources, Y.L.; data curation, Y.L.; writing—original draft preparation, Y.L.; writing—review and editing, Y.L.; visualization, Y.L.; supervision, Y.L.; project administration, Y.L.; funding acquisition, Y.L. and Z.C. All authors have read and agreed to the published version of the manuscript.

Funding: This research was funded by the National Natural Science Foundation of China, grant number 72171120, 71701098, 72001138 and Qing Lan Project of Jiangsu Province in China, grant number 2021.

Institutional Review Board Statement: Not applicable.

Informed Consent Statement: Not applicable.

Data Availability Statement: Data are contained within the article.

Conflicts of Interest: The authors declare no conflict of interest.

References

1. Ouyang, L.; Chen, J.; Park, C.; Ma, Y.; Jin, J. Bayesian closed-loop robust process design considering model uncertainty and data quality. *IISE Trans.* **2020**, *52*, 288–300. [CrossRef]
2. Ouyang, L.; Zheng, W.; Zhu, Y.; Zhou, X. An interval probability-based FMEA model for risk assessment: A real-world case. *Qual. Reliab. Eng. Int.* **2020**, *36*, 125–143. [CrossRef]
3. Montgomery, D.C.; Mastrangelo, C.M. Some Statistical Process Control Methods for Autocorrelated Data. *J. Qual. Technol.* **1991**, *23*, 179–204. [CrossRef]
4. Maragah, H.D.; Woodall, W.H. The Effect of Autocorrelation on the Retrospective X-chart. *J. Stat. Comput. Simul.* **1992**, *40*, 29–42. [CrossRef]
5. Runger, G.C. Assignable Causes and Autocorrelation: Control Charts for Observations or Residuals? *J. Qual. Technol.* **2002**, *34*, 165–170. [CrossRef]

6.   Franco, B.C.; Celano, G.; Castagliola, P.; Costa, A.F.B.; Machado, M.A.G. A New Sampling Strategy for the Shewhart Control Chart Monitoring A Process with Wandering Mean. *Int. J. Prod. Res.* **2015**, *53*, 4231–4248. [CrossRef]
7.   Kim, J.; Jeong, M.K.; Elsayed, E.A. Monitoring Multistage Processes with Autocorrelated Observations. *Int. J. Prod. Res.* **2017**, *55*, 2385–2396. [CrossRef]
8.   Yang, H.H.; Huang, M.L.; Lai, C.M.; Jin, J.R. An Approach Combining Data Mining and Control Charts-Based Model for Fault Detection in Wind Turbines. *Renew. Energy* **2018**, *115*, 808–816. [CrossRef]
9.   Li, Y.; Pan, E.; Xiao, Y. On autoregressive model selection for the exponentially weighted moving average control chart of residuals in monitoring the mean of autocorrelated processes. *Qual. Reliab. Eng. Int.* **2020**, *36*, 2351–2369. [CrossRef]
10.   Montgomery, D.C. *Statistical Quality Control: A Modern Introduction*, 6th ed.; John Wiley & Sons, Inc.: New York, NY, USA, 2009.
11.   Vasilopoulos, A.V.; Stamboulis, A.P. Modification of Control Chart Limits in the Presence of Data Correlation. *J. Qual. Technol.* **1978**, *20*, 20–30. [CrossRef]
12.   Wardell, D.G.; Moskowitz, H.; Plante, R.D. Control Charts in the Presence of Data Autocorrelation. *Manag. Sci.* **1992**, *38*, 1084–1105. [CrossRef]
13.   Yashchin, E. Performance of CUSUM Control Schemes for Serially Correlated. *Technometrics* **1993**, *35*, 37–52. [CrossRef]
14.   Schmid, W. On the Run Length of a Shewhart Chart for Correlated Data. *Stat. Pap.* **1995**, *36*, 111–130. [CrossRef]
15.   Jiang, W.; Tsui, K.L.; Woodall, W.H. A New SPC Monitoring Method: The ARMA Chart. *Technometrics* **2000**, *42*, 399–410. [CrossRef]
16.   Lu, C.W.; Reynolds, M.R., Jr. CUSUM Charts for Monitoring an Autocorrelated Process. *J. Qual. Technol.* **2001**, *33*, 316–334. [CrossRef]
17.   Castagliola, P.; Tsung, F. Auto-correlated Statistical Process Control for Non-Normal Situations. *Qual. Reliab. Eng. Int.* **2005**, *21*, 131–161. [CrossRef]
18.   Osei-Aning, R.; Abbasi, S.A.; Riaz, M. Optimization Design of the CUSUM and EWMA Charts for Autocorrelated Processes. *Qual. Reliab. Eng. Int.* **2017**, *33*, 1827–1841. [CrossRef]
19.   Garza-Venegas, J.A.; Tercero-Gomez, V.G.; Ho, L.L.; Castagliola, P.; Celano, G. Effect of Autocorrelation Estimators on the Performance of the (X)over-bar Control Chart. *J. Stat. Comput. Simul.* **2018**, *88*, 2612–2630. [CrossRef]
20.   Ryan, T.P. Discussion of Some Statistical Process Control Methods for Autocorrelated Data by D.C. Montgomery and C.M. Mastrangelo. *J. Qual. Technol.* **1991**, *23*, 200–202. [CrossRef]
21.   Wardell, D.G.; Moskowitz, H.; Plante, R.D. Run-length Distributions of Special-cause Control Charts for Correlated Process. *Technometrics* **1994**, *36*, 3–27. [CrossRef]
22.   Mastrangelo, C.M.; Montgomery, D.C. SPC with Correlated Observations for the Chemical and Process Industries. *Int. J. Reliab. Qual. Saf. Eng.* **1995**, *11*, 79–89. [CrossRef]
23.   Zhang, N.F. Detection Capability of Residual Control Chart for Stationary Process Data. *J. Appl. Stat.* **1997**, *24*, 363–380. [CrossRef]
24.   Lu, C.W.; Reynolds, M.R., Jr. EWMA Control Charts for Monitoring the Mean of Autocorrelated Processes. *J. Qual. Technol.* **1999**, *31*, 166–188. [CrossRef]
25.   Davoodi, M.; Niaki, S.T.A. Estimating the Step Change Time of the Location Parameter in Multistage Processes Using MLE. *Qual. Reliab. Eng. Int.* **2012**, *28*, 843–855. [CrossRef]
26.   Perez-Rave, J.; Munoz-Giraldo, L.; Correa-Morales, J.C. Use of Control Charts with Regression Analysis for Autocorrelated Data in the Context of Logistic Financial Budgeting. *Comput. Ind. Eng.* **2017**, *112*, 71–83. [CrossRef]
27.   Pan, X.; Jarrett, J. Using Vector Autoregressive Residuals to Monitor Multivariate Processes in the Presence of Serial Correlation. *Int. J. Prod. Econ.* **2007**, *106*, 204–216. [CrossRef]
28.   Hwarng, H.B.; Wang, Y. Shift Detection and Source Identification in Multivariate Autocorrelated Processes. *Int. J. Prod. Res.* **2010**, *48*, 835–859. [CrossRef]
29.   Vanhatalo, E.; Kulahci, M. The Effect of Autocorrelation on the Hotelling T-2 Control Chart. *Qual. Reliab. Eng. Int.* **2015**, *31*, 1779–1796. [CrossRef]
30.   Leoni, R.C.; Machado, G.; Aparecida, M.; Costa, B.; Fernando, A. The T-2 Chart with Mixed Samples to Control Bivariate Autocorrelated Processes. *Int. J. Prod. Res.* **2016**, *54*, 3294–3310. [CrossRef]
31.   Pan, J.N.; Li, C.I.; Wu, J.J. A New Approach to Detecting the Process Changes for Multistage Systems. *Expert Syst. Appl.* **2016**, *62*, 293–301. [CrossRef]
32.   Yang, S.F.; Yang, C.M. An Approach to Controlling Two Dependent Process Steps with Autocorrelated Observations. *Int. J. Adv. Manuf. Technol.* **2006**, *29*, 170–177. [CrossRef]
33.   Li, Y.; Zio, E.; Pan, E. An MEWMA-Based Segmental Multivariate Hidden Markov Model for Degradation Assessment and Prediction. *Proc. Inst. Mech. Eng. Part O J. Risk Reliab.* **2021**, *235*, 831–844. [CrossRef]
34.   Xia, T.; Sun, B.; Chen, Z.; Pan, E.; Wang, H.; Xi, L. Opportunistic maintenance policy integrating leasing profit and capacity balancing in service-oriented manufacturing. *Reliab. Eng. Syst. Saf.* **2021**, *205*, 107233. [CrossRef]
35.   Xia, T.; Dong, Y.; Pan, E.; Zheng, M.; Wang, H.; Xi, L. Fleet-level opportunistic maintenance for large-scale wind farms integrating real-time prognostic updating. *Renew. Energy* **2021**, *163*, 1444–1454. [CrossRef]
36.   Xia, T.; Zhuo, P.; Xiao, L.; Du, S.; Wang, D.; Xi, L. Multi-stage Fault Diagnosis Framework for Rolling Bearing Based on OHF Elman AdaBoost-Bagging Algorithm. *Neurocomputing* **2021**, *433*, 237–251. [CrossRef]

37. Tang, D.; Yu, J.; Chen, X.; Makis, V. An optimal condition-based maintenance policy for a degrading system subject to the competing risks of soft and hard failure. *Comput. Ind. Eng.* **2015**, *83*, 100–110. [CrossRef]
38. Tang, D.; Makis, V.; Jafari, L.; Yu, J. Optimal maintenance policy and residual life estimation for a slowly degrading system subject to condition monitoring. *Reliab. Eng. Syst. Saf.* **2015**, *134*, 198–207. [CrossRef]
39. Chiu, C.C.; Chen, M.K.; Lee, K.M.S. Shifts Recognition in Correlated Process Data Using a Neural Network. *Int. J. Syst. Sci.* **2001**, *32*, 137–143. [CrossRef]
40. Arkat, J.; Niaki, S.T.A.; Abbasi, B. Artificial Neural Networks in Applying MCUSUM Residuals Charts for AR(1) Processes. *Appl. Math. Comput.* **2007**, *189*, 1889–1901. [CrossRef]
41. Pacella, M.; Semeraro, Q. Using Recurrent Neural Networks to Detect Changes in Autocorrelated Processes for Quality Monitoring. *Comput. Ind. Eng.* **2007**, *52*, 502–520. [CrossRef]
42. Camargo, M.E.; Priesnitz, W.; Russo, S.L.; Dullius, A.I.D. Control Charts for Monitoring Autocorrelated Processes Based on Neural Networks Model. In Proceedings of the International Conference on Computers and Industrial Engineering, Troyes, France, 6–9 July 2009; pp. 1881–1884.
43. Yang, H.H.; Huang, M.L.; Yang, S.W. Integrating Auto-Associative Neural Networks with Hotelling T-2 Control Charts for Wind Turbine Fault Detection. *Energies* **2015**, *8*, 12100–12115. [CrossRef]
44. Rai, A.; Upadhyay, S.H. The Use of MD-CUMSUM and NARX Neural Network for Anticipating the Remaining Useful Life of Bearings. *Measurement* **2017**, *111*, 397–410. [CrossRef]
45. Lee, S.; Li, L.; Ni, J. Online Degradation Assessment and Adaptive Fault Detection Using Modified Hidden Markov Model. *J. Manuf. Sci. Eng.-Trans. ASME* **2010**, *132*, 021010. [CrossRef]
46. Alshraideh, H.; Runger, G. Process Monitoring Using Hidden Markov Models. *Qual. Reliab. Eng. Int.* **2014**, *30*, 1379–1387. [CrossRef]
47. Ross, S.M. *Introduction to Probability Models*, 11th ed.; Posts & Telecom Press: Beijing, China, 2015.
48. Chen, Z.; Xia, T.B.; Li, Y.; Pan, E.S. Degradation Modeling and Classification of Mixed Populations Using Segmental Continuous Hidden Markov Models. *Qual. Reliab. Eng. Int.* **2018**, *34*, 807–823. [CrossRef]
49. Guo, X.J.; Liu, S.F.; Wu, L.F.; Gao, Y.B.; Yang, Y.J. A Multi-variable Grey Model with a Self-memory Component and Its Application on Engineering Prediction. *Eng. Appl. Artif. Intell.* **2015**, *42*, 82–93. [CrossRef]

Article

An Integrated Approach-Based FMECA for Risk Assessment: Application to Offshore Wind Turbine Pitch System

Zhen Wang [1],*, Rongxi Wang [2],*, Wei Deng [1] and Yong Zhao [1]

[1] Xi'an Thermal Power Research Institute Co., Ltd., Xi'an 710054, China; dengwei@tpri.com.cn (W.D.); zhaoyong@tpri.com.cn (Y.Z.)

[2] The State Key Laboratory for Manufacturing Systems Engineering, Xi'an Jiaotong University, Xi'an 710049, China

* Correspondence: wangzhen@tpri.com.cn (Z.W.); rongxiwang@163.com (R.W.)

Abstract: Failure mode, effects and criticality analysis (FMECA) is a well-known reliability analysis tool for recognizing, evaluating and prioritizing the known or potential failures in system, design, and process. In conventional FMECA, the failure modes are evaluated by using three risk factors, severity (S), occurrence (O) and detectability (D), and their risk priorities are determined by multiplying the crisp values of risk factors to obtain their risk priority numbers (RPNs). However, the conventional RPN has been considerably criticized due to its various shortcomings. Although significant efforts have been made to enhance the performance of traditional FMECA, some drawbacks still exist and need to be addressed in the real application. In this paper, a new FMECA model for risk analysis is proposed by using an integrated approach, which introduces Z-number, Rough number, the Decision-making trial and evaluation laboratory (DEMATEL) method and the VIsekriterijumska optimizacija i KOmpromisno Resenje (VIKOR) method to FMECA to overcome its deficiencies in real application. The novelty of this paper in theory is that the proposed approach integrates the strong expressive ability of Z-numbers to vagueness and uncertainty information, the strong point of DEMATEL method in studying the dependence among failure modes, the advantage of rough numbers for aggregating experts' diversity evaluations, and the strength of VIKOR method to flexibly model multi-criteria decision-making problems. Based on the integrated approach, the proposed risk assessment model can favorably capture and aggregate FMECA team members' diversity evaluations and prioritize failure modes under different types of uncertainties with considering the failure propagation. In terms of application, the proposed approach was applied to the risk analysis of failure modes in offshore wind turbine pitch system, and it can also be used in many industrial fields for risk assessment and safety analysis.

Keywords: failure mode; effects and criticality analysis; Z-number; rough number; DEMATEL method; VIKOR method

Citation: Wang, Z.; Wang, R.; Deng, W.; Zhao, Y. An Integrated Approach-Based FMECA for Risk Assessment: Application to Offshore Wind Turbine Pitch System. *Energies* **2022**, *15*, 1858. https://doi.org/10.3390/en15051858

Academic Editor: Eugen Rusu

Received: 3 December 2021
Accepted: 30 December 2021
Published: 3 March 2022

Publisher's Note: MDPI stays neutral with regard to jurisdictional claims in published maps and institutional affiliations.

1. Introduction

Failure mode, effects and criticality analysis (FMECA), also known as failure mode and effects analysis (FMEA) when without referring to criticality analysis, is a risk and reliability analysis tool based on multidisciplinary team cooperation [1]. The FMEA method originates from the formal design methodology by NASA and first proposed in 1960s for solving their obvious reliability and safety requirements [2]. In many fields, it can be used to enhance the reliability and safety for a system by recognizing the various failure modes and analyzing their reasons and effects in the system and process during product design and manufacturing processes. The main task of FMEA is to evaluate the likelihood of the potential failure modes and their impact and severity to identify weaknesses and key projects in the system and then provide a basis for developing improved control measures. Differing from some other reliability management approaches, FMEA emphasizes taking precautions against failures rather than finding a solution after the failures happen [3], which can greatly reduce the frequency of occurrence of failure modes and avoid serious accidents. As a widely used methodology in safety and reliability analysis, FMEA has gained a widespread attention due to its visibility and simplicity, and up to now it has been extensively used in various industries [4–11].

In traditional FMECA, each failure mode identified in a system is evaluated by three risk factors of severity (S), occurrence (O) and detectability (D), and their risk priorities are determined by sorting their risk priority numbers (RPNs) [12], which is obtained by multiplying the values of S, O and D. Generally, S, O and D of failure modes are scored by experts and a number ranged from 1 to 10 is given for each of the three risk factors, usually the large the number, the worse the case is. Based on the values of RPNs, the risk priorities of failure modes are determined, which can help the analyst to pinpoint system inherent vulnerabilities. A failure mode with higher RPN is regarded as more important [13], which means it has greater harm to the system and will be given a higher risk priority. Thus for guaranteeing safety and reliability, some measures of prevention and improvement should be taken preferentially to the failure modes with high risk priority to avoid their occurrence. However, the crisp value of RPN has been highly criticized for various reasons [14–19], most of which are listed as follows:

1. The relative importance of the three risk factors has not been considered or are considered as equal importance, which may not consistent with the actual situation in many cases.
2. Multiplying the values of S, O and D in different groups may produce the same RPN value, but the hidden risk implications of each group can be completely different, which may lead to the limited resources and time being inappropriately allocated, or worse yet, some high risk failure modes being ignored.
3. The mean of computing RPN is debatable and hypersensitive to the variation of the values of risk factors. In some cases, a subtle alteration in the value of one risk factor may have a hugely different effect on RPN when the values of other risk factors are very large.
4. The evaluations for risk factors of S, O and D are usually given based on discrete ordinal scales of measure, on which the calculation of multiplication is meaningless because the obtained RPNs may be not continuous with many holes and heavily distributed ranged from 1 to 1000. In this case, the ranking results of failure modes are meaningless and even misleading.
5. The three risk factors are often hard to be determined precisely. The evaluations obtained from FMECA team members are expressed by linguistic items like high, moderate or low and so on.
6. FMECA team members may provide their evaluations in different way for the same risk factor due to their different expertise and backgrounds, and some of the assessment information may be vagueness and uncertain. In conventional FMECA, there is no means to describe the group judgment more comprehensively and explore the intrinsic link between different judgments [20].

In order to conquer the shortcomings mentioned above and enhance the applicability of traditional FMECA to real cases [3], much attention have been paid to its improvements and a variety of theories and methods have been introduced to FMECA. For example, fuzzy set has been introduced to FMECA for transforming the vagueness of experts' evaluation into a mathematical formula; information fusion method like Dempster–Shafer Theory and rough number, etc., are introduced to FMECA for aggregating different evaluations; multi-criteria decision making methods like the VIsekriterijumska optimizacija i KOmpromisno Resenje (VIKOR) method and Technique for Ordering Preference by Similarity to Ideal Solution (TOPSIS) method, etc., are introduced to FMECA for ranking failure modes. Some of the main theories and methods are presented in Table 1.

In studies of FMECA in wind turbines, some experts take the structures of different wind turbines, economic factors, costs and climatic regions into consideration. For example,

Mahmood et al. [21] developed a mathematical tool for risk and failure mode analysis of wind turbine systems (both onshore and offshore) by integrating the aspects of traditional FMEA and some economic considerations such as power production losses, and the costs of logistics and transportation. Samet et al. [22] proposed a FMECA methodology with considering different weather conditions or climatic regions and different wind turbine design types such as direct-drive model and geared-drive model. Nacef et al. [23] developed a hybrid cost-FMEA by integrating cost factors to assess the criticality, these costs vary from replacement costs to expected failure costs.

Table 1. Some of the theories and methods used in FMECA.

Categories	Theories and Methods	Roles in FMECA
Artificial intelligence techniques	Fuzzy rule-base system [24]	It can be used to deal with the drawback 5. in FMECA by transforming the vagueness of the evaluation of failure modes into a mathematical formula, which has the decision making ability by ranking the failure modes using fuzzy rules.
	Evidential reasoning method (ER) [25]	It can be used to deal with the drawbacks 5. and 6. in FMECA by modeling the diversity and uncertainty of experts' evaluations, which enables experts to evaluate failure modes in an independent way and aggregating their evaluations in a rigorous yet nonlinear rather than simple addition or multiplication manner.
	Rough number [26]	It can be used to deal with the drawback 6. in FMECA by aggregating the vague and uncertain evaluations of failure modes, which can reduce the subjectively of experts' opinion in aggregation process and make the decision-making more objective.
	Dempster–Shafer Theory (DST) [27]	It can be used to deal with the drawbacks 5. and 6. in FMECA by aggregating different types of subjective and uncertain evaluations using Dempster's rule, which has the ability of representing and handling various uncertainty information using belief structure.
	D number [28]	It can be used to deal with the drawbacks 5. and 6. in FMECA by representing and aggregating the experts' evaluations with cognitive uncertainty and imprecision for failure modes, which is capable of efficiently expressing various types of uncertainty.
Multi-criteria decision making (MCDM)	Analytic hierarchy process (AHP) method [29]	It can be used to deal with the drawbacks 2., 3. and 4. in FMECA, which determines the risk priorities of failure modes by using eigen vectors for synthesizing a series of paired comparison evaluations based on the evaluation of failure modes in a hierarchical way.
	TOPSIS method [30]	It can be used to deal with the drawbacks 2., 3. and 4. in FMECA, which determines the risk priorities of failure modes by comparing the Euclid distances simultaneously from the best evaluation value and from the worst evaluation value.
	VIKOR method [16]	It can be used to deal with the drawbacks 2., 3. and 4. in FMECA, which determines the risk priorities of failure modes by using a compromise solution of maximizing the group utility of the majority, and meanwhile minimizing the individual regret of the opponent.
	Decision making trial and evaluation laboratory (DEMATEL) method [31]	It can be used to deal with the drawbacks 2., 3., 4. and 7. in FMECA, which determines the risk priorities of failure modes by studying the dependence among failure modes in FMECA process using the graph theory and matrix tools.
	Grey theory method [32]	It can be used to deal with the drawbacks 2., 3. and 4. in FMECA, which determines the risk priorities of failure modes by calculating the grey relational coefficient between all comparability sequences and the reference sequence of the ideal target sequence and negative ideal target sequence).

Although many theories and methods have been introduced to FMECA to eliminate the defects of the traditional FMECA, the representation of expert's judgments on the evaluation of failure modes, the aggregation of experts' diversified evaluation information, and the determination of risk priorities of failure modes are still open issues, especially in terms of the defect of without considering the dependencies among different failure modes. In this paper, an integrated approach-based risk assessment model for FMECA was proposed to the existing defects, which integrates the strong expressive ability of Z-numbers to vagueness and uncertainty information, the strong point of the DEMATEL method in studying dependence among failure modes, the advantage of rough numbers in aggregating experts' diversity evaluation information, and the merit of VIKOR evaluation structure in flexibly modeling multi-criteria decision-making. Based on the integrated approach, the proposed risk assessment model can well capture and aggregate FMECA team members' diversity evaluations and prioritize failure modes under different types of uncertainties with considering the failure propagation. The rest of this paper is organized as follows. Some existing improvement methods to traditional FMECA are introduced in Section 2. Section 3 introduces the proposed new risk assessment model for FMECA using Z-number, rough number, DEMATEL method and VIKOR method. An illustrative case and the comparison and discussion for the proposed FMECA approach are respectively provided in Sections 4 and 5. Section 6 concludes the paper with a summary.

2. Literature Review

In the recent decades, scholars and researchers have done a lot of significant work to the improvements of FMECA. Among these improvement methods we can find they are mainly focusing on the following four aspects.

In term of the defect of traditional FMECA without considering the weights of risk factors, Hua et al. [33] introduced fuzzy analytic hierarchy process (FAHP) approach to FMECA for determining the weights of risk factors. Liu et al. [13] introduced a subjective weight and objective weight for risk factors by integrating fuzzy analytic hierarchy process (FAHP) and entropy method. Bozdag et al. [34] proposed a new fuzzy FMECA approach based on IT2 fuzzy sets for obtaining the uncertainty both in intrapersonal and interpersonal, which considers the optimal weights of risk factors and synthetizes them by using an ordered weighted averaging operator based on-cut. Liu et al. [35] introduced fuzzy digraph and matrix approach to FMECA for developing a new FMECA model with considering the relative weights of risk factors expressed by linguistic terms and transformed to fuzzy numbers, which determines the risk priorities of failure mode using their risk priority indexes that computed based on the formed corresponding fuzzy risk matrixes for failure modes. Zhou et al. [36] proposed a new generalized evidential FMECA (GEFMECA) model to handle the uncertain risk factors comprised of not only the conventional risk factors, but also the other incomplete risk factors. Based on the generalized evidence theory, the relative weights among all risk factors are well addressed. Liu et al. [37] proposed an integrated FMECA approach for the improvement of its performance based on the interval-valued intuitionistic fuzzy sets (IVIFSs) and multi-attributive border approximation area comparison (MABAC) method, in which the linear programming model is developed for obtaining the optimal weights of risk factors even if the weight information among risk factors is incompletely known.

In view of the defect that the evaluations obtained from FMECA team members are expressed in a linguistic way which are difficult to be converted directly and correctly into numerical value. To handle this case, fuzzy set theory and its improvement methods were introduced to FMECA by many researchers, which can be well used to transform the linguistic item into a mathematical formula and improve the decision making ability for FMECA in real application. Bowles and Peláez [2] first introduced fuzzy set theory into FMECA and proposed a technique based on fuzzy logic to prioritize failure modes in a system FMECA, which enables analysts to evaluate the failure modes using the linguistic terms directly and provides a more flexible structure to combine the parameters of risk factors. For dealing with the drawbacks of traditional fuzzy logic (i.e., rule-based) methods used in FMECA, Yang et al. [38] proposed a fuzzy rule-based Bayesian reasoning approach for the prioritization of failure modes. Jee et al. [39] presented a new fuzzy inference system (FIS)-based risk assessment model for FMECA to prioritizing failure modes, in which a new two-stage method is introduced for reducing the number of fuzzy rules which need to be gathered. By integrating FMECA and fuzzy linguistic scale method, Gajanand et al. [40] proposed a new strategy for the reliability-centered maintenance, in which the failure modes are prioritized by using the weighted Euclidean distance and centroid defuzzification based on fuzzy logic. Tooranloo et al. [41] proposed a new model for FMECA based on intuitionistic fuzzy approach, which evaluates failure modes under vague concepts and insufficient data. Jian et al. [42] proposed a new risk evaluation approach for failure mode analysis in FMECA by integrating intuitionistic fuzzy sets (IFSs) and evidence theory. In their method, linguistic items and intuitionistic fuzzy numbers are used to evaluate the risk factors of failure modes and then the evaluations are transformed into basic probability assignment functions based on evidence theory. Jiang et al. [43] assessed the risk factors of failure modes using fuzzy membership degree in their proposed fuzzy evidential method for FMECA, and ranked the failure modes by fusing the feature information of risk factors with D–S theory of evidence.

Aiming at the controversial mathematical formula for RPN calculation and the ranking problem of failure modes, many researchers have viewed the risk ranking problem of failure modes as a multiple criteria decision-making (MCDM) issue [16], and a lot of MCDM methods such as Analytical Hierarchy Process (AHP), technique for ordering preference by similarity to ideal solution (TOPSIS), Preference Ranking Organization Method for Enrichment Evaluation (PROMETHEE), grey theory, and VIsekriterijumska optimizacija i KOmpromisno Resenje (VIKOR) are introduced to FMECA. For example, Aydogan [44] introduced an integrated approach by using the rough AHP and fuzzy TOPSIS method for the performance analysis of organizations under fuzzy environment. Song et al. [20], taking advantage of the merit of rough set theory in manipulating uncertainty and the strength of the TOPSIS method in modeling multi-criteria decision making, proposed a new risk assessment model for FMECA. Liu et al. [45] introduced an intuitionistic fuzzy hybrid TOPSIS method to FMECA for determining the risk priorities of failure modes. Silvia et al. [46] proposed an maintenance approach based on by combining reliability analysis and MCDM method to optimize maintenance activities of complex systems, in which the AHP is used for weight evaluation of criteria and fuzzy TOPSIS method is responsible for risk ranking of failure modes identified in the system. Vahdani et al. [47] integrated fuzzy belief structure and TOPSIS method in FMECA to describe expert knowledge and rank failure modes in risk analysis. Zhou et al. [48] introduced grey theory and fuzzy theory into FMECA for the failure prediction of tanker equipment, in which the risk priorities of failure modes are determined by two criteria of the fuzzy risk priority numbers (FRPNs) obtained by fuzzy set theory and the grey relational coefficient obtained by grey theory. Liu et al. [28] introduced a new FMECA approach based on grey relational projection method (GRP) and D numbers for determining the risk priority orders of failure modes. Liu et al. [49] developed a framework for FMECA by integrating cloud model and PROMETHEE method for handling the representation of diversified risk evaluations of FMECA team members and the determination of the risk priorities of failure modes. Mandal et al. [50] presented a methodology utilizing VIKOR approach for ranking the human errors. Baloch et al. [51] integrated fuzzy VIKOR method and data envelopment analysis method into FMECA for determining the rankings of potential manners and selecting the most important impairment manner.

For better capturing and aggregating different experts' diversity evaluations which are difficult to be handled by traditional FMECA, evidential reasoning and Dempster–Shafer (D–S) Theory are introduced to FMECA in many literatures. Chin et al. [25] proposed an FMECA approach based on group-based evidential reasoning (ER) for capturing experts' diversity evaluations and prioritizing failure modes in the situation of various uncertainty. Liu et al. [52] proposed an improvement approach for FMECA based on fuzzy evidential reasoning (FER) and grey theory to solve the two shortcomings of traditional FMECA with respect to the acquirement and aggregation of different experts' evaluations and the determination of the risk priorities of failure modes. Liu et al. [53] proposed a new risk assessment model for the prioritization of failure modes in FMECA based on FER and belief rule-based (BRB) method. In their method, FER method is utilized to capture and aggregate the diversified, uncertain evaluations provided by experts and the relationships of nonlinear and uncertainty between risk factors and corresponding risk level are modeled by BRB method. Du et al. [54] proposed a new method in fuzzy FMECA based on evidential reasoning (ER) and TOPSIS method for precisely determining and aggregating the risk factors. Li et al. [55] proposed a new method by integrating D–S Theory, DEMATEL, and IFS method to prioritize alternatives and make risk assessment for system FMECA. Yang et al. [27] introduced D–S Theory to FMECA for analyzing different failure modes in the rotor blades of an aircraft engine under multiple evaluation sources with uncertainty. Su et al. [56] aiming at the method of Yang et al. proposed a modification for dealing with the combination of conflicting evidence by using the Gaussian distribution-based uncertain reasoning method to reconstruct the basic belief assignments (BBAs) with considering the weight of each expert. Shi et al. [57] proposed a aggregation method for aggregating hybrid preference information based on IFS and D–S Theory, which determines the weight of each expert based on the conflict degree that is obtained by computing the conflict coefficient with Jousselme distance [58]. Jiang et al. [59] proposed a modified method for improving the performance of evidence theory used in FMECA, which reassigns the basic belief assignments by considering the reliability coefficients obtained based on evidence distance to reduce the conflicts among expert's opinion [60].

3. Proposed FMECA Approach

3.1. Methodologies

In this section, Z-number, rough number, DEMATEL method and VIKOR method are briefly introduced. These methods will be used in the proposed risk assessment model.

(1)　Z-number

Z-number, a 2-tuple fuzzy numbers that includes the restriction of the evaluation and the reliability of the judgment, was first introduced by Zadeh in the year of 2011 for overcoming the limitation that fuzzy numbers does not consider the reliability of the information [61]. The idea of a Z-number is providing a mode for calculation with numbers that has partial reliability in the evaluation [62]. A Z-number can be utilized to express the information of an uncertain judgement in the form of two fuzzy numbers that the first fuzzy number indicates the fuzzy restriction and the second fuzzy number represents an idea of confidence, reliability, and probability. Thus, Z-number is more efficient than fuzzy number in describing the knowledge of human judgment since it describes both the restraint and reliability. Due to the powerful ability in modeling uncertain information in real world, Z-number has gained attention by some researchers and efforts have been made to apply Z-number to various situations such as in computing with words (CWW) [63] and decision making problems [64].

A Z-number can be denoted as $Z = (A, R)$ where the first component is the fuzzy restriction for the evaluation of objects and the second component is the reliability of the first component. In Z-number, A and R are described in natural language using linguistic terms and presented in a fuzzy number form such as triangular or trapezoidal fuzzy numbers [61]. For example, in risk analysis, the severity of a failure mode is very high, with a confidence of very sure, then the Z-number for evaluating the failure mode can be written as $Z = $ (Very high, Very sure).

(2)　Rough number

Rough set theory as a mathematical tool for dealing with the imprecision, uncertainty and vagueness knowledge [65] has been extensively applied in the fields of knowledge discovery, data mining, decision analysis and pattern recognition. By using its lower approximation and upper approximation, rough set theory can fully express and describe the ambiguity and randomness of uncertain information and can lessen the information loss of aggregation process to a certain extent. Based on rough set theory, rough number is developed by Zhai et al. [66] for managing customers' subjective judgments and determining their boundary intervals. By introducing rough number to FMECA, the evaluations of experts in FMECA can be transformed to rough numbers by calculating their lower approximation and upper approximation on the basis of original data without any requirement of auxiliary information. Since it can effectively extract experts' actual opinion and reduce their subjectively in decision-making [67], in this section, rough number is applied to aggregate the evaluations of experts.

(3)   DEMATEL method

Decision-making trial and evaluation laboratory (DEMATEL) method was first proposed in 1973 to solve the fragmented and antagonistic issues of world societies [68]. It is a method of system analysis using the structural modeling technique to find the influence relation among complex elements. DEMATEL is a tool of based on the graph theory and matrix, which constructs the direct influence matrix by means of the logical relation among various elements in the system and calculates the effect degree and cause degree of each element to other elements. Because of its ability to pragmatically visualize complicated causal relationships [69], DEMATEL can be used as an effective tool in studying the interdependence among elements in a complex systems and can be well used to identify the dependence among failure modes in FMECA process.

(4)   VIKOR method

The VIKOR (VIsekriterijumska optimizacija i KOmpromisno Resenje) method was first proposed by Opricovic [70] to rank and select the optimum solution among a set of choices under different units criteria. As one of the MCDM method, VIKOR ranks alternatives based on the multicriteria ranking index by calculating the particular measure of "closeness" to the "ideal" solution [71]. It is an effective method in the field of multicriteria decision making especially in the case where the decision makers may not have enough knowledge to express their preferences at the beginning of system design [72]. Comparing to other MCDM methods, VIKOR helps decision makers reach a feasible decision closest to the ideal by proposing a compromised solution with an advantage rate. Moreover, it is facile to conduct without any parameter settings. Thus, VIKOR has been extensively applied to practical decision making issues.

3.2. The Proposed Risk Assessment Model for FMECA

In this paper, a new risk assessment model for FMECA by integrating Z-number, Rough number, DEMATEL method and VIKOR method is proposed. In the proposed approach, Z-number is introduced to express experts' judgements on the evaluation of failure modes, which has a strong ability to describe the knowledge of human beings by using a 2-tuple fuzzy numbers and can be well used in representing vagueness and uncertainty information. Rough number is applied to aggregate different types of evaluations transformed by the given 2-tuple fuzzy numbers of experts and manipulate the subjectivity and vagueness in the evaluation process. Based on its flexible boundary interval, the epistemic uncertainty of evaluations can be generally represented and the different sources of uncertainty can be effectively tackled in aggregation process. DEMATEL method is introduced to calculate the effect degree and cause degree of each failure mode by constructing the direct influence matrix of failure modes, which is a very effective tool to study the relationship among various failure modes in complex systems. Finally, VIKOR method is utilized to determine the risk priorities of failure modes under a compromise way, which can help experts achieving a reasonable ranking results on the basis of maximizing the group utility for the "majority" and minimizing the individual regret for the "opponent".

The framework of the proposed approach is depicted in Figure 1, which comprises four different stages. The first stage is to evaluate the failure modes by using Z-number, the second stage is the aggregation of different experts' evaluations based on rough number, the third stage is to determine the dependency among failure modes on the basis of historical failure data, and the fourth stage is to rank the failure modes using VIKOR method. The four stages are explained in detail as follows.

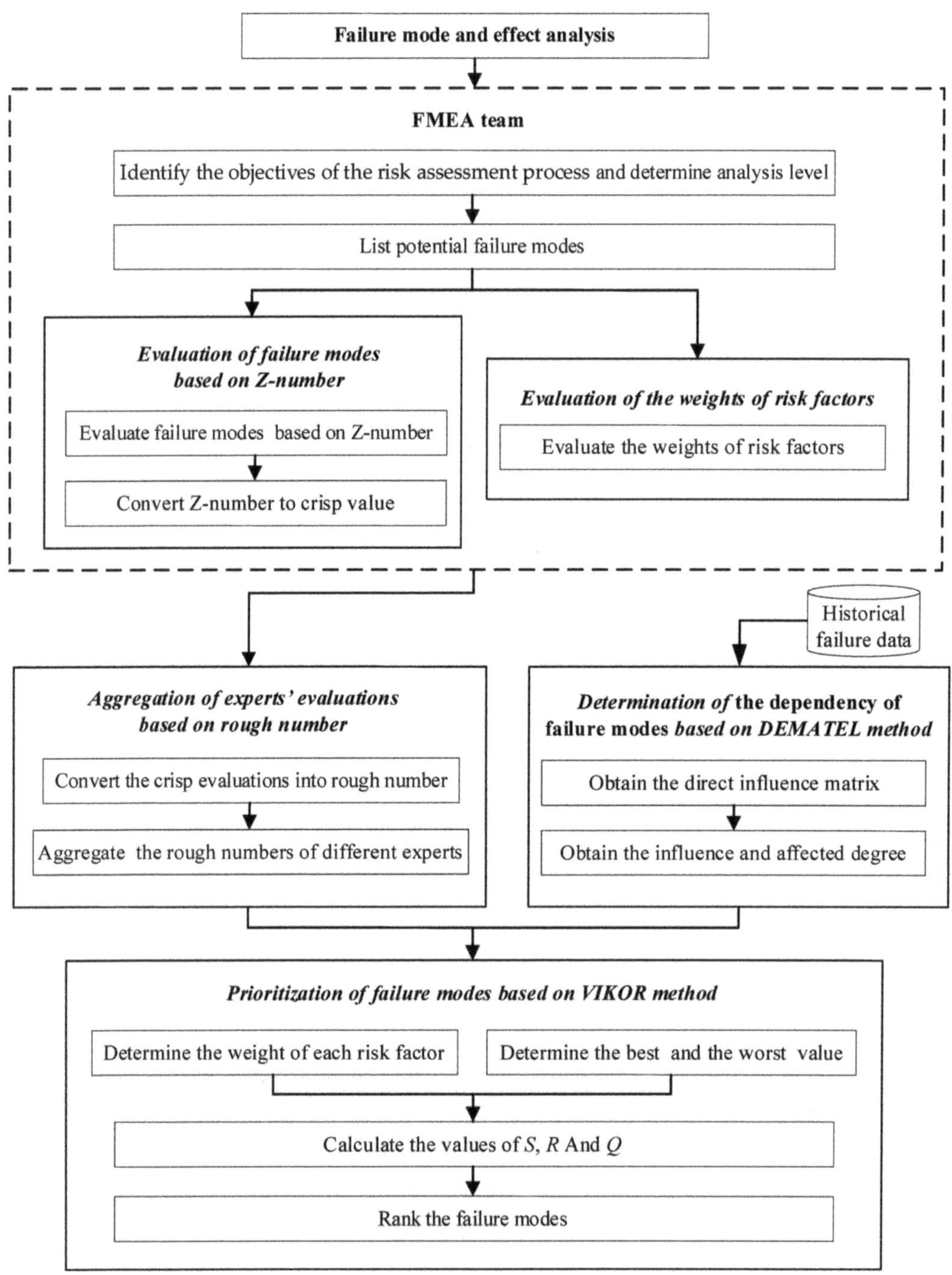

Figure 1. The framework of the proposed FMECA approach.

Step 1: Identify the objectives of the risk assessment process and determine the analysis level.

Step 2: Establish the FMECA team, list the potential failure modes and describe a finite set of relevant risk factors.

Suppose there are m failure modes in FMECA needed to be ranked according to the evaluations of failure modes and K experts are responsible for the evaluation with respect to the risk factors of severity, occurrence, detectability and failure propagation of failure modes.

Step 3: Evaluate the identified failure modes based on Z-number

In FMECA, failure modes are usually evaluated using linguistic variables such as very high, high, moderate, low, and very low, these evaluations are usually expressed in a fuzzy and imprecise way. In this section, failure modes are evaluated by using Z-number, which can not only express the evaluation of failure modes in a fuzzy and imprecise way, but also consider the confidence and reliability of the evaluations. In our work, failure modes are first evaluated according to Table 2, then the given linguistic terms are converted to fuzzy number according to Table 3. The transferred evaluations for failure mode in the form of 2-tuple fuzzy numbers are expressed as

$$Z = (A, B) = \{(\alpha_1, \alpha_2, \alpha_3), (\beta_1, \beta_2, \beta_3)\} \tag{1}$$

Table 2. Evaluation criterion for S, O, and D and the corresponding linguistic terms.

Severity	Occurrence	Detectability	Linguistic Variables
Failure is hazardous and causes system failure	Extremely high: Failure almost inevitable	Design control cannot detect failures	Extremely high (EH)
Failure involves hazardous outcomes	Very high	Very remote chance to detect failures	Very high (VH)
System is inoperable with loss of primary function	Repeated failures	Remote chance to detect failures	Relatively high (RH)
System performance is severely affected but functions	High	Very low chance to detect failures	High (H)
System performance is degraded, of which the comfort or convince functions may not operate	Moderately high	Low chance to detect failures	Moderately high (MH)
Moderate effect on system performance and system requires repair	Moderate	Moderate chance to detect failures	Moderate (M)
Small effect on system performance and system does not require repair	Relatively low	Good chance to detect failures	Relatively low (RL)
Minor effect on system performance	Low	High chance to detect failures	Low (L)
Very minor effect on system performance	Remote	Very high chance to detect failures	Very low (VL)
No effect	Nearly impossible	Design control will almost certainly detect failures	None (N)

Table 3. The relationship between linguistic terms and fuzzy numbers.

Linguistic Variables		Fuzzy Number
Fuzzy Restriction (A)	**Idea of Confidence (B)**	
High (EH)	Exactly Sure (ES)	(8.4, 10, 10)
Very High (VH)	Very Sure (VS)	(7.2, 8.4, 9.6)
High (H)	Sure (S)	(6, 7.2, 8.4)
Relatively High (RH)	Relatively Sure (RS)	(4.8, 6, 7.2)
Moderately High (MH)	Not Sure (NS)	(3.6, 4.8, 6)
Moderate (M)	Uncertain (U)	(2.4, 3.6, 4.8)
Relatively Low (RL)	Relatively Uncertain (RU)	(1.2, 2.4, 3.6)
Low (L)	Very Uncertain (VU)	(0, 1.2, 2.4)
Very low (VL)	Exactly Uncertainty (EU)	(0, 0, 1.2)

Step 4: Convert Z-numbers into crisp number.

For effectively aggregating the evaluations of experts, the Z-number form evaluations should be defuzzified to obtain a crisp value. The crisp value of evaluations can be obtained by

$$v = \frac{\int x\mu_B(x)dx}{\int 10\mu_B(x)dx} \cdot \frac{(\alpha_1 + 4 \times \alpha_2 + \alpha_3)}{6}, \tag{2}$$

$$\mu_B(x) = \begin{cases} 0, x \in (-\infty, \beta_1) \\ \frac{x-\beta_1}{\beta_2-\beta_1}, x \in [\beta_1, \beta_2] \\ \frac{\beta_3-x}{\beta_3-\beta_2}, x \in [\beta_2, \beta_3] \\ 0, x \in (\beta_3, +\infty) \end{cases} \tag{3}$$

where $\int$ is an algebraic integration, $\mu_B(x)$ is the membership function of triangular fuzzy number $(\beta_1, \beta_2, \beta_3)$.

Step 5: Aggregate the evaluations of K experts for each failure mode by using rough number.

For failure mode i $(i = 1, 2, \cdots, m)$ with respect to risk factor j $(j = S, O, D)$, the evaluations is denoted as $V_{ij} = \left\{ v_{ij}^1, v_{ij}^2, \cdots, v_{ij}^K \right\}$. The first step in the aggregation process is to obtain the lower approximation and upper approximation of $v_{ij}^k (k = 1, 2, \cdots, K)$ by the following equations:

$$\underline{Apr}(v_{ij}^k) = \cup\left\{ v_{ij}^t \in V_{ij} / v_{ij}^t \leq v_{ij}^k \right\}, \tag{4}$$

$$\overline{Apr}(v_{ij}^k) = \cup\left\{ v_{ij}^t \in V_{ij} / v_{ij}^t \geq v_{ij}^k \right\}. \tag{5}$$

Based on the lower approximation and upper approximation of v_{ij}^k, the lower limit and upper limit of v_{ij}^k are obtained by

$$\underline{Lim}(v_{ij}^k) = \frac{1}{M_L}\sum v_{ij}^t \left| v_{ij}^t \in \underline{Apr}(v_{ij}^k) \right., \tag{6}$$

$$\overline{Lim}(v_{ij}^k) = \frac{1}{M_U}\sum v_{ij}^t \left| v_{ij}^t \in \overline{Apr}(v_{ij}^k) \right. \tag{7}$$

where M_L is the number of elements contained in $\underline{Apr}(v_{ij}^k)$, and M_U is the number of elements contained in $\overline{Apr}(v_{ij}^k)$.

Then the rough number of v_{ij}^k is obtained by its corresponding lower limit and upper limit, namely

$$RN(v_{ij}^k) = [\underline{Lim}(v_{ij}^k), \overline{Lim}(v_{ij}^k)]. \tag{8}$$

The interval between $\underline{Lim}(v_{ij}^k)$ and $\overline{Lim}(v_{ij}^k)$ is the rough boundary interval denoted as

$$RBnd(v_{ij}^k) = \overline{Lim}(v_{ij}^k) - \underline{Lim}(v_{ij}^k). \tag{9}$$

With the obtained rough numbers of $v_{ij}^k (k = 1, 2, \cdots, K)$, the rough sequence $RS(V_{ij})$ of V_{ij} can be obtained by

$$RS(V_{ij}) = \left\{ [v_{ij}^L, v_{ij}^U]_1, [v_{ij}^L, v_{ij}^U]_2, \cdots, [v_{ij}^L, v_{ij}^U]_K \right\}. \tag{10}$$

Thus the rough number of the evaluation for failure mode i with respect to risk factor j (V_{ij}) is obtained by averaging the rough sequence, that is

$$RN(V_{ij}) = [v_{ij}^L, v_{ij}^U] = \frac{1}{K}\left([v_{ij}^L, v_{ij}^U]_1 + [v_{ij}^L, v_{ij}^U]_2 + \cdots + [v_{ij}^L, v_{ij}^U]_K \right). \tag{11}$$

Then the aggregated evaluation matrix EM for failure modes with respect to S, O and D is given as:

$$EM = \begin{vmatrix} [v_{1S}^L, v_{1S}^U] & [v_{1O}^L, v_{1O}^U] & [v_{1D}^L, v_{1D}^U] \\ \vdots & \vdots & \vdots \\ [v_{mS}^L, v_{mS}^U] & [v_{mO}^L, v_{mO}^U] & [v_{mD}^L, v_{mD}^U] \end{vmatrix} \tag{12}$$

In this step, DEMATEL method is applied to obtain the effect degree (R) and the cause degree (C) of each failure mode. The first procedure in DEMATEL method is to obtain the direct effect degree between any two failure modes, referred to as a_{ij} $(i = 1, 2, \cdots, m; j = 1, 2, \cdots, m)$, which can be obtained by making statistical analysis for the historical failure data or the expertise and experience of experts. The value of a_{ij} represents the degree that failure mode j influenced by failure mode i, and the value of a_{ji} represents the degree that failure mode i influenced by failure mode j. In general, a_{ij} is not equal to a_{ji}. Specifically, $a_{ij} = 0$ if $i = j$. For m failure modes in FMECA, the direct relation matrix among these failure modes can be expressed as

$$A = \begin{vmatrix} 0 & a_{1m} & \cdots & a_{1m} \\ a_{21} & 0 & \cdots & a_{2m} \\ \vdots & \vdots & \ddots & \vdots \\ a_{m1} & a_{m2} & \cdots & 0 \end{vmatrix}. \tag{13}$$

The initial direct relation matrix A can be normalized by using the following equations [73]

$$D = A \times S, \tag{14}$$

$$S = Min \left[\frac{1}{\max_{1 \leq i \leq m} \sum_{j=1}^{m} a_{ij}}, \frac{1}{\max_{1 \leq j \leq m} \sum_{i=1}^{m} a_{ij}} \right] \tag{15}$$

where the value of each element in matrix D ranges from 0 to 1.

Then the total relation matrix is obtained by the following equation

$$T = D(I - D)^{-1} = [t_{ij}]_{m \times m} \tag{16}$$

where I is the identity matrix.

The sums of rows and of columns in the total relation matrix T are the effect degree (R) and the cause degree (C) of failure modes, respectively, which are obtained by using the following equations

$$R = (r_1, r_2, \cdots, r_m) = \left[\sum_{j=1}^{m} t_{ij} \right]_{m \times 1}, \tag{17}$$

$$C = (c_1, c_2, \cdots, c_m) = \left[\sum_{i=1}^{m} t_{ij} \right]_{1 \times m} \tag{18}$$

where r_i in vector R is the sum of ith row of matrix T, which represents both the direct and indirect effects of failure mode i acting on the other failure modes, and c_j in vector C is the sum of jth column of matrix T, which represents both the direct and indirect effects of failure mode j caused by other failure modes.

Step 7: Obtain the ultimate decision making matrix for failure modes.

The effect degree (R) and the cause degree (C) of each failure mode are regarded as the risk factor for assessing the risk priority of failure mode, namely five risk factors as severity, occurrence, detectability, effect degree and cause degree are chosen in the proposed FMECA approach for the prioritization of failure modes. Thus the ultimate decision making matrix for failure modes is given as:

$$DM = \begin{vmatrix} [v_{1S}^L, v_{1S}^U] & [v_{1O}^L, v_{1O}^U] & [v_{1D}^L, v_{1D}^U] & v_{1R} & v_{1C} \\ \vdots & \vdots & \vdots & \vdots & \vdots \\ [v_{mS}^L, v_{mS}^U] & [v_{mO}^L, v_{mO}^U] & [v_{mD}^L, v_{mD}^U] & v_{mR} & v_{mC} \end{vmatrix} \tag{19}$$

where $v_{iR} = r_i$ and $v_{iC} = c_i$ are the effect degree and the cause degree of failure mode i respectively.

Step 8: Determine the weight of each risk factor.

As similar as the evaluations of failure modes, the relative weights among risk factors need to be assessed by experts and aggregated using rough number, the rough numbers for the weights of risk factors are expressed as

$$W = [\ [w_S^L, w_S^U] \quad [w_O^L, w_O^U] \quad [w_D^L, w_D^U] \quad [w_R^L, y_R^U] \quad [w_C^L, w_C^U]\ ]. \tag{20}$$

Step 9: Determine the risk rankings of failure modes using VIKOR method.

In this step, VIKOR method is applied to determine the risk rankings of failure modes. Firstly the weights of risk factors in rough number form need to be converted to crisp value by the following equation:

$$w_j = \lambda \left(1 - \frac{w_j^U - w_j^L}{2(\beta - \alpha)}\right) + (1 - \lambda)\frac{w_j^U + w_j^L}{2(\beta - \alpha)}, j = S, O, D, R, C \tag{21}$$

where w_j is weight of risk factor j, w_j^U and w_j^L are the lower limit and upper limit of the rough number of risk factor j, $\beta = \max_j w_j^U, \alpha = \min_j w_j^L$ λ, is a discount factor, expressing the effect degree of rough boundary interval imposing on the weight of risk factor. $0 \leq \lambda \leq 1$, and the greater the value of λ, the more effect is imposing on the weight of risk factor, here suppose $\lambda = 0.5$.

The normalized weight of each risk factor is obtained by using the following equation:

$$w_j' = \frac{w_j}{\sum\limits_{l=1}^{5} w_l}, j = S, O, D, R, C. \tag{22}$$

In VIKOR method, the first step is to determine the optimal and the worst value of each risk factor in DM, which is determined by

$$v_j^* = \begin{cases} \max\limits_i v_{ij}^U, j = S, O, D \\ \max\limits_i v_{ij}, j = R, C \end{cases}, \tag{23}$$

$$v_j^- = \begin{cases} \min\limits_i v_{ij}^L, j = S, O, D \\ \min\limits_i v_{ij}, j = R, C \end{cases}. \tag{24}$$

Based on v_j^* and v_j^-, the values of S_i and R_i can be calculated by the following relations

$$S_i = \sum_{j=1}^{3} w_j' \frac{\left\{(v_{ij}^L - v_j^*)^2 + (v_{ij}^U - v_j^*)^2\right\}^{1/2}}{\left|v_j^* - v_j^-\right|} + \sum_{j=4}^{5} w_j' \frac{\left|v_{ij} - v_j^*\right|}{\left|v_j^* - v_j^-\right|}, j = S, O, D, R, C, \tag{25}$$

$$R_i = \max_j \left(w_j' \frac{\left\{(v_{ij}^L - v_j^*)^2 + (v_{ij}^U - v_j^*)^2\right\}^{1/2}}{\left|v_j^* - v_j^-\right|}, w_j' \frac{\left|v_{ij} - v_j^*\right|}{\left|v_j^* - v_j^-\right|}\right), j = S, O, D, R, C. \tag{26}$$

Then the values of Q_i $(i = 1, 2, \cdots, m)$ are determined by

$$Q_i = v\frac{S_i - S^*}{S^- - S^*} + (1 - v)\frac{R_i - R^*}{R^- - R^*} \tag{27}$$

where S^* is the minimum value of S_i and S^- is the maximum value of S_i, R^* is the minimum value of R_i and R^- is the maximum value of R_i, v is the weight of the strategy of "the majority of criteria" (or "the maximum group utility"), whereas $1 - v$ is the weight of the individual regret. Here suppose $v = 0.5$.

Based on the values of S, R and Q, the failure modes can be prioritized with three ranking lists. Moreover, VIKOR method proposes a compromise solution, the failure mode $(FM^{(1)})$, which is the best ranked by the measure Q (Minimum) if the following two conditions are satisfied:

C1. Acceptable advantage: $Q(FM^{(2)}) - Q(FM^{(1)}) \geq 1/(m - 1)$, where $FM^{(2)}$ is the failure mode with second position in the ranking list by Q.

C2. Acceptable stability in decision making: The failure mode $FM^{(1)}$ must also be the best ranked by S or/and R. This compromise solution is stable within a decision making process, which could be: "voting by majority rule" (when $v > 0.5$ is needed), or "by consensus" $v \approx 0.5$, or "with veto" $(v < 0.5)$.

If one of the conditions is not satisfied, then a set of compromise solutions is proposed, which consists of:

- Failure mode $FM^{(1)}$ and $FM^{(2)}$ if only the condition **C2** is not satisfied,

  or

- Failure mode $FM^{(1)}, FM^{(2)}, \ldots, FM^{(M)}$ if the condition **C1** is not satisfied.

$FM^{(M)}$ is determined by the relation: $Q(FM^{(M)}) - Q(FM^{(1)}) < 1/(m - 1)$ for maximum M.

4. Case Study: Application to the Risk Analysis of the Failure Modes in Offshore Wind Turbine Pitch System

In this section, the proposed approach is used to the risk prioritization of the failure modes in offshore wind turbine pitch system. There are seven main malfunctions of the pitch system, namely, pitch bearing failure, pitch gearbox failure, pitch motor failure, pitch actuator failure, backup power and charger failure, encoder and limit switch failure, and control module failure. Each of the malfunctions could cause the pitch system failure and eventually result in the turbine shutdown. In order to ensure the operation quality and safety of the pitch system, it is necessary to analyze the malfunctions, excavate the potential failure reasons, and identify the weak links and dangerous source of the system.

In this case, four experts with different backgrounds and professional knowledge were invited to identify and evaluate the potential failure modes of pitch system. They are from wind turbine manufacturer, pitch system manufacturer, wind farm and the operation and maintenance enterprise for wind turbine respectively, and all of them have rich experience and knowledge about the fault analysis and diagnosis of pitch system. Based on the analysis of the historical data of pitch system in a wind farm subordinate to Huaneng Group and the experts 'experience knowledge, twenty-four failure modes which are able to cause the seven kinds of malfunctions were identified, and the four experts are responsible for evaluating the severity, occurrence, detectability and failure propagation of these failure modes.

For identifying the weak links and dangerous source of the system, the identified failure modes should to be prioritized based on the values of their severity, occurrence, detectability, effect degree, and cause degree. The twenty-four failure modes and the corresponding code are given in Table 4, and the propagation relationship among different failure modes are provided in Figure 2, which reveals the dependency of the failure modes participated in the failure propagation.

Table 4. The potential failure modes and their code.

Code	Failure Modes	Code	Failure Modes
FM1	Bearing internal component failure	FM13	IGBT damage
FM2	Crack or fracture of bolt connected with hub	FM14	Backup battery or capacitor failure
FM3	Gear failure of the gearbox	FM15	Charger failure
FM4	Bearing failure of the gearbox	FM16	Pitch angle A/B encoder failure
FM5	Oil spill of the gearbox	FM17	Pitch angle limit switch failure
FM6	Short circuit and open circuit of motor winding	FM18	Blade angle failure
FM7	Bearing failure of the motor	FM19	Hardware failure of controller module (PLC failure)
FM8	Motor brake failure	FM20	Power conversion module failure
FM9	Motor fan failure	FM21	Switches/contactors/relays failure
FM10	Motor wiring and interface problems	FM22	heater and cooling fan failure
FM11	Motor overload	FM23	Input/output line failure
FM12	Communication failure	FM24	Pitch safety chain module failure

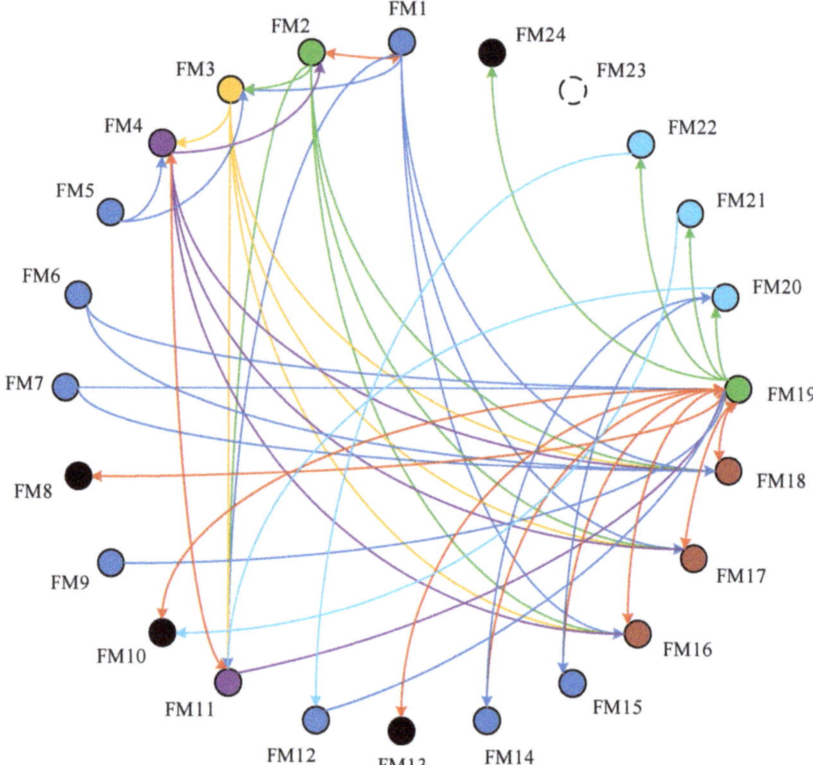

Figure 2. The propagation relationship among different failure modes.

The direct relation matrix among the failure modes obtained based on the historical failure data is given as follows:

$$
A = \begin{bmatrix}
0 & 0.3 & 0.2 & 0 & 0 & 0 & 0 & 0 & 0 & 0 & 0.2 & 0 & 0 & 0 & 0 & 0.1 & 0.1 & 0.1 & 0 & 0 & 0 & 0 & 0 & 0 \\
0.4 & 0 & 0.2 & 0 & 0 & 0 & 0 & 0 & 0 & 0 & 0.2 & 0 & 0 & 0 & 0 & 0.1 & 0.1 & 0.1 & 0 & 0 & 0 & 0 & 0 & 0 \\
0 & 0 & 0 & 0.3 & 0 & 0 & 0 & 0 & 0 & 0 & 0.4 & 0 & 0 & 0 & 0 & 0.3 & 0.3 & 0.3 & 0 & 0 & 0 & 0 & 0 & 0 \\
0 & 0 & 0.3 & 0 & 0 & 0 & 0 & 0 & 0 & 0 & 0.2 & 0 & 0 & 0 & 0 & 0.3 & 0.3 & 0.3 & 0 & 0 & 0 & 0 & 0 & 0 \\
0 & 0 & 0.4 & 0.4 & 0 & 0 & 0 & 0 & 0 & 0 & 0 & 0 & 0 & 0 & 0 & 0 & 0 & 0 & 0 & 0 & 0 & 0 & 0 & 0 \\
0 & 0 & 0 & 0 & 0 & 0 & 0 & 0 & 0 & 0 & 0 & 0 & 0 & 0 & 0 & 0 & 0 & 0.2 & 0.3 & 0 & 0 & 0 & 0 & 0 \\
0 & 0 & 0 & 0 & 0 & 0 & 0 & 0 & 0 & 0 & 0 & 0 & 0 & 0 & 0 & 0 & 0 & 0.4 & 0.1 & 0 & 0 & 0 & 0 & 0 \\
0 & 0 & 0 & 0 & 0 & 0 & 0 & 0 & 0 & 0 & 0 & 0 & 0 & 0 & 0 & 0 & 0 & 0 & 0.2 & 0 & 0 & 0 & 0 & 0 \\
0 & 0 & 0 & 0 & 0 & 0 & 0 & 0 & 0 & 0 & 0 & 0 & 0 & 0 & 0 & 0 & 0 & 0 & 0.2 & 0 & 0 & 0 & 0 & 0 \\
0 & 0 & 0 & 0 & 0 & 0 & 0 & 0 & 0 & 0 & 0 & 0 & 0 & 0 & 0 & 0 & 0 & 0 & 0.2 & 0 & 0 & 0 & 0 & 0 \\
0 & 0 & 0 & 0 & 0 & 0 & 0 & 0 & 0 & 0 & 0 & 0 & 0 & 0 & 0 & 0 & 0 & 0 & 0.2 & 0 & 0 & 0 & 0 & 0 \\
0 & 0 & 0 & 0 & 0 & 0 & 0 & 0 & 0 & 0 & 0 & 0 & 0 & 0 & 0 & 0 & 0 & 0 & 0.3 & 0 & 0 & 0 & 0 & 0 \\
0 & 0 & 0 & 0 & 0 & 0 & 0 & 0 & 0 & 0 & 0 & 0 & 0 & 0 & 0 & 0 & 0 & 0 & 0.4 & 0 & 0 & 0 & 0 & 0 \\
0 & 0 & 0 & 0 & 0 & 0 & 0 & 0 & 0 & 0 & 0 & 0 & 0 & 0 & 0 & 0 & 0 & 0 & 0.3 & 0.5 & 0 & 0 & 0 & 0 \\
0 & 0 & 0 & 0 & 0 & 0 & 0 & 0 & 0 & 0 & 0 & 0 & 0 & 0 & 0 & 0 & 0 & 0 & 0.2 & 0.4 & 0 & 0 & 0 & 0 \\
0 & 0 & 0 & 0 & 0 & 0 & 0 & 0 & 0 & 0 & 0 & 0 & 0 & 0 & 0 & 0 & 0 & 0 & 0.1 & 0 & 0 & 0 & 0 & 0 \\
0 & 0 & 0 & 0 & 0 & 0 & 0 & 0 & 0 & 0 & 0 & 0 & 0 & 0 & 0 & 0 & 0 & 0 & 0.1 & 0 & 0 & 0 & 0 & 0 \\
0 & 0 & 0 & 0 & 0 & 0 & 0 & 0 & 0 & 0 & 0 & 0 & 0 & 0 & 0 & 0 & 0 & 0 & 0.1 & 0 & 0 & 0 & 0 & 0 \\
0 & 0 & 0 & 0 & 0 & 0 & 0 & 0.5 & 0 & 0 & 0 & 0 & 0.3 & 0 & 0 & 0.2 & 0.2 & 0.2 & 0 & 0.3 & 0.3 & 0.2 & 0 & 0.2 \\
0 & 0 & 0 & 0 & 0 & 0 & 0 & 0 & 0 & 0 & 0.1 & 0 & 0 & 0.2 & 0.2 & 0 & 0 & 0 & 0 & 0 & 0 & 0 & 0 & 0 \\
0 & 0 & 0 & 0 & 0 & 0 & 0 & 0 & 0 & 0.4 & 0 & 0 & 0 & 0 & 0 & 0 & 0 & 0 & 0 & 0 & 0 & 0 & 0 & 0 \\
0 & 0 & 0 & 0 & 0 & 0 & 0 & 0 & 0 & 0 & 0 & 0.4 & 0 & 0 & 0 & 0 & 0 & 0 & 0 & 0 & 0 & 0 & 0 & 0 \\
0 & 0 & 0 & 0 & 0 & 0 & 0 & 0 & 0 & 0 & 0 & 0 & 0 & 0 & 0 & 0 & 0 & 0 & 0 & 0 & 0 & 0 & 0 & 0 \\
0 & 0 & 0 & 0 & 0 & 0 & 0 & 0 & 0 & 0 & 0 & 0 & 0 & 0 & 0 & 0 & 0 & 0 & 0 & 0 & 0 & 0 & 0 & 0
\end{bmatrix}
$$

According to Table 2, the twenty-four potential failure modes were evaluated with respect to severity, occurrence and detectability, and the evaluations for these failure modes were transformed to Z-numbers according to Table 3. The evaluations given by expert 1 and the corresponding Z-numbers for these evaluations are presented in Tables 5 and 6, respectively. For the sake of space, the other three experts' evaluation information are provided in Appendix A. It is necessary to mention that in our work, the weights of importance of experts are considered as equal. Since each of them has his/her good points, it is difficult to assign a subjective weight to each expert. After converting the Z-numbers into the crisp values, the evaluations (in the form of crisp value) given by the four experts were aggregated by using rough number. The aggregation results are presented in Table 7.

Table 5. The assessment on S, O and D of the twenty-four potential failure modes given by expert 1.

Code	Severity	Occurrence	Detectability	Code	Severity	Occurrence	Detectability
FM1	(VH, VS)	(RL, NS)	(M, S)	FM13	(MH, RS)	(MH, NS)	(L, NS)
FM2	(RH, S)	(RL, NS)	(RH, S)	FM14	(L, NS)	(RH, RS)	(L, RS)
FM3	(MH, S)	(M, RS)	(M, NS)	FM15	(L, NS)	(L, NS)	(RL, RS)
FM4	(MH, S)	(L, NS)	(M, NS)	FM16	(VL, RS)	(RH, U)	(RL, U)
FM5	(L, NS)	(MH, NS)	(VL, S)	FM17	(VL, NS)	(MH, U)	(RL, NS)
FM6	(MH, RS)	(RH, U)	(L, RS)	FM18	(VL, U)	(L, NS)	(RL, U)
FM7	(RL, NS)	(M, NS)	(M, NS)	FM19	(RL, RS)	(L, RS)	(L, RS)
FM8	(RL, RS)	(M, U)	(RL, S)	FM20	(RL, NS)	(VL, RS)	(L, RS)
FM9	(L, U)	(MH, NS)	(RL, S)	FM21	(L, NS)	(M, U)	(L, NS)
FM10	(L, RS)	(M, NS)	(RL, NS)	FM22	(VL, RS)	(L, NS)	(L, RS)
FM11	(M, NS)	(RL, NS)	(L, RS)	FM23	(L, RS)	(RL, RS)	(L, U)

According to the direct relation matrix among the failure modes, the effect degree and the cause degree of each failure mode were obtained by using DEMATEL method. In this paper, the effect degree and the cause degree are considered as two risk factors, which reveal the correlation strength between each failure mode and the other failure modes. The greater the effect degree of a failure mode, the more likely the failure mode will lead to other failures/faults to happen, meaning it has a higher severity. The greater the cause degree of a failure mode, the more likely the failure mode can be caused by other failure modes, meaning it has a higher probability of occurrence. The effect degrees and the cause degrees of failure modes are presented in Table 7, thus the ultimate decision matrix for the twenty-four potential failure modes with respect to five risk factors is formed. Similarly, the evaluations of the weights of risk factors were aggregated and presented in Table 8.

Table 6. The transformed Z-numbers of the twenty-four potential failure modes given by expert 1.

Code	Severity	Occurrence	Detectability
FM1	{(7.2, 8.4, 9.6),(7.2, 8.4, 9.6)}	{(1.2, 2.4, 3.6), (3.6, 4.8, 6)}	{(2.4, 3.6, 4.8), (4.8, 6, 7.2)}
FM2	{(4.8, 6, 7.2), (6, 7.2, 8.4)}	{(1.2, 2.4, 3.6), (3.6, 4.8, 6)}	{(4.8, 6, 7.2),(6, 7.2, 8.4)}
FM3	{(3.6, 4.8, 6), (6, 7.2, 8.4)}	{(2.4, 3.6, 4.8), (4.8, 6, 7.2)}	{(2.4, 3.6, 4.8), (3.6, 4.8, 6)}
FM4	{(3.6, 4.8, 6), (6, 7.2, 8.4)}	{(0, 1.2, 2.4), (3.6, 4.8, 6)}	{(2.4, 3.6, 4.8), (3.6, 4.8, 6)}
FM5	{(0, 1.2, 2.4), (3.6, 4.8, 6)}	{(3.6, 4.8, 6), (3.6, 4.8, 6)}	{(0, 0, 1.2), (6, 7.2, 8.4)}
FM6	{(3.6, 4.8, 6), (4.8, 6, 7.2)}	{(6, 7.2, 8.4), (2.4, 3.6, 4.8)}	{(0, 1.2, 2.4), (4.8, 6, 7.2)}
FM7	{(1.2, 2.4, 3.6), (3.6, 4.8, 6)}	{(2.4, 3.6, 4.8), (3.6, 4.8, 6)}	{(2.4, 3.6, 4.8), (3.6, 4.8, 6)}
FM8	{(1.2, 2.4, 3.6), (4.8, 6, 7.2)}	{(2.4, 3.6, 4.8), (2.4, 3.6, 4.8)}	{(1.2, 2.4, 3.6), (6, 7.2, 8.4)}
FM9	{(0, 1.2, 2.4), (2.4, 3.6, 4.8)}	{(3.6, 4.8, 6), (3.6, 4.8, 6)}	{(1.2, 2.4, 3.6), (6, 7.2, 8.4)}
FM10	{(0, 1.2, 2.4), (4.8, 6, 7.2)}	{(2.4, 3.6, 4.8), (3.6, 4.8, 6)}	{(1.2, 2.4, 3.6), (3.6, 4.8, 6)}
FM11	{(2.4, 3.6, 4.8), (3.6, 4.8, 6)}	{(1.2, 2.4, 3.6), (3.6, 4.8, 6)}	{(0, 1.2, 2.4), (4.8, 6, 7.2)}
FM12	{(0, 1.2, 2.4), (4.8, 6, 7.2)}	{(4.8, 6, 7.2), (2.4, 3.6, 4.8)}	{(1.2, 2.4, 3.6), (2.4, 3.6, 4.8)}
FM13	{(3.6, 4.8, 6), (4.8, 6, 7.2)}	{(3.6, 4.8, 6), (3.6, 4.8, 6)}	{(0, 1.2, 2.4), (3.6, 4.8, 6)}
FM14	{(0, 1.2, 2.4), (3.6, 4.8, 6)}	{(6, 7.2, 8.4), (4.8, 6, 7.2)}	{(0, 1.2, 2.4), (4.8, 6, 7.2)}
FM15	{(0, 1.2, 2.4), (3.6, 4.8, 6)}	{(0, 1.2, 2.4), (3.6, 4.8, 6)}	{(1.2, 2.4, 3.6), (4.8, 6, 7.2)}
FM16	{(0, 0, 1.2), (4.8, 6, 7.2)}	{(4.8, 6, 7.2), (2.4, 3.6, 4.8)}	{(1.2, 2.4, 3.6), (2.4, 3.6, 4.8)}
FM17	{(0, 0, 1.2), (3.6, 4.8, 6)}	{(3.6, 4.8, 6), (2.4, 3.6, 4.8)}	{(1.2, 2.4, 3.6), (3.6, 4.8, 6)}
FM18	{(0, 0, 1.2), (2.4, 3.6, 4.8)}	{(0, 1.2, 2.4), (3.6, 4.8, 6)}	{(1.2, 2.4, 3.6), (2.4, 3.6, 4.8)}
FM19	{(1.2, 2.4, 3.6), (4.8, 6, 7.2)}	{(0, 1.2, 2.4), (4.8, 6, 7.2)}	{(0, 1.2, 2.4), (4.8, 6, 7.2)}
FM20	{(1.2, 2.4, 3.6), (3.6, 4.8, 6)}	{(0, 0, 1.2), (4.8, 6, 7.2)}	{(0, 1.2, 2.4), (4.8, 6, 7.2)}
FM21	{(0, 1.2, 2.4), (3.6, 4.8, 6)}	{(2.4, 3.6, 4.8), (2.4, 3.6, 4.8)}	{(0, 1.2, 2.4), (3.6, 4.8, 6)}
FM22	{(0, 0, 1.2), (4.8, 6, 7.2)}	{(0, 1.2, 2.4), (3.6, 4.8, 6)}	{(0, 1.2, 2.4), (4.8, 6, 7.2)}
FM23	{(0, 1.2, 2.4), (4.8, 6, 7.2)}	{(1.2, 2.4, 3.6), (4.8, 6, 7.2)}	{(0, 1.2, 2.4), (2.4, 3.6, 4.8)}
FM24	{(1.2, 2.4, 3.6), (3.6, 4.8, 6)}	{(0, 0, 1.2), (2.4, 3.6, 4.8)}	{(0, 1.2, 2.4), (2.4, 3.6, 4.8)}

Table 7. Decision matrix for the twenty-four failure modes.

Code	Severity	Occurrence	Detectability	Effect Degree	Cause Degree
FM1	[5.08, 6.58]	[1.3, 1.58]	[1.61, 2.78]	0.69	0.17
FM2	[4.01, 4.28]	[1.19, 1.4]	[2.25, 3.73]	0.74	0.14
FM3	[2.63, 3.37]	[1.38, 2.09]	[1.38, 1.88]	0.80	0.52
FM4	[2.63, 3.37]	[0.63, 1.1]	[1.37, 1.65]	0.74	0.34
FM5	[0.59, 0.7]	[1.96, 2.2]	[0.24, 0.48]	0.54	0
FM6	[2.48, 3.75]	[3.33, 4.97]	[0.83, 1.19]	0.37	0
FM7	[1.66, 2.56]	[1.2, 1.91]	[1.27, 1.55]	0.30	0
FM8	[1.91, 3.38]	[1.33, 2.07]	[0.96, 1.69]	0.20	0.47
FM9	[0.54, 0.9]	[1.94, 2.7]	[0.93, 1.46]	0.20	0.00
FM10	[0.59, 1.21]	[1.05, 1.89]	[0.65, 1.02]	0.20	0.20
FM11	[1.49, 1.97]	[1.3, 1.58]	[0.83, 1.19]	0.20	0.67
FM12	[0.67, 1.81]	[2.61, 4.07]	[0.86, 1.3]	0.27	0.19
FM13	[2.12, 2.7]	[1.98, 2.25]	[0.63, 0.9]	0.33	0.31
FM14	[0.61, 1.33]	[3.74, 4.94]	[1.14, 1.74]	0.52	0.21
FM15	[0.56, 0.98]	[0.46, 0.7]	[0.77, 1.04]	0.41	0.16
FM16	[0.2, 0.92]	[2.37, 3.02]	[0.78, 1.26]	0.13	0.72
FM17	[0.32, 1.63]	[1.55, 1.91]	[0.94, 1.6]	0.13	0.72
FM18	[0.18, 0.91]	[0.58, 0.8]	[0.96, 1.37]	0.13	0.96
FM19	[1.06, 1.39]	[0.63, 0.9]	[0.7, 1.06]	1.20	1.59
FM20	[0.72, 1.07]	[0.09, 0.27]	[0.69, 0.98]	0.38	0.71
FM21	[0.7, 1.4]	[0.89, 1.08]	[0.65, 1.08]	0.18	0.31
FM22	[0.34, 0.83]	[0.47, 0.54]	[0.83, 1.19]	0.19	0.24
FM23	[0.83, 1.28]	[0.49, 0.97]	[0.5, 0.83]	0	0
FM24	[0.98, 1.94]	[0.07, 0.17]	[0.58, 0.95]	0	0.24

Table 8. The evaluations and weights for risk factors.

Experts	S	O	D	R	C
Expert 1	0.4	0.3	0.3	0.3	0.3
Expert 2	0.5	0.3	0.2	0.2	0.2
Expert 3	0.35	0.35	0.3	0.3	0.2
Expert 4	0.4	0.35	0.35	0.3	0.2
Aggregation results	[0.38, 0.44]	[0.31, 0.34]	[0.26, 0.32]	[0.26, 0.30]	[0.21, 0.24]
weights	0.27	0.21	0.19	0.18	0.15

After the aggregation process, VIKOR method was applied to sort the risks of the failure modes based on the decision matrix. The risk priorities of the twenty-four failure modes were determined by calculating the measure of closeness to the weighted vectors of positive ideal point. In the stage of VIKOR method, the optimal and the worst value of each risk factor were determined by Equations (22) and (23), and the values of S, R and Q for all failure modes were computed by using Equations (24)–(26) and presented in Table 9. A failure mode would be closer to the optimal values as the corresponding measure values approaches to zero. Thus, the failure modes can be prioritized or ranked according as the values of S, R, and Q in descending order. In order to make the ranking results better accepted by decision-makers, VIKOR method provides a compromise solution as illustrated in Section 3.2.

Table 9. The values and rankings of S, R and Q for all failure modes.

Code	S		R		Q	
	Value	Ranking	Value	Ranking	Value	Ranking
FM1	0.614	1	0.214	2	0.013	1
FM2	0.655	2	0.223	4	0.079	2
FM3	0.736	3	0.215	3	0.128	3
FM4	0.822	7	0.249	6	0.322	5
FM5	1.038	16	0.354	21	0.871	20
FM6	0.765	4	0.210	1	0.140	4
FM7	0.940	11	0.268	8	0.494	8
FM8	0.881	8	0.239	5	0.343	6
FM9	1.008	15	0.350	20	0.829	17
FM10	1.058	18	0.339	18	0.840	18
FM11	0.951	12	0.290	9	0.576	10
FM12	0.905	10	0.320	12	0.636	12
FM13	0.902	9	0.249	7	0.397	7
FM14	0.798	6	0.335	15	0.586	11
FM15	1.084	21	0.347	19	0.890	21
FM16	0.951	13	0.360	23	0.809	15
FM17	0.967	14	0.337	16	0.747	13
FM18	1.038	17	0.361	24	0.893	22
FM19	0.794	5	0.320	11	0.531	9
FM20	1.059	19	0.339	17	0.841	19
FM21	1.067	20	0.331	14	0.820	16
FM22	1.117	22	0.358	22	0.957	24
FM23	1.154	24	0.330	13	0.898	23
FM24	1.137	23	0.307	10	0.805	14

By comparing the risk rankings of the twenty-four failure modes, we see that in the pitch system the weakest link from a reliability standpoint is the pitch bearing, whose failure modes are ranked first and second in all failure modes identified in pitch system, and followed by the pitch gearbox and pitch motor. The failure of pitch bearing may lead to blade pitch to be out of sync or cannot pitch, causing impeller aerodynamic imbalance and fan speeding, which can result in the failure of safely starting and stopping the turbine, and bring about the blade rupture and other accidents. The pitch gearbox is also the key component that affects the reliability of the pitch system, whose failures of gear and bearing in the gearbox are ranked third and fifth in all failure modes. Through statistical analysis of historical fault data of pitch system, we found that the failure of pitch bearing and pitch gear accounts for 71% of the failure of the whole pitch system, which reveals that attention should be paid to these failure modes, and necessary measures and controls should be taken to lessen the possibility of their occurrence. There are many types of failure of pitch motor, among which the most serious failure modes are short circuit and open circuit of motor winding and motor brake failure, ranked fourth and sixth in all failure modes, respectively. It can also be seen that the failures of bearings, gear and other mechanical components have higher rankings, while the failures of switch, line and other electrical components have relatively lower rankings. This is because mechanical failures are difficult to be detected in time, and electrical failures are easy to be detected according to the abnormal current and voltage signals. Thus, in the reliability design of pitch system, higher reliability should be allocated to mechanical components, and in order to identify the failures of the mechanical components such as bearings and gears early before accidents to ensure the reliable operation of wind turbine, it is necessary to study the on-line monitoring technology for the mechanical failures of pitch system.

From the above analysis, we see that the ranking results of the twenty-four potential failure modes are in accordance with the practical engineering background, which proves the effectiveness of the proposed approach in practical application.

5. Comparison and Discussion

To further demonstrate the validity and availability of the proposed approach, three comparable method of traditional FMECA, fuzzy TOPSIS and combination weighting-based fuzzy VIKOR were also applied in the case study. The ranking results of the three methods are given in Table 10 and compared with that of the proposed FMECA approach. Based on the rankings in Table 10, it can be seen that the four approaches have a certain degree of similarity on the overall ranking trends of the twenty-four failure modes. For example, FM1 is recognized as the most critical failure mode in the four approaches since it has the highest or second-highest risk ranking. In each of approach, the top four ranked failure modes all contain FM1, FM2, and FM3, and the lowest ranked failure mode is all FM22. Moreover, failure mode FM15, FM17 and FM21 have very similar rankings in the four approaches. However, there are also some failure modes whose rankings of are very different in the four approaches, such as FM4, FM5, FM9, FM11, FM13, FM14, FM16, FM18, FM19, FM20 and FM24. The reasons contributing to the different rankings are analyzed as follows.

First, the weights of risk factors are different in the four approaches. The traditional FMECA reckons the weights of risk factors as equal, which is not reasonable in actual case. Since under the hypothesis of equal weights, some risk factors may be overestimated and others may be underestimated. In the fuzzy TOPSIS, fuzzy VIKOR and the proposed approach, such equal weight assumption is abandoned by determining the real weights of risk factors based on evaluations of experts. In the three kinds of approaches, the weights of risk factors are evaluated by experts using linguistic items. Meanwhile, the fuzzy TOPSIS determines the weight of risk factors by fuzzy AHP method, in which the weight of risk factors is ($w_S = 0.41, w_O = 0.31, w_D = 0.28$). The fuzzy VIKOR determines the weight of risk factor based on a combined weighting method integrated by fuzzy AHP and entropy method, in which the weight of risk factors is ($w_S = 0.4, w_O = 0.38, w_D = 0.22$). The proposed approach determines the weight of risk factors based on rough number and Equations (20) and (21), in which the weight of risk factors is ($w_S = 0.27, w_O = 0.21, w_D = 0.19, w_R = 0.18, w_C = 0.15$). Take the FM14 as an example, although experts evaluate FM14 with respect to occurrence with high value, they put relatively low importance on occurrence. Thus, FM14 gets relatively high ranking in the traditional FMECA compared to the rankings in the other three approaches, since occurrence is overestimated when regarding the weights of severity, occurrence and detectability as equal in traditional FMECA.

Table 10. The values and rankings of S, R and Q for all failure modes.

Code	Traditional FMECA		Fuzzy TOPSIS		Fuzzy VIKOR		Proposed Approach	
	RPN	Ranking	CC	Ranking	Q	Ranking	Q	Ranking
FM1	115.28	1	0.211	1	0.194	2	0.013	1
FM2	110.91	2	0.192	3	0.258	3	0.079	2
FM3	74.38	4	0.177	4	0.260	4	0.128	3
FM4	48.13	10	0.159	9	0.528	9	0.322	5
FM5	18.56	17	0.104	19	0.822	17	0.871	20
FM6	105.00	3	0.209	2	0.018	1	0.140	4
FM7	56.00	7	0.157	10	0.427	7	0.494	8
FM8	52.59	9	0.175	5	0.348	6	0.343	6
FM9	32.81	14	0.128	14	0.737	13	0.829	17
FM10	25.00	15	0.125	15	0.799	16	0.840	18
FM11	35.00	12	0.128	13	0.506	8	0.576	10
FM12	52.94	8	0.160	8	0.584	11	0.636	12
FM13	56.11	6	0.151	11	0.271	5	0.397	7
FM14	60.00	5	0.166	6	0.577	10	0.586	11
FM15	11.39	21	0.076	23	0.952	23	0.890	21
FM16	35.44	11	0.160	7	0.743	14	0.809	15
FM17	33.75	13	0.145	12	0.723	12	0.747	13
FM18	14.22	18	0.113	17	0.941	22	0.893	22
FM19	13.92	20	0.086	20	0.846	18	0.531	9
FM20	10.00	23	0.083	22	0.937	21	0.841	19
FM21	24.06	16	0.121	16	0.771	15	0.820	16
FM22	8.59	24	0.065	24	0.987	24	0.957	24
FM23	14.06	19	0.085	21	0.887	19	0.898	23
FM24	10.16	22	0.104	18	0.908	20	0.805	14

The second reason is the different representation and aggregation method for experts' evaluation information in the four approaches. As we know, FMECA is a team collaboration behavior which cannot be implemented alone on an individual basis [25]. On one hand, traditional FMECA, fuzzy TOPSIS and fuzzy VIKOR aggregate different experts' evaluations by average method. The aggregation results by this method are largely influenced by expert's opinion with subjectively and uncertainly. In fact, because of the different experience and backgrounds of experts, the evaluations of experts may be different and diverse, and some of which may be vague, imprecise and uncertain. In the proposed approach, the evaluations of different experts were aggregated by rough number, which could effectively aggregate the diversity evaluations and reduce the subjectivity and uncertainly in aggregation process. On the other hand, traditional FMECA, fuzzy TOPSIS and fuzzy VIKOR evaluate failure modes in the form of crisp number or triangular fuzzy number. Although fuzzy numbers are able to deal with the human vagueness evaluation to some extent, it does not consider the reliability of the restricted evaluation. In the proposed approach, the limitations of fuzzy number are overcome by Z-number, which describe the evaluations of failure modes by using 2-tuple fuzzy numbers. Compared to fuzzy number, Z-number has a stronger ability to express vague and uncertain information.

The third reason is the different ranking mechanism for failure modes in the four approaches. The traditional FMECA ranks the failure modes by multiplying the values of S, O, and D, which is questionable as mentioned in Introduction section. While the fuzzy TOPSIS, fuzzy VIKOR and the proposed approach take the ranking problem of failure modes as a multiple criteria decision-making (MCDM) issue and rank the failure modes by TOPSIS and VIKOR method. One difference between TOPSIS and VIKOR method is the different mechanism of aggregation function for ranking in the two methods. The aggregation function of VIKOR method represents the distance from the optimal values [72] with the ranking index of aggregating all risk factors, the weights of risk factors, and the balance between group and individual satisfaction. While the aggregation function of TOPSIS method represents the distances from the optimal value and from the worst value, which introduces the ranking index by summing these distances without considering their relative importance. The other difference between these two methods is the different means of normalization. The VIKOR method utilizes linear normalization method for normalizing, while the TOPSIS method uses vector normalization method. Moreover, the VIKOR method proposes a compromise solution with an advantage rate.

The last and most critical reason is that the traditional FMECA, fuzzy TOPSIS and fuzzy TOPSIS do not consider the propagation effect among failure modes, while the proposed approach considers. The failure propagation takes into account how a failure of a component could spread within a system, leading other components to failure. In fact, the practical impact to system reliability of the failure propagation is to increase the severity and occurrence of failure modes which can cause the occurrence of other failure modes or be affected by other failure modes. It can be seen that a very different ranking of failure mode is found in FM19 between the proposed approach and the other three approaches, which is ranked as twentieth in the traditional FMECA and fuzzy TOPSIS, eighteenth in the fuzzy VIKOR and ninth in the proposed approach. The remarkably different ranking for FM19 result from its high effect degree and cause degree. The high effect degree indicates that FM19 has a large possibility of causing other failures, which means its severity can be increased through the failure propagation. The high cause degree indicates that FM19 is more likely to be caused by other failures, which means its occurrence can be increased through the failure propagation. Thus, due to consideration of the failure propagation, the ranking of FM19 has greatly increased in the proposed approach compared to the ranking in the other three approaches, and so does the other failure modes such as FM3, FM4, FM20, etc.

6. Conclusions

Although FMECA has been extensively used in many fields for risk analysis, there are still some flaws that limit its performance of application in actual case, especially in terms of the issues of the representation of expert's opinions on the evaluation of failure modes, the aggregation of experts' diversity evaluations, and the determination of risk priorities of failure modes. In this paper, a new risk assessment model is proposed by using an integrated approach, which integrates the strong expressive ability of Z-numbers to vagueness and uncertainty information, the strong point of DEMATEL method in studying the dependence among failure modes, the advantage of rough numbers for aggregating experts' diversity evaluations, and the strength of VIKOR method to flexibly model multi-criteria decision-making problems. Based on the integrated approach, the proposed risk assessment model has the follow advantage features compared to the traditional FMECA and its variant:

1. The proposed model can well describe the judgements of experts on the evaluation of failure modes by using 2-tuple fuzzy numbers (Z-number) that the first fuzzy number represents the fuzzy restriction of the evaluation and the second fuzzy number represents its confidence or reliability.
2. The proposed model can effectively aggregate the diversity evaluations of experts by using rough number, which can reduce the subjectivity and uncertainty of evaluations in aggregation process and help to inspect the consistency of experts' perspective in decision making.
3. The proposed model takes the dependency among failure modes into consideration to identify the effect degree and cause degree of each failure mode by using DEMATEL method, which can recognize the potential high risk failure modes by analyzing the effects of failure propagation.
4. The proposed model determines the risk priorities of failure modes by using VIKOR method, which ranks the failure modes in a compromise way and helps experts in

FMECA team to reach a feasible ranking results based on maximizing the group utility for the "majority" and minimizing the individual regret for the "opponent".

To validate the performance of application in real case of the proposed FMECA approach and verify its effectiveness, the proposed risk assessment model is applied to the risk analysis of the failure modes in offshore wind turbine pitch system. By analyzing the ranking results of the twenty-four potential failure modes, we see that the proposed FMECA approach can be well used in real case, especially in the situations that the evaluations of experts are vague and uncertain and the failure modes are interacted with each other. Through the comparison with other approaches, we see that the ranking results obtained by proposed approach are more rational and more consistent with the actual results.

As a recommendation for future research, it is suggested that the evaluations of different experts for failure modes should be aggregated in the form of Z-number without converting the Z-numbers into crisp value, and some efficient fusion approaches should be excavated and applied to aggregation process. Moreover, the complexity of the proposed approach needs to be optimized to make it more applicable in practice. Moreover, in future work, the proposed model will be applied for risk management decision making in other fields of quality and reliability engineering to further verify its effectiveness.

Author Contributions: Conceptualization, Z.W. and R.W.; methodology, Z.W.; software, R.W.; validation, W.D. and Y.Z.; formal analysis, R.W.; investigation, Z.W.; resources, Y.Z.; data curation, W.D.; writing—original draft preparation, Z.W.; writing—review and editing, Z.W. and R.W.; visualization, W.D.; supervision, Y.Z.; project administration, W.D.; funding acquisition, Y.Z. and R.W. All authors have read and agreed to the published version of the manuscript.

Funding: This research was funded by [the science and technology project of China Huaneng Group Co., Ltd., Beijing 100031, China] grant number [HNKJ20-H72-02].

Acknowledgments: The authors gratefully acknowledge the valuable cooperation of the Clean energy branch of Huaneng (Zhejiang) Energy Development Co., Ltd., Hangzhou 310005, China in accomplishing this research project.

Nomenclature

v	Crisp value of evaluation
$\mu_B(x)$	Membership function of a triangular fuzzy number
$(\alpha_1, \alpha_2, \alpha_3)$	Triangular fuzzy number for fuzzy restriction in Z-number
$(\beta_1, \beta_2, \beta_3)$	Triangular fuzzy number for idea of confidence in Z-number
$Apr(v)$	Lower approximation of v
$\overline{Apr}(v)$	Upper approximation of v
$\underline{Lim}(v)$	Lower limit of v
$\overline{Lim}(v)$	Upper limit of v
M_L	Number of elements contained in $\underline{Apr}(v)$
M_U	Number of elements contained in $\overline{Apr}(v)$
$RN(v)$	Rough number of v
$Bnd(v)$	Boundary region of v
a_{ij}	Degree that a failure mode influenced by another failure mode
R	Effect degree
C	Cause degree
w_j	Weight of criterion (or risk factor)
v_j^*	Optimal value of risk factor
v_j^*	Optimal value of risk factor

Appendix A

Table A1. The assessment on *S*, *O* and *D* of the twenty-four potential failure modes given by expert 2.

Code	Severity	Occurrence	Detectability	Code	Severity	Occurrence	Detectability
FM1	(RH, S)	(M, NS)	(RH, RS)	FM13	(M, RS)	(MH, NS)	(RL, NS)
FM2	(RH, RS)	(RL, NS)	(M, U)	FM14	(M, RS)	(RH, NS)	(RL, RS)
FM3	(MH, RS)	(MH, U)	(M, RS)	FM15	(RL, RS)	(L, NS)	(L, RS)
FM4	(MH, RS)	(M, U)	(M, NS)	FM16	(M, NS)	(RH, U)	(M, NS)
FM5	(L, NS)	(RH, U)	(VL, RS)	FM17	(RH, RS)	(M, U)	(L, NS)
FM6	(RH, S)	(RH, NS)	(RL, NS)	FM18	(M, NS)	(RL, U)	(M, NS)
FM7	(MH, RS)	(M, U)	(M, U)	FM19	(RL, RS)	(L, NS)	(RL, RS)
FM8	(H, RS)	(MH, NS)	(L, RS)	FM20	(RL, NS)	(L, U)	(RL, NS)
FM9	(RL, NS)	(RH, NS)	(L, RS)	FM21	(M, NS)	(RL, U)	(M, NS)
FM10	(M, NS)	(M, U)	(RL, NS)	FM22	(L, NS)	(L, NS)	(RL, RS)
FM11	(M, NS)	(M, NS)	(RL, RS)	FM23	(RL, RS)	(L, NS)	(RL, NS)
FM12	(MH, RS)	(RH, U)	(L, NS)	FM24	(MH, RS)	(VL, U)	(RL, NS)

Table A2. The assessment on *S*, *O* and *D* of the twenty-four potential failure modes given by expert 3.

Code	Severity	Occurrence	Detectability	Code	Severity	Occurrence	Detectability
FM1	(EH, S)	(M, NS)	(RL, NS)	FM13	(RH, NS)	(RH, U)	(L, RS)
FM2	(H, RS)	(M, NS)	(MH, RS)	FM14	(L, U)	(EH, U)	(RL, RS)
FM3	(RH, RS)	(MH, NS)	(RL, RS)	FM15	(L, U)	(RL, U)	(L, RS)
FM4	(RH, RS)	(RL, NS)	(RL, RS)	FM16	(VL, NS)	(VH, U)	(L, NS)
FM5	(RL, U)	(RH, U)	(L, NS)	FM17	(VL, NS)	(RH, U)	(M, RS)
FM6	(RH, RS)	(EH, NS)	(L, RS)	FM18	(VL, NS)	(RL, U)	(RL, NS)
FM7	(MH, RS)	(MH, NS)	(RL, RS)	FM19	(RL, RS)	(RL, NS)	(L, RS)
FM8	(MH, S)	(MH, NS)	(L, RS)	FM20	(RL, U)	(VL, U)	(L, NS)
FM9	(RL, U)	(RH, NS)	(L, S)	FM21	(RL, RS)	(M, RU)	(L, NS)
FM10	(L, NS)	(MH, NS)	(L, NS)	FM22	(RL, NS)	(L, U)	(L, RS)
FM11	(MH, NS)	(M, NS)	(L, RS)	FM23	(RL, RS)	(RL, RU)	(L, NS)
FM12	(L, U)	(VH, NS)	(RL, RS)	FM24	(RL, NS)	(L, RU)	(L, NS)

Table A3. The assessment on *S*, *O* and *D* of the twenty-four potential failure modes given by expert 4.

Code	Severity	Occurrence	Detectability	Code	Severity	Occurrence	Detectability
FM1	(VH, RS)	(RL, NS)	(M, NS)	FM13	(M, NS)	(MH, U)	(L, NS)
FM2	(RH, RS)	(RL, NS)	(RH, RS)	FM14	(L, NS)	(EH, RS)	(M, RS)
FM3	(M, RS)	(RL, U)	(RL, NS)	FM15	(L, NS)	(L, RU)	(L, RS)
FM4	(M, RS)	(L, U)	(RL, NS)	FM16	(VL, NS)	(H, NS)	(RL, U)
FM5	(L, NS)	(MH, U)	(L, NS)	FM17	(VL, NS)	(MH, U)	(RL, NS)
FM6	(M, NS)	(EH, RS)	(RL, RS)	FM18	(VL, U)	(L, U)	(RL, U)
FM7	(RL, RS)	(RL, U)	(RL, NS)	FM19	(L, NS)	(L, NS)	(L, NS)
FM8	(RL, RS)	(RL, U)	(M, RS)	FM20	(L, U)	(VL, RU)	(RL, U)
FM9	(L, U)	(M, U)	(RL, RS)	FM21	(L, U)	(RL, U)	(L, NS)
FM10	(L, U)	(RL, RU)	(L, U)	FM22	(L, U)	(L, U)	(RL, NS)
FM11	(RL, NS)	(RL, NS)	(RL, NS)	FM23	(L, NS)	(L, RU)	(L, U)
FM12	(L, NS)	(VH, RS)	(RL, RS)	FM24	(L, NS)	(VL, RU)	(RL, U)

References

1. Liu, H.C.; You, J.X.; Li, P.; Su, Q. Failure Mode and Effect Analysis Under Uncertainty: An Integrated Multiple Criteria Decision Making Approach. *IEEE Trans. Reliab.* **2016**, *65*, 1380–1392. [CrossRef]
2. Bowles, J.B.; Pelaez, C.E. Fuzzy logic prioritization of failures in a system failure mode, effects and criticality analysis. *Reliab. Eng. Syst. Saf.* **1995**, *50*, 203–213. [CrossRef]
3. Liu, H.C. *FMEA Using Uncertainty Theories and MCDM Methods*; Springer: Singapore, 2016.

4. Jong, C.H.; Tay, M.K.; Lim, C.P. Application of the fuzzy Failure Mode and Effect Analysis methodology to edible bird nest processing. *Comput. Electron. Agric.* **2013**, *96*, 90–108. [CrossRef]
5. Kurt, L.; Ozilgen, S. Failure mode and effect analysis for dairy product manufacturing: Practical safety improvement action plan with cases from Turkey. *Saf. Sci.* **2013**, *55*, 195–206. [CrossRef]
6. Lin, Q.L.; Wang, D.J.; Lin, W.G.; Liu, H.C. Human reliability assessment for medical devices based on failure mode and effects analysis and fuzzy linguistic theory. *Saf. Sci.* **2014**, *62*, 248–256. [CrossRef]
7. Abrahamsen, H.B.; Abrahamsen, E.B.; Høyland, S. On the need for revising healthcare failure mode and effect analysis for assessing potential for patient harm in healthcare processes. *Reliab. Eng. Syst. Saf.* **2016**, *155*, 160–168. [CrossRef]
8. Lijesh, K.P.; Muzakkir, S.M.; Hirani, H. Failure mode and effect analysis of passive magnetic bearing. *Eng. Fail. Anal.* **2016**, *62*, 1–20. [CrossRef]
9. Saulino, M.F.; Patel, T.; Fisher, S.P. The Application of Failure Modes and Effects Analysis Methodology to Intrathecal Drug Delivery for Pain Management. *Neuromodulation* **2017**, *20*, 177–186. [CrossRef]
10. Teixeira, F.C.; de Almeida, C.E.; Saiful, H.M. Failure mode and effects analysis based risk profile assessment for stereotactic radiosurgery programs at three cancer centers in Brazil. *Med. Phys.* **2016**, *43*, 171. [CrossRef]
11. Chai, K.C.; Jong, C.H.; Tay, K.M.; Lim, C.P. A perceptual computing-based method to prioritize failure modes in failure mode and effect analysis and its application to edible bird nest farming. *Appl. Soft Comput.* **2016**, *49*, 734–747. [CrossRef]
12. Liu, H.C.; Fan, X.J.; Li, P.; Chen, Y.Z. Evaluating the risk of failure modes with extended MULTIMOORA method under fuzzy environment. *Eng. Appl. Artif. Intell.* **2014**, *34*, 168–177. [CrossRef]
13. Liu, H.C.; You, J.X.; You, X.Y.; Shan, M.M. A novel approach for failure mode and effects analysis using combination weighting and fuzzy VIKOR method. *Appl. Soft Comput.* **2015**, *28*, 579–588. [CrossRef]
14. Liu, H.C.; Liu, L.; Liu, N. Risk evaluation approaches in failure mode and effects analysis: A literature review. *Expert Syst. Appl.* **2013**, *40*, 828–838. [CrossRef]
15. Certa, A.; Hopps, F.; Inghilleri, R.; Fata, C.M.L. A Dempster-Shafer Theory-based approach to the Failure Mode, Effects and Criticality Analysis (FMECA) under epistemic uncertainty: Application to the propulsion system of a fishing vessel. *Reliab. Eng. Syst. Saf.* **2017**, *159*, 69–79. [CrossRef]
16. Mohsen, O.; Fereshteh, N. An extended VIKOR method based on entropy measure for the failure modes risk assessment—A case study of the geothermal power plant (GPP). *Saf. Sci.* **2017**, *92*, 160–172. [CrossRef]
17. Huang, J.; Li, Z.; Liu, H.C. New approach for failure mode and effect analysis using linguistic distribution assessments and TODIM method. *Reliab. Eng. Syst. Saf.* **2017**, *167*, 302–309. [CrossRef]
18. Kerk, Y.W.; Tay, K.M.; Lim, C.P. An analytical interval fuzzy inference system for risk evaluation and prioritization in failure mode and effect analysis. *IEEE Syst. J.* **2017**, *11*, 1589–1600. [CrossRef]
19. Certa, A.; Enea, M.; Galante, G.M.; Fata, C.M.L. ELECTRE TRI-based approach to the failure modes classification on the basis of risk parameters: An alternative to the Risk Priority Number. *Comput. Ind. Eng.* **2017**, *108*, 100–110. [CrossRef]
20. Song, W.; Ming, X.; Wu, Z.; Zhu, B. A rough TOPSIS Approach for Failure Mode and Effects Analysis in Uncertain Environments. *Qual. Reliab. Eng. Int.* **2014**, *30*, 473–486. [CrossRef]
21. Shafiee, M.; Dinmohammadi, F. An FMEA-Based Risk Assessment Approach for Wind Turbine Systems: A Comparative Study of Onshore and Offshore. *Energies* **2014**, *7*, 619–642. [CrossRef]
22. Ozturk, S.; Fthenakis, V.; Faulstich, S. Failure Modes, Effects and Criticality Analysis for Wind Turbines Considering Climatic Regions and Comparing Geared and Direct Drive Wind Turbines. *Energies* **2018**, *11*, 2317.
23. Tazi, N.; Châtelet, E.; Bouzidi, Y. Using a Hybrid Cost-FMEA Analysis for Wind Turbine Reliability Analysis. *Energies* **2017**, *10*, 276. [CrossRef]
24. Kai, M.T.; Lim, C.P. Fuzzy FMEA with a guided rules reduction system for prioritization of failures. *Int. J. Qual. Reliab. Manag.* **2006**, *23*, 1047–1066.
25. Chin, K.S.; Wang, Y.M.; Gary, K.K.P.; Yang, J.B. Failure mode and effects analysis using a group-based evidential reasoning approach. *Comput. Oper. Res.* **2009**, *36*, 1768–1779. [CrossRef]
26. Huang, G.; Xiao, L.; Zhang, G. Risk evaluation model for failure mode and effect analysis using intuitionistic fuzzy rough number approach. *Soft Comput.* **2021**, *25*, 4875–4897. [CrossRef]
27. Yang, J.; Huang, H.Z.; He, L.P.; Zhu, S.P.; Wen, D. Risk evaluation in failure mode and effects analysis of aircraft turbine rotor blades using Dempster–Shafer evidence theory under uncertainty. *Eng. Fail. Anal.* **2011**, *18*, 2084–2092. [CrossRef]
28. Liu, H.C.; You, J.X.; Fan, X.J.; Lin, Q.L. Failure mode and effects analysis using D numbers and grey relational projection method. *Expert Syst. Appl.* **2014**, *41*, 4670–4679. [CrossRef]
29. Aslani, R.K.; Feili, H.R.; Javanshir, H. A hybrid of fuzzy FMEA-AHP to determine factors affecting alternator failure causes. *Manag. Sci. Lett.* **2014**, *4*, 1981–1984. [CrossRef]
30. Salmon, C.; Mobin, M.; Roshani, A. TOPSIS as a Method to Populate Risk Matrix Axes. In Proceedings of the Industrial and Systems Engineering Research Conference, Nashville, TN, USA, 30 May–2 June 2015.
31. Vafadarnikjoo, A.; Mobin, M.; Firouzabadi, A.K. An Intuitionistic Fuzzy-Based DEMATEL to Rank Risks of Construction Projects. In Proceedings of the International Conference on Industrial Engineering and Operations Management, Detroit, MI, USA, 23–25 September 2016.

32. Skeete, A.; Mobin, M. Aviation Technical Publication Content Management System Selection Using Integrated Fuzzy-Grey MCDM Method. In Proceedings of the Industrial and Systems Engineering Research Conference, Nashville, TN, USA, 30 May–2 June 2015.
33. Hu, A.H.; Hsu, C.W.; Kuo, T.C.; Wu, W.C. Risk evaluation of green components to hazardous substance using FMEA and FAHP. *Expert Syst. Appl.* **2009**, *36*, 7142–7147. [CrossRef]
34. Bozdag, E.; Asan, U.; Soyer, A.; Serdarasan, S. Risk prioritization in Failure Mode and Effects Analysis using interval type-2 fuzzy sets. *Expert Syst. Appl.* **2015**, *42*, 4000–4015. [CrossRef]
35. Liu, H.C.; Chen, Y.Z.; You, J.X.; Li, H. Risk evaluation in failure mode and effects analysis using fuzzy digraph and matrix approach. *J. Intell. Manuf.* **2016**, *27*, 805–816. [CrossRef]
36. Zhou, D.; Tang, Y.; Jiang, W. A Modified Model of Failure Mode and Effects Analysis Based on Generalized Evidence Theory. *Math. Probl. Eng.* **2016**, *2016*, 4512383. [CrossRef]
37. Liu, H.C.; You, J.X.; Duan, C.Y. An integrated approach for failure mode and effect analysis under interval-valued intuitionistic fuzzy environment. *Int. J. Prod. Econ.* **2017**, *207*, 163–172. [CrossRef]
38. Yang, Z.; Bonsall, S.; Wang, J. Fuzzy Rule-Based Bayesian Reasoning Approach for Prioritization of Failures in FMEA. *IEEE Trans. Reliab.* **2008**, *57*, 517–528. [CrossRef]
39. Jee, T.L.; Tay, K.M.; Lim, C.P. A new two-stage fuzzy inference system-based approach to prioritize failures in failure mode and effect analysis. *IEEE Trans. Reliab.* **2015**, *64*, 869–877. [CrossRef]
40. Gupta, G.; Mishra, R.P. A Failure Mode Effect and Criticality Analysis of Conventional Milling Machine Using Fuzzy Logic: Case Study of RCM. *Qual. Reliab. Eng. Int.* **2016**, *33*, 347–356. [CrossRef]
41. Tooranloo, H.S.; Ayatollah, A.S. A model for failure mode and effects analysis based on intuitionistic fuzzy approach. *Appl. Soft Comput.* **2016**, *49*, 238–247. [CrossRef]
42. Guo, J. A risk assessment approach for failure mode and effects analysis based on intuitionistic fuzzy sets and evidence theory. *J. Intell. Fuzzy Syst.* **2016**, *30*, 869–881. [CrossRef]
43. Jiang, W.; Xie, C.; Zhuang, M.; Tang, Y. Failure Mode and Effects Analysis based on a novel fuzzy evidential method. *Appl. Soft Comput.* **2017**, *57*, 672–683. [CrossRef]
44. Aydogan, E.K. Performance measurement model for Turkish aviation firms using the rough-AHP and TOPSIS methods under fuzzy environment. *Expert Syst. Appl.* **2011**, *38*, 3992–3998. [CrossRef]
45. Liu, H.C.; You, J.X.; Shan, M.M.; Shao, L.N. Failure mode and effects analysis using intuitionistic fuzzy hybrid TOPSIS approach. *Soft Comput.* **2015**, *19*, 1085–1098. [CrossRef]
46. Carpitella, S.; Certa, A.; Izquierdo, J.; La Fata, C.M. A combined multi-criteria approach to support FMECA analyses: A real-world case. *Reliab. Eng. Syst. Saf.* **2018**, *169*, 394–402. [CrossRef]
47. Vahdani, B.; Salimi, M.; Charkhchian, M. A new FMEA method by integrating fuzzy belief structure and TOPSIS to improve risk evaluation process. *Int. J. Adv. Manuf. Technol.* **2015**, *77*, 357–368. [CrossRef]
48. Zhou, Q.; Thai, V.V. Fuzzy and grey theories in failure mode and effect analysis for tanker equipment failure prediction. *Saf. Sci.* **2016**, *83*, 74–79. [CrossRef]
49. Liu, H.C.; Li, Z.; Song, W.; Su, Q. Failure Mode and Effect Analysis Using Cloud Model Theory and PROMETHEE Method. *IEEE Trans. Reliab.* **2017**, *66*, 1058–1072. [CrossRef]
50. Mandal, S.; Singh, K.; Behera, R.K.; Sahu, S.K.; Raj, N.; Maiti, J. Human error identification and risk prioritization in overhead crane operations using HTA, SHERPA and fuzzy VIKOR method. *Expert Syst. Appl.* **2015**, *42*, 7195–7206. [CrossRef]
51. Baloch, A.U.; Mohammadian, H. Fuzzy failure modes and effects analysis by using fuzzy Vikor and Data Envelopment Analysis-based fuzzy AHP. *Int. J. Adv. Appl. Sci.* **2016**, *3*, 23–30. [CrossRef]
52. Liu, H.C.; Liu, L.; Bian, Q.H.; Lin, Q.L.; Dong, N.; Xu, P.C. Failure mode and effects analysis using fuzzy evidential reasoning approach and grey theory. *Expert Syst. Appl.* **2011**, *38*, 4403–4415. [CrossRef]
53. Liu, H.C.; Liu, L.; Lin, Q.L. Fuzzy Failure Mode and Effects Analysis Using Fuzzy Evidential Reasoning and Belief Rule-Based Methodology. *IEEE Trans. Reliab.* **2013**, *62*, 23–36. [CrossRef]
54. Du, Y.; Mo, H.; Deng, X.; Sadiq, R.; Deng, Y. A new method in failure mode and effects analysis based on evidential reasoning. *Int. J. Syst. Assur. Eng. Manag.* **2014**, *5*, 1–10. [CrossRef]
55. Li, Y.; Deng, Y.; Kang, B. A Risk Assessment Methodology Based on Evaluation of Group Intuitionistic Fuzzy Set. *J. Inf. Comput. Sci.* **2012**, *9*, 1855–1862.
56. Su, X.; Deng, Y.; Mahadevan, S.; Bao, Q. An improved method for risk evaluation in failure modes and effects analysis of aircraft engine rotor blades. *Eng. Fail. Anal.* **2012**, *26*, 164–174. [CrossRef]
57. Shi, C.; Cheng, Y.M.; Pan, Q. Method to aggregate hybrid preference information based on intuitionistic fuzzy and evidence theory. *Control Decis.* **2012**, *27*, 1163–1168.
58. Jousselme, A.L.; Grenier, D.; Bossé, É. A new distance between two bodies of evidence. *Inf. Fusion* **2001**, *2*, 91–101. [CrossRef]
59. Jiang, W.; Xie, C.; Wei, B.; Zhou, D. A modified method for risk evaluation in failure modes and effects analysis of aircraft turbine rotor blades. *Adv. Mech. Eng.* **2016**, *8*, 579. [CrossRef]
60. Yang, Y.; Han, D. A new distance-based total uncertainty measure in the theory of belief functions. *Knowl. Based Syst.* **2016**, *94*, 114–123. [CrossRef]

61. Mohamad, D.; Shaharani, S.A.; Kamis, N.H. A Z-number-based decision making procedure with ranking fuzzy numbers method. In Proceedings of the International Conference on Quantitative Sciences & Its Applications, Langkawi Kedah, Malaysia, 12–14 August 2014; pp. 160–166.
62. Azadeh, A.; Saberi, M.; Atashbar, N.Z.; Chang, E.; Pazhoheshfar, P. Z-AHP: A Z-number extension of fuzzy analytical hierarchy process. In Proceedings of the IEEE International Conference on Digital Ecosystems and Technologies, Menlo Park, CA, USA, 24–26 July 2013; pp. 141–147.
63. Zadeh, L.A. A Note on Z-numbers. *Inf. Sci.* **2011**, *181*, 2923–2932. [CrossRef]
64. Salari, M.; Bagherpour, M.; Wang, J. A novel earned value management model using Z-number. *Int. J. Appl. Decis. Sci.* **2014**, *7*, 97–119. [CrossRef]
65. He, Y.H.; Wang, L.B.; He, Z.Z.; Xie, M. A fuzzy TOPSIS and Rough Set based approach for mechanism analysis of product infant failure. *Eng. Appl. Artif. Intell.* **2015**, *47*, 25–37. [CrossRef]
66. Zhai, L.Y.; Khoo, L.P.; Zhong, Z.W. A rough set enhanced fuzzy approach to quality function deployment. *Int. J. Adv. Manuf. Technol.* **2008**, *37*, 613–624. [CrossRef]
67. Zhu, G.N.; Hu, J.; Qi, J.; Gu, C.C.; Peng, Y.H. An integrated AHP and VIKOR for design concept evaluation based on rough number. *Adv. Eng. Inform.* **2015**, *29*, 408–418. [CrossRef]
68. Wu, H.-H.; Chang, S.-Y. A case study of using DEMATEL method to identify critical factors in green supply chain management. *Appl. Math. Comput.* **2015**, *256*, 394–403. [CrossRef]
69. Chang, B.; Chang, C.-W.; Wu, C.-H. Fuzzy DEMATEL method for developing supplier selection criteria. *Expert Syst. Appl.* **2011**, *38*, 1850–1858. [CrossRef]
70. Opricovic, S.; Tzeng, G.H. Extended VIKOR method in comparison with outranking methods. *Eur. J. Oper. Res.* **2007**, *178*, 514–529. [CrossRef]
71. Liou, J.J.; Tsai, C.-Y.; Lin, R.-H.; Tzeng, G.-H. A modified VIKOR multiple-criteria decision method for improving domestic airlines service quality. *J. Air Transp. Manag.* **2011**, *17*, 57–61. [CrossRef]
72. Opricovic, S.; Tzeng, G.H. Compromise solution by MCDM methods: A comparative analysis of VIKOR and TOPSIS. *Eur. J. Oper. Res.* **2004**, *156*, 445–455. [CrossRef]
73. Kumar, A.; Dixit, G. Evaluating critical barriers to implementation of WEEE management using DEMATEL approach. *Resour. Conserv. Recycl.* **2018**, *131*, 101–121. [CrossRef]

 *energies*

Article

Energy–Carbon Emissions Nexus Causal Model towards Low-Carbon Products in Future Transport-Manufacturing Industries

Olukorede Tijani Adenuga *, Khumbulani Mpofu and Ragosebo Kgaugelo Modise

Department of Industrial Engineering, Tshwane University of Technology, Pretoria-Campus, Pretoria 0183, South Africa
* Correspondence: olukorede.adenuga@gmail.com; Tel.: +27-(061)-6318731

Abstract: Climate change is progressing faster than previously envisioned. Efforts to mitigate the challenges of greenhouse gas emissions by countries through the establishment of the Intergovernmental Panel on Climate Change has resulted in continuous environmental improvements in the energy efficiency and carbon emission signatures of products. In this paper, an energy–carbon emissions nexus causal model was applied using the Leontief Input–Output mathematical model for low-carbon products in future transport-manufacturing industries., The relationship between energy savings, energy efficiency, and the carbon intensity of products for the carbon emissions signature of the future transport manufacturing in South Africa was established. The interrelationship between the variables resulted in a 29% improvement in the total energy intensity of the vehicle body part products, 7.22% in the cumulative energy savings, and 16.25% in the energy efficiency. The scope that has been examined in this paper will be interesting to agencies of government, researchers, policymakers, business owners, and practicing engineers in future transport manufacturing and could serve as a fundamental guideline for future studies in these areas.

Keywords: energy–carbon emissions nexus; causal model; low-carbon product; future transport-manufacturing industries

Citation: Adenuga, O.T.; Mpofu, K.; Modise, R.K. Energy–Carbon Emissions Nexus Causal Model towards Low-Carbon Products in Future Transport-Manufacturing Industries. *Energies* **2022**, *15*, 6322. https://doi.org/10.3390/en15176322

Academic Editors: Ershun Pan, Rongxi Wang, Yupeng Li, Xi Gu and Tangbin Xia

Received: 10 August 2022
Accepted: 28 August 2022
Published: 30 August 2022

Publisher's Note: MDPI stays neutral with regard to jurisdictional claims in published maps and institutional affiliations.

1. Introduction

Today's fast-paced manufacturing industry is increasingly characterized by technology. The manufacturing sector accounts for about 33% of the primary energy use and 38% of the CO_2 emissions globally [1]. Concerning transport manufacturing, energy purchases have a major impact on the production costs and, ultimately, on the industry's competitiveness [2]. Energy efficiency becomes a driver for the manufacturing industry since it is historically one of the greatest energy consumers and carbon emitters in the world [3]. Author [4] argues that promoting efficiency without any curbs on the consumption will not tackle the problem of reducing CO_2 emissions. The targets for the reduction of CO_2 emissions have a great effect on the manufacturing industry. South African sectoral electricity, specifically for the industry sector, sits at 49% vs. a supply of 43.7% [5]. South Africa is committed to reducing emissions through the introduction of smart manufacturing as an essential route to meet the greenhouse gas (GHG) emissions target that was out in the agreements on climate change by the International Panel on Climate Change [6]. However, it is not only the introduction of smart manufacturing that will aid the achievement of the CO_2 emission target, but the main enabler of the commission of these technologies, namely the policies that recognize the energy efficiency and carbon emission reduction system with the benefits of digital technologies to overcome the barriers to the implementation and the accelerated technology deployment in South Africa. Du Plessis 2015 [7] presented a study that explored the nature and the extent of the various policy instruments and legislation that relate to energy efficiency.

Several publications have focused on the application of energy efficiency in the different sector's policies [8–10], general trading, decision making [11,12], ICT [13], and the building sector [7]. Other energy efficiency-related research includes the industry application in the supply chain [14], the impact on the South African energy crisis on emissions [15], carbon accounting [16], and predicting carbon emissions [17,18]. A review of the literature that is related to carbon emissions in most of the applications, in general, accounts for carbon emission [19], integrated energy efficiency, and the carbon emissions in the industry, focusing on the implication of improvement for emissions [20], conducted an experiential study and modeling of energy efficiency by [21], while the reviewed literature focused on a causal relationship of energy consumption, price, and emissions. The manufacturing sector consumes a high amount of energy and emits even more [22], and extensive research has been carried out on the energy efficiency of manufacturing, focusing on improvement [23], low-cost energy efficiency measures [24], energy efficiency development in manufacturing [25], and energy management that includes energy efficiency [26]. Further studies have focused on the energy efficiency of manufacturing systems and processes [3,27,28]. The research has observed that the literature that has been presented on energy efficiency has a high interest in modeling energy efficiency in manufacturing, including rail manufacturing [29–32] and carbon dioxide prediction [33]. Energy efficiency is recognized globally as a critical solution toward the reduction of energy consumption, while the management of global carbon dioxide emissions complements climate change policies, abates the costs of reducing carbon emissions, and the improves economic competitiveness. Built on the existing reviews on energy efficiency and carbon emissions, this paper considers the causal relationship between energy consumption, energy intensity, and carbon emission in future transport manufacturing. The research considers an investigative study that explores the implications of energy efficiency improvement for CO_2 emissions in the energy-intensive industries and includes the element of prediction, to advise the decision makers on low-carbon products. The outline of the paper is structured as follows: Section 2 reviews the systematic pieces of the literary techniques in transport manufacturing and the future transport sector regarding energy resource consumption and carbon dioxide emission. The model for the asymmetric energy–carbon emission and energy and carbon emissions efficiency regression-based approach are formulated in Section 3. Section 4 expounds on the results and the discussion. Section 5 presents a conclusion.

2. Literature Review

The transport-manufacturing sector is one of the most important sectors in the industry and is considered to shape the economic growth and job creation that supports policies that are related to energy consumption [34], however, the policies that support energy systems through digital technologies are rare. A new framework for driving analytical data to reduce energy consumption and carbon emission in energy-intensive manufacturing has been suggested [35]. Authors [36] investigated the causal relationship between energy resource consumption, energy prices, and carbon dioxide emission in the building sector to determine the effects of energy sources and prices on carbon emissions, thus, further research is required in the industrial and transportation sectors. A structured literature review technique was used in the collection of empirical evidence in a particular field for the study to assert the evidence of the co-benefits between energy consumption and carbon emission to determine the current state of knowledge. It required a critical assessment of the evidence and the identification of both the potential for energy and carbon efficiency, with direct economic savings, and the ability to summarize the findings. To achieve the objective, we have adopted the following three-step approach to identify the relevant research literature: search term, filtering approach, and information removal.

2.1. Future Transport Manufacturing

The report 'European Commission' has emphasized the importance of the transport sector for economic growth and has widely acknowledged that targeted innovations and targeted research activities are key factors for fostering competitiveness in the future transport sector [37]. Research that was conducted in United States has highlighted that transport manufacturing is the eighth largest industrial energy consumer; the energy expenditure increased by 20%, and the purchases of electricity went up by nearly 10% [2]. The modern/smart manufacturing industry is investing in new technologies, such as the Internet of Things (IoT), big data analytics, cloud computing, and cybersecurity, to cope with system complexity, to increase information visibility, to improve production performance, and to gain competitive advantages in the global market [8,38].

2.2. Energy Efficiency and Carbon Emission in Future Transport Manufacturing

The impact of the energy efficiency of emissions in the transport-manufacturing sector in South Africa can be seen from the recent technological improvements. Giampieri et al., 2019 [39] suggested that automotive manufacturers are facing economic and environmental pressure for the realization of a sustainable low carbon process, therefore, improved energy efficiency is necessary to decrease greenhouse gas emissions, and the carbon risks are mainly related to the emissions from the purchased electricity in Korean automobile manu-factures [40]. The application of a conceptual model of the wind turbine into the transport sector to produce energy for powering the car has also been suggested [41]. However, there have been studies on a causal relationship between energy consumption and CO_2 emission in building sectors [36].

2.3. Recent Studies on Energy and Carbon Emission

Botts et al., 2021 [42] developed a decision tool for the energy efficiency of a blower heater on a normalized basis, in terms of the performance and the cost. The energy consumption estimation in Indian refineries based on empirical-analytical-based panel data econometrics was postulated by [43], which concluded on the formulation of a policy to reduce the energy consumption. An energy planning and carbon dioxide estimation system dynamic model was presented for Nigeria's power sector by [44]. The model investigated ways to bridge the demand gaps and the electricity supply through the simulation of variables, real socio-economic factors, and the estimation of CO_2, in various performance scenarios. Sunde, 2020 [45] investigated the economic growth and the energy consumption of SADC countries using causality analysis to model the growth variables with the implied notion of an increased level of energy consumption leading to an economic output increase. Brahmana and Ono [46] justified the need for energy efficiency as a significant part of company performance in Japanese listed companies, which affects the market-based performance with significant impacts on the return on assets (accounting-book performance), thereby debunking the energy-efficiency paradox. Olanrewaju et al. developed a forecast model that was dependent on an artificial neural network to model the energy consumption between 2002 and 2009, which was based on the gross domestic product and the population [47]. It was discovered that an artificial neural network is a better modeling technique compared to regression analysis. The link between an energy-restricted environment and the emissions in South Africa was evaluated by [15], with findings of undeniable facts on the negative impact of the emissions that are caused by energy production. Gamede et al., [48] proposed a business model for intergrading the energy efficiency performance in the manufacturing industry by using a rail car case study that was recommended for energy service companies. Authors [49] proposed the energy efficiency analysis modeling system (EEAMS) in transport manufacturing, focusing on rail. The tool provides an estimate of the energy costs by using the rail car manufacturing plan load profiles as a case study to provide a consumer-oriented analysis to produce a first-cut energy-efficient program baseline cost.

This paper examined the use of the casual relationship between energy consumption, energy intensity, and carbon dioxide emission in the future transport-manufacturing sectors, respectively.

3. Methodology

In this section, the concepts and specifications of the mathematical model development and the measurement and verification integration model for energy efficiency and the carbon emission signatures approach proposed in this work are presented. The integration of the high volume of data used was in real-time in the determination of energy intensity on the realization of the low-carbon product. The proposed architecture of the framework was based on the Leontief Input–Output mathematical model approach for energy savings. The entire architecture's key components are presented here to understand how it works. The components that constitute the architecture are the physical layer, communication layer, middleware layer, database layer, and the application and management layer. The moving average over the past year, i.e., the last 12 values, was determined with specific functions for rolling statistics in Pandas in the determination of the time series techniques for the actual data-processing method. We collected Tier 2 automotive company data for 2015 to 2018 and examined the results of IoT-based energy monitoring devices using an asymmetric energy causal mathematical model for the analysis through decomposing into the following components: prediction, testing, and forecasting, for decision making and policy formulation. We used the vehicle body productions' energy intensity with the expected energy consumption and savings, with the expected energy consumption adjustment from the 2015 to 2016 energy performance data, to compare the results.

3.1. Asymmetric Energy–Carbon Emission Causal Mathematical Model

The asymmetry energy causal mathematical model is an order of the dependency variables characteristics of the energy variables and vehicle production parts from a heuristic assumption. The definition represents inferences from the statistical data of energy consumed in producing one unit (inputs) as an exogenous variable, which is determined by embodied energy intensity per unit as an output variable.

Energy Analysis—Input and Output Theory

The following equilibrium structure predicts the future transport manufacturing energy consumption over carbon and energy intensity, thereby entailing probabilistic independence or dependence of the process variables. The adoption of the asymmetric energy–carbon emission causal mathematical models, which are presented in Figure 1, is the heuristic dependency determines the observed probabilistic correlations among variables, or the outcomes of the lowest influence variables. We then manipulate the variables within the equations to produce asymmetric causal equations as follows:

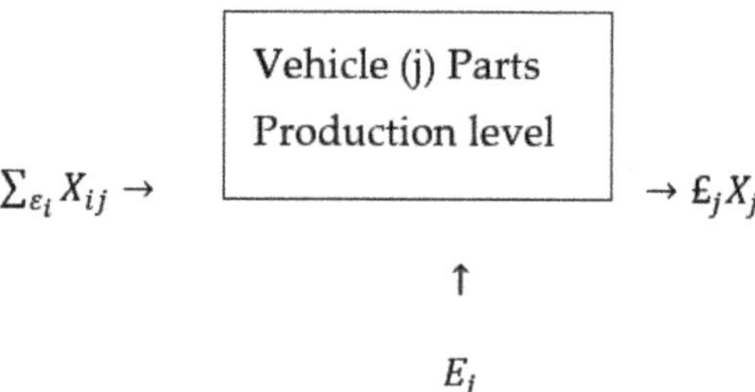

Figure 1. Causal graph embodied energy intensity in an input and output equilibrium structure.

X_{ij} = transaction from sector i to sector j, X_j = total output of sector j, $\pounds_j$ = embodied energy intensity per unit of X_j (amount of energy consumed in producing one unit), and E_j = energy consumed to restate the demand of the reporting periods under a common set of conditions. E_s = energy savings in Equation (1), β_{peu} = baseline energy usage for a period of use, ρ_{peu} = reported energy usage for a period of use, and A = sum of the adjustment as Equation (2)

$$E_S = \beta_{peu} - \rho_{peu} \pm A \tag{1}$$

$$A = 1 - \left(1 - R^2\right)^{n-1}/n - p_p^{-1} \tag{2}$$

Carbon Emissions converted from Energy

Evaluating the carbon footprint relative to the energy consumption during the production of vehicle bodies using applicable emission factors $(7.0555 * 10^{-4})$ to estimate the carbon emissions signature (CESTM) for the energy mix for a particular year.

$$\text{Carbon emission } [\text{kgtCO}_2)] = \text{\pounds}_j * \text{CESTM } [\text{kg tCO}_2 / \text{GJ}] \tag{3}$$

The annualized non-baseload CO_2 output emission rate is used to convert reductions in hours into avoided units of carbon dioxide emissions. The average carbon intensity factor of $0.4 \text{ kgCO}_2/\text{GJ}$ is adopted for the equivalent of emission reductions from energy efficiency programs that are assumed to affect non-baseload generation (power plants that are brought online, as necessary to meet demand).

$$\text{Energy}_{\text{consumption}} = f\left(E_c, \text{CO}_2, V_{op}, E_i,\right)\text{CO}_2 \tag{4}$$

$$\text{Energy}_{\text{consumption}} = f(E_c, V_b, E_i, N_c) \tag{5}$$

E_c = electricity consumed, V_b = vehicle bodies, E_i = energy intensity, N_c = normalization coefficient in CUSUM savings (895.56) to produce parts body.

Carbon Emission Savings

$$\text{CO}_2 \text{ savings} = f\left(E_s, V_{op}, E_i,\right) \tag{6}$$

Considering the variables required in total internal requirements within the model, the equation was X = AX, that is, the total input is equal to the total output. X = AX + D (I − A) X = D, where I is a 3 by 3 identity matrix X = (I−A) − 1D, all consumption is within the industries. There is no external demand.

Producing P_1 units of E_s required $E_{s_{p1}}$ units of CO_2 savings; producing P_2 units of V_{op} required $V_{op_{p1}}$ units of CO_2 savings; producing P_3 units of E_i required $E_{i_{p1}}$ units of CO_2 savings.

Definition 1.

$$\begin{bmatrix} P_1 \\ P_2 \\ P_3 \end{bmatrix} = P_1 \begin{bmatrix} E_{s_{p1}} \\ E_{s_{p2}} \\ E_{s_{p3}} \end{bmatrix} + \begin{bmatrix} V_{op_{p1}} \\ V_{op_{p2}} \\ V_{op_{p3}} \end{bmatrix} + \begin{bmatrix} E_{i_{p1}} \\ E_{i_{p2}} \\ E_{i_{p3}} \end{bmatrix} * \begin{pmatrix} P_1 \\ P_2 \\ P_3 \end{pmatrix} \tag{7}$$

Total required units of CO_2 savings

$$E_{s_{p1}} + V_{op_{p1}} + E_{i_{p1}} \tag{8}$$

Definition 2. *The matrix M is the consumption matrix.*
Applying the Leontief Input–Output Model implies the following:

$$\overline{P} = M_{\overline{P}} + \overline{d} \tag{9}$$

The consumption matrix is made up of consumption vectors. The jth column is the jth consumption vector and contains the necessary input required from each of the sectors for sector j to produce one CO_2-saving. Vector $\overline{P}$ = production, vector $\overline{d}$ = external demand, and vector $M_{\overline{P}}$ = internal demand.

The case is open if $\overline{d} = 0$ and closed if $\overline{d} = \overline{0}$; since cases of $\overline{d} = 0$ are rare, the case of $\overline{d} = \overline{0}$ and $I - M$ are as follows:

$$\overline{P} = (I - M)^{-1}\,\overline{d} \tag{10}$$

$$\overline{P} = \left(I + M + M^2 + M^3 + - - - -\right)^{-1}\overline{d} \tag{11}$$

3.2. Energy and Carbon Emissions Efficiency Regression-Based Approach

The adoption of the regression-based approach (RBA) for this study requires defining the boundary through the production operations, including energy savings commitment. The international performance measurement and verification protocol (IPMVP) framework is used in the determination of energy intensity on the realization of the low-carbon product. The compared consumption was measured before and after implementation to make suitable adjustments following the measurement and verification integration. The processes were carried out as follows:

Step 1. Secondary energy data from the electricity generated were obtained for produce vehicles in automotive plants for a 3-year–period was obtained through the quantitative method.

Step 2. Baseline year was established using 2015 and 2018 data for tracking the energy performance to capture the energy savings and energy intensity of vehicle production bodies to determine the energy improvement in energy intensity.

In **step 3.** Relevant variables were determined, including output units for the production. The considered variables of production variability, product variability, feedstock quality, and quantity were dependent on the processes and outputs. Energy performance was normalized using a regression approach for these variables to track the energy consumption and baseline year consumption. Before the normalization, variables were identified by observations using the best technical judgments, in this case, the production level and units. Energy performance indicator (EnPI) tools were used to automate the process of evaluating all possible variables for a given year.

Step 4. Gathering of energy consumption data was performed for the baseline year and subsequent year for annual reporting. In this study, energy analysis was carried out using the input and output theory for equating energy consumption as the total of energy sources for vehicle part production, excluding feedstock in million thermal units (MMBtu), as presented in Equation (1). Electricity was valued by the primary energy required for the generation, transmission, and distribution of energy. Monthly energy consumption was collected on production data and relevant variable data for the regression model.

Step 5. Using regression analysis to normalize the data, the techniques were used to estimate the dependence of actual energy consumption as a dependent variable (kWh per unit of vehicles) for a given period and the production levels of parts as an independent variable, while controlling other variables simultaneously. The regression linear model in Equation (2) was used based on the strength of estimating energy savings through the measurement and verification of projects when the variations in operation conditions included the input data, as follows:

$$E_c = m_1 x_1 + m_2 x_2 + m_3 x_3 + b \tag{12}$$

where m_1, m_2, m_3 = kWh per unit of vehicles, x_1, x_2, x_3 = independent variables, and b = energy use when x_1, x_2, x_3 are 0 (kWh per month).

The linear equation is developed to model the energy consumption for a given period when an independent variable is set to zero using a known set of conditions. Comparing the actual energy consumed to the modeled energy consumption can help to estimate the energy performance improvements of the produced parts. Equation (2) presents the energy savings.

Step 6. Requires the determination of energy intensity from the baseline year. Equation (3) presents the total improvements in energy intensity depending on the regression method analysis, as follows:

$$E_i = \sum (e_{l1} * e_{c1}) + (e_{12} * e_{c2}) + (e_{1n} * e_{cn})/(e_{c1} + e_{c2} - e_{cn}) * 100\% \tag{13}$$

where E_i = total improvement in energy intensity, $e_{l1,}$ = modeled baseline energy use per product, e_{c1} = actual energy consumption per product, and n represents the number of products.

4. Results and Discussions

A balance between the manufacturing sector's expansion and energy efficiency can be achieved through interventions such as cogeneration plants and energy-efficient measures in enterprises, which will give the businesses the chance to become ready for the industrial policy consequences to be incorporated. The predicted energy consumption with carbon emissions quantifies the case company's future energy needs and serves as a benchmark for the implementation of some important energy consumers' small-scale renewable technology needs. When applying Equation (12), the analysis provides a direct energy and carbon relationship, which implies larger tax burdens.

A detailed analysis of the model projects the future climate policy on the vehicle production energy econometric as an industry response to climate policy over a medium to long-term time scale. Figure 2 shows the energy intensity for the vehicle bodies' production model based on the data, will help the industry to become more competitive in lower energy consumption and carbon intensity for production without a greenhouse gas emission constraint.

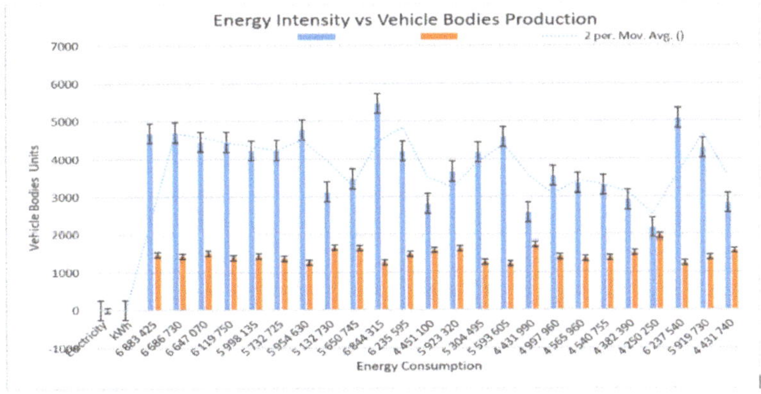

Figure 2. Energy intensity for the vehicle bodies production model based on the collected data.

This paper applied univariate TS analysis as a regression model based on 1086 observations of data with a 385 minimum sample size of vehicle body production energy intensity, as shown in Figure 3. Figure 4 presents a graphic representation of the energy consumption (MWh) and CO_2 emissions (Mt), while Figure 5 is the energy (MWh) and CO_2 emission (Mt) profiles.

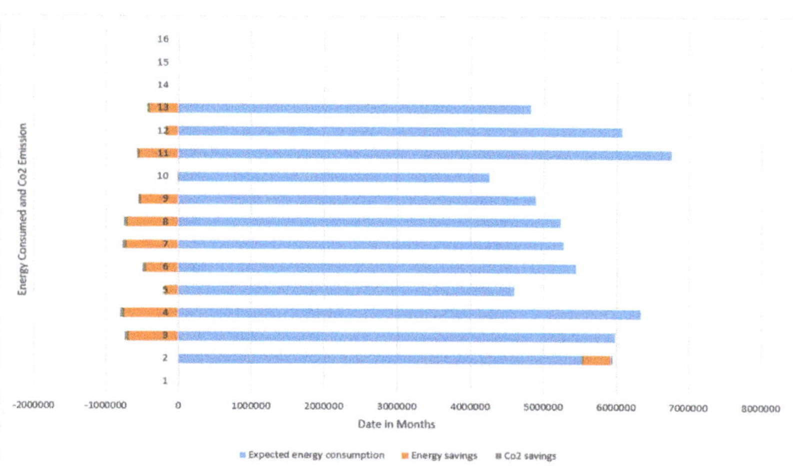

Figure 3. Analysis of the energy performance indicator based on energy intensity of the manufactured product.

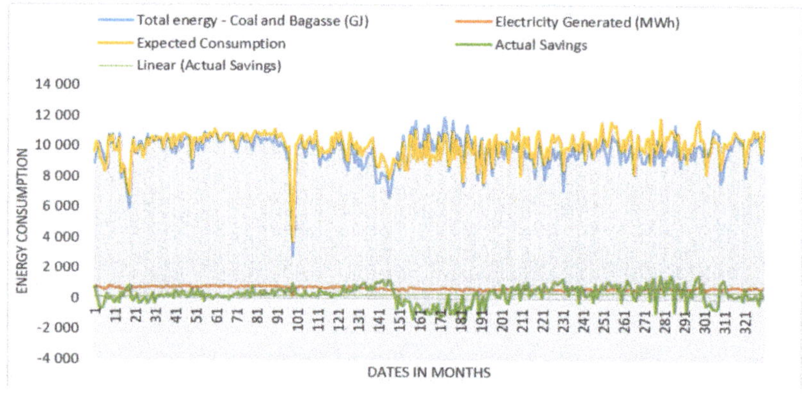

Figure 4. Graphic representation of the energy consumption (MWh) and CO_2 emissions (Mt).

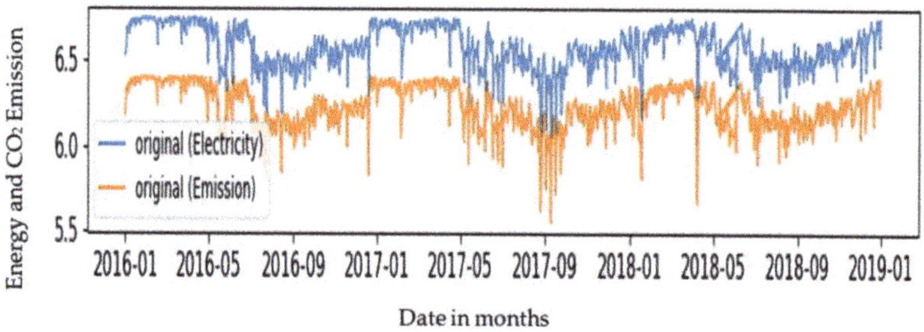

Figure 5. Graphic representation of the energy (MWh) and CO_2 emission (Mt) profiles.

Figure 6 shows that there is a significant positive trend that exhibits less than the 10% critical value margin of error, with a 95% confidence and correlation coefficient as the fraction of the total variation in the regression of the results.

Figure 6. Significant positive trend that exhibits less than the 10% critical value margin of error with 95% confidence.

Table 1 presents the test statistics and the results for the vehicle body production, the energy intensity with the expected energy consumption, and the savings. Although the variation in the standard deviation is small, the mean is increasing with time, and this is not a stationary series, as presented in Figure 7. It shows that there is seasonality that exhibits less than the 10% critical value margin of error with 95% confidence. in energy (MWh) and CO_2 emission (Mt) profiles that exhibits less than the 10% critical value margin of error, with a 95% confidence and correlation coefficient as the fraction of the total variation in regressing the results.

Table 1. Test statistics and results for vehicle body production, carbon intensity with expected energy consumption, and energy savings.

Test Statistics	Test Results	Smaller than the 1% Critical Value	Less than the 10% Critical Value	Lower than the 1% Critical Value
Test statistics	−3.602351	−9.397422	−7.793783	−7.590148
p-value	0.005716	6.319220	7.815122	2.544113
#Lags used	6.000000	1.50000	8.000000	1.300000
Number of observations used	1086.000	1.066000	1.084000	7.0400000
Critical value (1%)	−3.436386	−3.436499	−3.436397	−3.439673
Critical value (5%)	−2.864205	−2.864255	−2.864210	−2.865654
Critical value (10%)	−2.568189	−2.568216	−2.568192	−2.568961

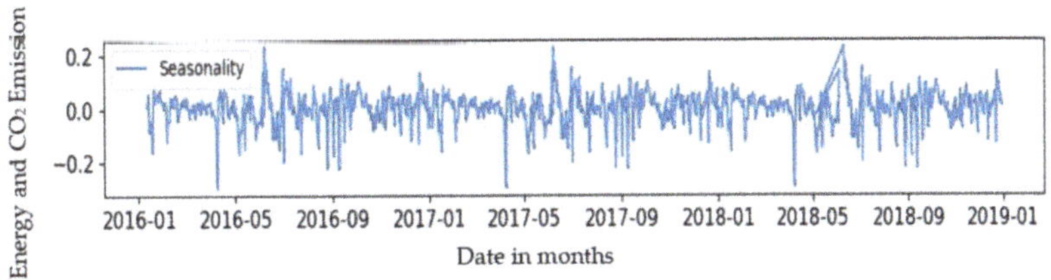

Figure 7. Seasonality that exhibits less than the 10% critical value margin of error with 95% confidence.

The test statistic is less than the critical values. It is important to note that the signed values should be compared and not the absolute values. TS has even smaller variations in the mean and the standard deviation in magnitude. The test statistic is smaller than the 1% critical value, which is better than in the previous case. In this case, there are no missing values as all the values from the beginning are given as weights. It will not work with the previous values. The mean and the standard deviations have small variations with time. The result is less than the 10% critical value, thus, the TS is stationary with 95% confidence. We can take second or third-order differences, which might obtain better results in certain applications.

Figure 8 is the representation of the simulation of the test and training root mean square error (RMSE) values of five years of energy and CO_2 emission prediction using the Dickey–Fuller test statistic to determine heuristically the RMSE as the normalized distance between the vector of the predicted values and the vector of the observed values. The test RMSE was evaluated on unused test data, which use the energy-carbon dioxide causal regression model that was fitted to the training data as a measure of how well fitted the model is, but not simply how well the training RMSE fits the data that were used to train the model. This study proves the expert's prediction that CO_2 emissions from transport manufacturing are the largest cause of climate change that is expanding at the quickest rate. At the same time, the expansion of CO_2 intensity may result in more greenhouse gas (GHG) pollution from the increased energy use. The test statistic is smaller than the 1% critical value, which is better than in the previous case.

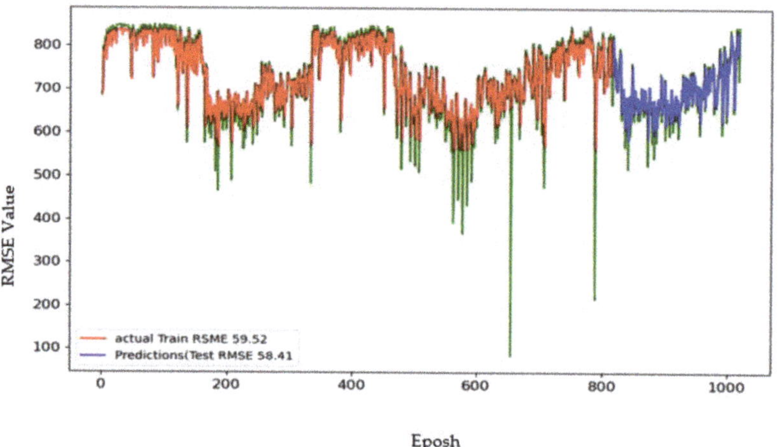

Figure 8. The actual trained RMSE and prediction test RMSE.

5. Conclusions

In this paper, we employed a statistical approach to establish the baseline demand for future transport manufacturing to anticipate the electricity demand and CO_2 emissions for that industry. To increase the forecast accuracy, we used the errors to train the causal model for the energy–carbon nexus. To accurately estimate the CO_2 emissions for effective mitigation and reduction by climate change targets, it is essential to identify the flaws in previous data using the notion of demand intelligence. The model used here will aid in lowering CO_2 emissions, which is necessary for ongoing technological advancement, for investing in cutting-edge energy and resource efficiency, for setting up programs to lower greenhouse gas emissions, and for contributing to climate science research. The contribution of this study is threefold. Firstly, we have demonstrated the energy-efficiency paradox in manufacturing, concerning whether the energy efficiency and carbon emissions signature of the manufactured products. Secondly, we have contributed to the literature by extending the understanding of the low carbon emission of products with an approach towards adoption in a production facility. Thirdly, we have carefully studied the energy performance by testing the moving average over the past year, i.e., the last 12 values were determined with specific functions for rolling statistics in Pandas. The energy efficiency of production industries and the carbon emissions signature of a manufactured product is bringing a revolution to manufacturing industries by using big data in industry 4.0 technologies. This study has evaluated research that has been conducted in future transport-manufacturing literature, establishing that minimal research has been conducted on the relationship between energy efficiency, energy savings, and carbon intensity for the future transport manufacturing industries. The interrelationship between the examined variables resulted in a 29% improvement in the total energy intensity in the vehicle body part products, 7.22% in the cumulative energy savings, and 16.25% in the energy efficiency. At a micro level, industries' adoption of energy efficiency in terms of fuel is still limited in responses to climate change campaigns, while EE marginal abatement cost could provide an insight into incentives for industries to exploit investments in EE through this study. The study will encourage companies to optimize their profits through more cash in hand, due to the energy savings, the energy intensity reduction, and a low carbon footprint of products. As evidenced in the reviewed literature, the harmful effects of GHG pollution and local air pollution on the environment and human health is growing to pose a threat to development, if no changes are made to investment strategies and legislation to enforce cleaner environmental practices, due to a perceived threat to their profits. The scope that has been examined in this paper will be interesting to agencies of government, researchers, policymakers, business owners, and practicing engineers in future transport manufacturing, and it could serve as a fundamental guideline for future studies in these areas. Future research will concentrate on the incorporation of energy and carbon emission prediction efficiency into a data monitoring device for online and mobile applications for the manufacturing of future transportation systems.

Author Contributions: Conceptualization, R.K.M. and O.T.A.; methodology, R.K.M.; software, O.T.A.; validation, R.K.M., O.T.A. and K.M.; formal analysis, R.K.M. and O.T.A.; investigation, O.T.A.; resources, R.K.M. and O.T.A.; data, O.T.A.; editing, K.M.; visualization, O.T.A.; supervision, O.T.A. and K.M.; project administration, R.K.M.; funding acquisition, K.M. All authors have read and agreed to the published version of the manuscript.

Funding: This research was funded by the National Research Foundation (NRF), grant number 123575, and the APC was funded by the Research Chair in Future Transport Manufacturing Technologies.

Acknowledgments: The researchers acknowledge the support and assistance of the Industrial Engineering Department of Tshwane University of Technology, Gibela Rail, and the National Research Foundation (123575) of South Africa for their financial and material assistance in executing this research project. The opinions that are presented in this paper are those of the authors and not the funders.

Conflicts of Interest: The authors declare no conflict of interest.

References

1. Kearney, D. *EIA's Outlook Through 2035 from the Annual Energy Outlook 2010*; Annual Energy Review; Energy Information Administration: Washington, DC, USA, 2010.
2. *Technology Roadmap for Energy Reduction in Automotive Manufacturing*; USDOE: Washington, DC, USA, 2008. [CrossRef]
3. Apostolos, F.; Alexios, P.; Georgios, P. Energy Efficiency of Manufacturing Processes: A Critical Review. *Procedia CIRP* **2013**, *7*, 628–633. [CrossRef]
4. Herring, H. Energy efficiency—A critical view. *Energy* **2006**, *31*, 10–20. [CrossRef]
5. Industrial Development Corporation (IDC). Developing a vibrant ESCO market: Prospects for South Africa's energy efficiency future. Federal Ministry for Economic Cooperation and Development. 2015. Available online: https://www.IDC.co.za/wp-content/uploads/2018/11/ESCO_Market_Study_Report-Developing_a_vibrant_ESCO_Market.png (accessed on 2 October 2021).
6. Adenuga, O.T.; Mpofu, K.; Modise, K.R. An approach for enhancing optimal resource recovery from different classes of waste in South Africa: Selection of appropriate waste to energy technology. *Sustain. Futures* **2020**, *2*, 100033. [CrossRef]
7. du Plessis, W. Energy efficiency and the law: A multidisciplinary approach. *S. Afr. J. Sci.* **2015**, *111*, 1–8. [CrossRef]
8. Kluczek, A. An energy-led sustainability assessment of production systems—An approach for improving energy efficiency performance. *Int. J. Prod. Econ.* **2019**, *216*, 190–203.
9. Malinauskaite, J.; Jouhara, H.; Ahmad, L.; Milani, M.; Montorsi, L.; Venturelli, M. Energy efficiency in industry: EU and national policies in Italy and the UK. *Energy* **2019**, *172*, 255–269.
10. Thollander, P.; Danestig, M.; Rohdin, P. Energy policies for increased industrial energy efficiency: Evaluation of a local energy programme for manufacturing SMEs. *Energy Policy* **2007**, *35*, 5774–5783. [CrossRef]
11. Bonilla-Campos, I.; Nieto, N.; del Portillo-Valdes, L.; Egilegor, B.; Manzanedo, J.; Gaztañaga, H. Energy efficiency assessment: Process modelling and waste heat recovery analysis. *Energy Convers. Manag.* **2019**, *196*, 1180–1192. [CrossRef]
12. du Can, S.D.L.R.; Pudleiner, D.; Pielli, K. Energy efficiency as a means to expand energy access: A Uganda roadmap. *Energy Policy* **2018**, *120*, 354–364.
13. Chavez, K.M.G. Energy Efficiency in Wireless Access Networks: Measurements, Models and Algorithms. Ph.D. Dissertation, University of Trento, Trento, Italy, 2013. In School in Information and Communication Technologies.

14. Marchi, B.; Zanoni, S. Supply Chain Management for Improved Energy Efficiency: Review and Opportunities. *Energies* **2017**, *10*, 1618. [CrossRef]

15. Pretorius, I.; Piketh, S.J.; Burger, R.P. The impact of the South African energy crisis on emissions. *Air Pollut. XXIII* **2015**, *198*, 255–264.

16. Meng, Z.; Wang, H.; Wang, B. Empirical Analysis of Carbon Emission Accounting and Influencing Factors of Energy Consumption in China. *Int. J. Environ. Res. Public Health* **2018**, *15*, 2467. [CrossRef] [PubMed]

17. Modise, R.K.; Mpofu, K.; Adenuga, O.T. Energy and Carbon Emission Efficiency Prediction: Applications in Future Transport Manufacturing. *Energies* **2021**, *14*, 8466. [CrossRef]

18. Phiri, H.; Kunda, D. An Approach for Predicting CO_2 Emissions using Data Mining Techniques. *Int. J. Comput. Appl.* **2017**, *172*, 7–10.

19. Lee, K.H. Carbon accounting for supply chain management in the automobile industry. *J. Clean. Prod.* **2012**, *36*, 83–93. [CrossRef]

20. Siltonen, S. *Implication of Energy Efficiency Improvement for CO_2 Emmission in Energy-Intensive Industry in Energy Technology*; Aalto University: Espoo, Finland, 2010; p. 51.

21. Dufour, T.; Hoang, H.M.; Oignet, J.; Osswald, V.; Fournaison, L.; Delahaye, A. Experimental and modelling study of energy efficiency of CO_2 hydrate slurry in a coil heat exchanger. *Appl. Energy* **2019**, *242*, 492–505. [CrossRef]

22. Nota, G.; Nota, F.D.; Peluso, D.; Lazo, T. Energy Efficiency in Industry 4.0: The Case of Batch Production Processes. *Sustainability* **2020**, *12*, 6631. [CrossRef]

23. Benedetti, M.; Vittorio, C.; Vito, I. Improving Energy Efficiency in Manufacturing Systems—Literature Review and Analysis of the Impact on the Energy Network of Consolidated Practices and Upcoming Opportunities. In *Energy Efficiency Improvements in Smart Grid Components*; InTech: Singapore, 2015.

24. Schleich, J.; Fleiter, T. Effectiveness of energy audits in small business organizations. *Resour. Energy Econ.* **2019**, *56*, 59–70. [CrossRef]

25. Martínez, C.P. Energy efficiency developments in the manufacturing industries of Germany and Colombia, 1998–2005. *Energy Sustain. Dev.* **2009**, *13*, 189–201. [CrossRef]

26. Javied, T.; Rackow, T.; Franke, J. Implementing Energy Management System to Increase Energy Efficiency in Manufacturing Companies. *Procedia CIRP* **2015**, *26*, 156–161. [CrossRef]

27. Flick, D.; Ji, L.; Dehning, P.; Thiede, S. Energy Efficiency Evaluation of Manufacturing Systems by Considering Relevant Influencing Factors. *Procedia CIRP* **2017**, *63*, 586–591. [CrossRef]

28. Diaz, C.J.L.; Ocampo-Martinez, C. Energy efficiency in discrete-manufacturing systems: Insights, trends, and control strategies. *J. Manuf. Syst.* **2019**, *52*, 131–145. [CrossRef]

29. Adenuga, O.T.; Mpofu, K.; Ramatsetse, B.I. Exploring energy efficiency prediction method for Industry 4.0: A reconfigurable vibrating screen case study. *Procedia Manuf.* **2020**, *51*, 243–250. [CrossRef]

30. May, G.; Kiritsis, D. Business Model for Energy Efficiency in Manufacturing. *Procedia CIRP* **2017**, *61*, 410–415. [CrossRef]

31. Mawson, V.J.; Hughes, B.R. The development of modelling tools to improve energy efficiency in manufacturing processes and systems. *J. Manuf. Syst.* **2019**, *51*, 95–105. [CrossRef]

32. Heo, Y.; Zavala, V.M. Gaussian process modeling for measurement and verification of building energy savings. *Energy Build.* **2012**, *53*, 7–18. [CrossRef]

33. Saleh, C.; Dzakiyullah, N.R.; Nugroho, J.B. Carbon dioxide emission prediction using support vector machine. *IOP Conf. Ser. Mater. Sci. Eng.* **2016**, *114*, 012148. [CrossRef]

34. Fahmy-Abdullah, M.; Ismail, R.; Sulaiman, N.; Talib, B.A. Technical Efficiency in Transport Manufacturing Firms: Evidence from Malaysia. *Asian Acad. Manag. J.* **2017**, *22*, 57–77. [CrossRef]

35. Zhang, Y.; Ma, S.; Yang, H.; Lv, J.; Liu, Y. A big data driven analytical framework for energy-intensive manufacturing industries. *J. Clean. Prod.* **2018**, *197*, 57–72. [CrossRef]

36. Lee, S.; Chong, W.O. Causal relationships of energy consumption, price, and CO_2 emissions in the U.S. building sector. *Resour. Conserv. Recycl.* **2016**, *107*, 220–226. [CrossRef]

37. Aggelakakis, A.; Bernandino, J.; Boile, M.; Christidis, P.; Condeco, A.; Krail, M.; Papanikolaou, A.; Reichenbach, M.; Schippl, J. *Future of Transport Industry Report 2015*; Institution for Prospective Technological Studies Luxembourg: Luxembourg, 2015; p. 43.

38. Yang, H.; Soundar, K.; Satish, T.S.; Fugee, T. The internet of things for smart manufacturing: A review. *IISE Trans.* **2019**, *51*, 1190–1216. [CrossRef]

39. Giampieri, A.; Ling-Chin, J.; Taylor, W.; Smallbone, A.; Roskilly, A.P. Moving towards low-carbon manufacturing in the UK automotive industry. *Energy Procedia* **2019**, *158*, 3381–3386. [CrossRef]

40. Lee, G.; Sul, S.K.; Kim, J. Energy-saving method of parallel mechanism by redundant actuation. *Int. J. Precis. Eng. Manuf.-Green Technol.* **2015**, *2*, 345–351. [CrossRef]

41. Rubio, F.; Llopis-Albert, C. Analysis of the Use of a Wind Turbine as an Energy Recovery Device in Transport Systems. *Mathematics* **2021**, *9*, 2265. [CrossRef]

42. Botts, A.; Gopalakrishnan, B.; Nimbarte, A.; Currie, K.R.; Karki, V. Energy efficiency of blower heater non-purge compressed air dryers. *Int. J. Energy Technol. Policy* **2021**, *17*, 239. [CrossRef]

43. Jignesh, M.J.; Narendra, N.D.; Pratik, M. Estimating the energy consumption of Indian refineries: An empirical analysis based on panel data econometrics. *Int. J. Energy Technol. Policy (IJETP)* **2021**, *17*, 275–298.

44. Shari, B.E.; Moumouni, Y. A system dynamic modelling for energy planning and carbon dioxide estimation of the Nigerian power sector. *Int. J. Energy Technol. Policy* **2020**, *16*, 470. [CrossRef]
45. Sunde, T. Energy consumption and economic growth modelling in SADC countries: An application of the VAR Granger causality analysis. *Int. J. Energy Technol. Policy* **2020**, *16*, 41. [CrossRef]
46. Brahmana, R.K.; Ono, H. Energy efficiency and company performance in Japanese listed companies. *Int. J. Energy Technol. Policy* **2020**, *16*, 24. [CrossRef]
47. Olanrewaju, O.A.; Munda, J.L.; Jimoh, A.A. Understanding the Impacts of GDP and Population in South Africa's Energy Consumption. In Proceedings of the IASTED International Conference Modelling and Simulation (AfricaMS 2014), Gaborone, Botswan, 1–3 September 2014.
48. Gamede, G.B.; Mpofu, K.; Adenuga, O.T. Business model for integrating energy efficiency performance in manufacturing industries: Rail car case study. *Procedia CIRP* **2019**, *81*, 1441–1647. [CrossRef]
49. Adenuga, O.T.; Mpofu, K.; Ramatsetse, B.I. Energy efficiency analysis modelling system for manufacturing in the context of industry 4. *Procedia CIRP* **2019**, *80*, 735–740. [CrossRef]

Article

An Accuracy Prediction Method of the RV Reducer to Be Assembled Considering Dendritic Weighting Function

Shousong Jin, Yanxi Chen, Yiping Shao and Yaliang Wang *

School of Mechanical Engineering, Zhejiang University of Technology, Hangzhou 310023, China
* Correspondence: wangyaliang@zjut.edu.cn

Abstract: There are many factors affecting the assembly quality of rotate vector reducer, and the assembly quality is unstable. Matching is an assembly method that can obtain high-precision products or avoid a large number of secondary rejects. Selecting suitable parts to assemble together can improve the transmission accuracy of the reducer. In the actual assembly of the reducer, the success rate of one-time selection of parts is low, and "trial and error assembly" will lead to a waste of labor, time cost, and errors accumulation. In view of this situation, a dendritic neural network prediction model based on mass production and practical engineering applications has been established. The size parameters of the parts that affected transmission error of the reducer were selected as influencing factors for input. The key performance index of reducer was transmission error as output index. After data standardization preprocessing, a quality prediction model was established to predict the transmission error. The experimental results show that the dendritic neural network model can realize the regression prediction of reducer mass and has good prediction accuracy and generalization capability. The proposed method can provide help for the selection of parts in the assembly process of the RV reducer.

Keywords: RV reducer; assembly quality; dendrites; neural network; transmission accuracy

Citation: Jin, S.; Chen, Y.; Shao, Y.; Wang, Y. An Accuracy Prediction Method of the RV Reducer to Be Assembled Considering Dendritic Weighting Function. *Energies* **2022**, *15*, 7069. https://doi.org/10.3390/en15197069

Academic Editors: Xi Gu, Tangbin Xia, Ershun Pan, Rongxi Wang and Yupeng Li

Received: 31 August 2022
Accepted: 21 September 2022
Published: 26 September 2022

Publisher's Note: MDPI stays neutral with regard to jurisdictional claims in published maps and institutional affiliations.

1. Introduction

Rotate vector (RV) reducer has the advantages of small size, compact structure and large transmission ratio [1]. The quality of its assembly determines the performance, production cost, and production efficiency of the product. It is mainly used in robot joints with high precision and large load. At present, foreign countries already have a relatively complete theoretical system for RV reducers. Domestic RV reducers have been developed for many years without major breakthroughs in accuracy. The dynamic transmission error of RV reducers depends on the manufacturing error of each component, assembly errors, and elastic deformation. However, the material properties of parts are easy to determine and the manufacturing accuracy is difficult to ensure. In this case, companies generally measure parts and assemble parts to improve the dynamic transmission accuracy of the reducer.

In 2007, Kannan SM et al. used particle swarm algorithm to obtain the best combination of parts [2] and successfully made the assembly deviation less than the sum of the tolerances of the parts. Gentilini et al. established a finite element model; this method can predict and show the final shape of the assembly [3]. There is no need for physical assembly in future practice, reducing the time and cost of product quality inspection. S.Khodaygan et al. proposed to estimate the tolerance assembly and the reliability of the mechanical assembly to meet the quality requirements through the Bayesian modeling [4], which can formulate accurate assembly functions for complex mechanical assemblies. At present, artificial neural networks are widely used in the field of speech recognition, computer vision, and bioinformatics, etc. In recent years, some scholars have used neural networks to develop assembly models, which avoids the complicated operation and heavy workload of traditional methods to solve the accuracy. Steinberg Fabian [5] utilized a gradient boosting classifier to identify assembly start delayers. Stark Rainerr et al. selected three approaches based upon regression, natural language processing, and clustering into intelligent services in the digital production process to reduce the time for resolving upcoming issues in assembly [6]. It has certain guiding significance for improving the assembly quality.

The current neuron model relies on the McCulloch–Pitts structure. This neuron uses the weight between the synapses, and then obtains the output through the activation function. It does not consider the local interaction played by the dendritic structure as a part of the neuron and is only used to transmit signals. Experiments had shown that dendrites play an important role in information processing in biological neurons, which can nonlinearly integrate postsynaptic signals and filter out irrelevant background information [7–9]. The dendrites receive signals (action potentials) from upstream neurons. After the information is integrated and processed, it is transmitted to the neuron soma. The powerful information processing capabilities of dendrites enable individual neurons to process thousands of different synaptic inputs unequally [10]. Inspired by biological phenomena, Yuki Todo et al. proposed a single dendritic neuron with synaptic, dendritic, membranous, and somatic layers [11]. A synapse performs a sigmoid non-linear operation on its input, the nodes in the synaptic layer contain the initial weights and thresholds of the dendritic neural network, each branch receives a signal and performs a multiplication operation at its synapse, the membrane sums the results of all branches, ultimately transmitting the signal to the cell body. When the threshold is exceeded, cells send signals to other neurons through the axon. At present, this model has been widely used, such as computer-aided medical diagnosis [12] and morphological hardware implementation [13]. Gang et al. believed that the dendritic neuron model proposed by Yuki Todo and Gao [14] is not enough to be expressive, can only express first-order input, and is not conducive to computer operation [15]. Considering that the definition of multivalued logic and the integration of potentials in biological neurons can be described in a multiplicative form, a dendrite (DD) containing only matrix multiplication and Hadamard product was proposed to simulate the function of data interaction processing. Gang [16] developed a Taylor series using the dendritic network proposed by him and constructed a relational spectrum model to analyze the synergy and coupling of hand muscles, which is helpful for the design of prosthetic hands. In this article, the DD module is used to develop a prediction model to capture the internal structure dependence and increase the interaction between information, which provides a new idea for the prediction of reducer assembly accuracy.

2. Analysis of Quality Influencing Factors

Transmission error (TE) is an important index to evaluate the gear meshing mass, which is directly related to working accuracy, reliability, vibration noise, and service life of gear transmission [17]. In the actual manufacturing and installation process of the reducer, transmission errors are inevitable and the main sources include their own processing errors and assembly errors. According to the different specifications of the RV reducer, the transmission error is strictly limited within (1~1.5′).

The RV reducer has a sophisticated and complex structure, and its components mainly include input shaft, planetary gear, planet carrier, flange, crankshaft, pin teeth, etc. Figure 1 is a schematic diagram of the RV-40E reducer. Figure 1a shows the component assembly structure of the reducer. The transmission of the RV reducer is divided into two stages, which are the first-stage involute planetary gear transmission and the second-stage cycloidal pinwheel transmission [18,19]. The movement of the input shaft drives the planetary gear to achieve first-level deceleration. The planetary gear drives the crankshaft, then drives the swivel arm bearing, and transmits the power to the cycloidal pinwheels. Under the combined action of the swivel arm bearing and the pin teeth, the cycloid pinwheel produces a revolution motion that rotates around the central circular axis of the pin tooth and an autorotation motion that rotates around its own axis. The autorotation drives the flange and output planet carrier to achieve 1:1 speed ratio output to complete the second-stage of deceleration. Since the influence of the cycloidal pinwheel transmission on the transmission error is directly reflected on the output shaft, the first-stage planetary gear reduction mechanism is far from the output, and its transmission ratio is only the reciprocal of the second-stage cycloidal pinwheel transmission ratio. Therefore, the RV reducer transmission accuracy mainly depends on the second transmission. Figure 1b shows the schematic diagram of the transmission principle mechanism of the reducer [20,21]. The key parts involved in cycloidal pinwheel transmission include: cycloidal wheel, pinwheel, crankshaft, and crank bearing. However, in actual engineering assembly, the crank bearing clearance is usually adjusted to a small value, and the influence on the transmission error can be ignored. The influence of error factors of remaining key components on transmission error were analyzed, and the influence degree of each factor on the output shaft error were obtained. The parameters of each influencing factor $\theta = [\theta_1, \theta_2, \cdots, \theta_n]$ have errors $\Delta\theta = [\Delta\theta_1, \Delta\theta_2, \cdots, \Delta\theta_n]$. The transmission error of RV reducer can be recorded as $\varphi(\theta) = \varphi(\theta_1, \theta_2, \cdots, \theta_n)$, and each error sensitivity index can be defined as [22,23]:

$$S_N = \nabla\varphi(\theta) = \left[\frac{\partial\varphi}{\partial\theta_1}, \frac{\partial\varphi}{\partial\theta_2}, \cdots, \frac{\partial\varphi}{\partial\theta_n}\right] = [S_1, S_2, \cdots, S_n] \tag{1}$$

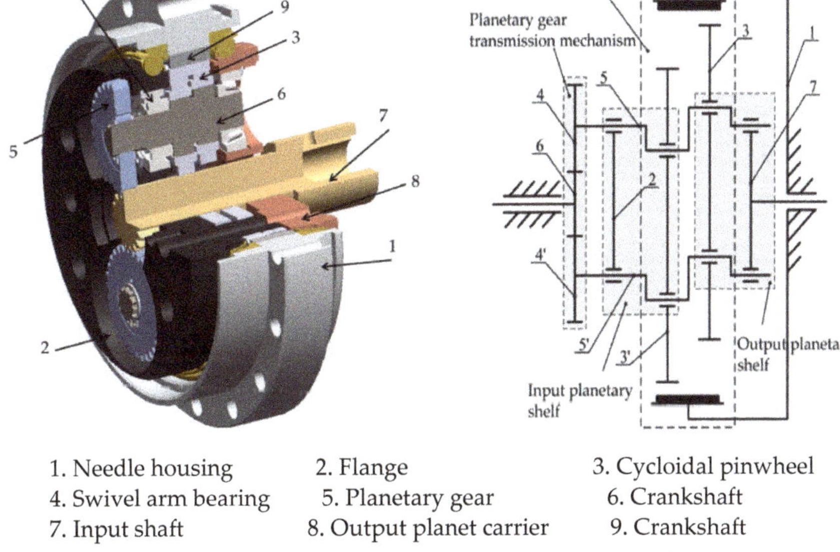

1. Needle housing   2. Flange   3. Cycloidal pinwheel
4. Swivel arm bearing   5. Planetary gear   6. Crankshaft
7. Input shaft   8. Output planet carrier   9. Crankshaft

(**a**)   (**b**)

Figure 1. Schematic diagram of RV-40E reducer structure: (**a**) structure diagram; and (**b**) schematic diagram.

This article takes the RV-40E type reducer as an example. Taking the center circle radius error of the needle tooth as reference error. The sensitivity index of the needle tooth center circle radius error is 1, the sensitivity index of other error parameters is compared with the reference error parameter. Through the calculation of sensitivity. It can be concluded that the factors that have a larger sensitivity index to the transmission error are: needle tooth center circle radius error δr_p, needle tooth pin radius error δr_{rp}, needle tooth pin hole circumferential position error δt_Σ, equidistant modification error $\delta\Delta r_p$, shift distance modification error $\delta\Delta r_{rp}$.

3. Development a Network Model

In the back propagation feed-forward neural network, each neuron takes the output of the node in the upper layer as the input. The input is activated by linear weighting and nonlinear function to obtain the output of this node, then pass the output to the node in the next layer. When using McCulloch–Pitts neurons to learn the relationship between input space and output space, it ignores the function of dendritic structure as a part of neurons to process information. For this phenomenon, this article uses the logic extractor DD proposed by gang to realize data interaction processing function and constructs a dendrite neural network (DDNN) model for the reducer transmission error prediction. In this model, the dendrites first to perform interactive preprocessing on the signals transmitted from upstream neurons, and then transmit them to the cell body for linear weighting. Finally, the axons are activated by nonlinear functions, then the signals are transmitted to downstream neurons. Compared with McCulloch–Pitts neuron, the neuron model constructed by DD module has lower computational complexity, stronger network generalization, and fitting capability.

3.1. Back Propagation Feed-Forward Neural Network

In the artificial neuron, $X = \{x_1, x_2, \cdots, x_n\}$ is feature input, n is the number of input features, x_i is the ith influencing factor. The input value of the neuron is transmitted by the connection of the weighting coefficient. The positive and negative values of the weighting coefficient simulate the excitation and inhibition of the synapse. The size indicates the strength of the connection, $\sum_{i=1}^{n} w_i x_i$ is the integration of all input signals by the cell body. The threshold b controls the activation of neurons. When the sum of the inputs exceeds the threshold, the neuron is activated and an output signal is generated to transmit information. f represents nonlinear activation function, the output $\hat{y}$ can be expressed as:

$$\hat{y} = f(W^{2,1}X + b) \tag{2}$$

where $W^{2,1}$ is the weight matrix from the first layer to the second layer, b is the bias.

Back propagation feed-forward (bp) neural network is a multi-layer feed-forward network trained according to the error back propagation algorithm, which is widely used at present. The network usually composed of three layers: input layer, hidden layer, and output layer. Its learning rule is gradient descent, which gradually adjusts weights and thresholds between layers through back propagation error signals to minimize the loss function of the network.

Firstly, define E as the loss function of the latter layer in back propagation and a as the learning rate. The weights from the input layer to the hidden layer is represented by v_j, and from the hidden layer to the output layer is represented by w_j. ins_p is the input of the pth neuron in the output layer, and ins_q is the input of the qth neuron in the hidden layer. The output vector of the output layer is o, the expected output vector is e, l represents the lth layer of the network model. The output error is:

$$E = \frac{1}{2}\sum_{p=1}^{l}\left(e_p - o_p\right)^2 = \frac{1}{2}\sum_{p=1}^{l}\left\{e_p - f\left[\sum_{q=0}^{m} w_{qp}f\left(\sum_{i=0}^{n} v_{jq}x_n\right)\right]\right\}^2 \tag{3}$$

It can be seen from the above formula that the size of the output error of the network is related to the weights v_j and w_j of each layer. The size $W = \{v_j, w_j\}$ of the output error E can be changed by adjusting the weights. The weight matrix update can be simplified as:

$$W^{2,1(\text{new})} = W^{2,1(\text{old})} - a\,\frac{\partial E}{\partial W^{2,1(\text{old})}} \tag{4}$$

3.2. Dynamic Adam Optimization Algorithm

For machine learning problems in high-dimensional parameter spaces or large datasets, Jimmy Ba et al. proposed a dynamic Adam optimization algorithm based on gradient optimization of the objective function with variable learning rate during iterations [24]. It continues the strengths of Adagrad and RMSprop. Independent adaptive learning rates are set for different parameters through the first and second moments of the gradient, which makes the algorithm converge faster. It improves problems, such as objective function fluctuations, and maintains prediction performance in non-dense gradient problems and unstable data. The first-order and second-order moment deviations are calculated as follows:

$$m_t = \beta_1 m_{t-1} + (1 - \beta_1) dW^{l,l-1} \tag{5}$$

$$v_t = \beta_2 m_{t-1} + (1 - \beta_2)(dW^{l,l-1})^2 \tag{6}$$

$$\begin{cases} \hat{m} = \dfrac{m_t}{1-\beta_1^t} \\[2mm] \hat{n} = \dfrac{n_t}{1-\beta_2^t} \end{cases} \tag{7}$$

where t is the number of iterations, m_t is the first-order moment vector, n_t is the second-order moment vector. $\hat{m}$ is the first-order moment deviation, $\hat{n}$ is the second-order moment deviation, β_1 is the first-order moment attenuation coefficient, β_2 is the second-order moment attenuation coefficient. This article takes 0.9 and 0.999, respectively. $W^{l,l-1}$ is the weight matrix from the $(l-1)$th layer to the lth layer in the DD module. The weight matrix using the Adam optimizer can be updated as:

$$W^{2,1(\text{new})} = W^{2,1(\text{old})} - a^{2,1} \frac{\hat{m}}{\sqrt{\hat{n}} + \varepsilon} \tag{8}$$

where $\varepsilon = 10^{-8}$ prevents the divisor from becoming 0.

3.3. Dendritic Unit

In biological nervous systems, dendrites had been shown to have logical operations [25]. The dendritic used in this article is the 0 product between the current input and the previous input. The Hadamard product can be used to establish the logical relationship between inputs, so DD is the expression of the logical relationship between features. DD can capture the logical relationship between features. The DD module can not only be used to extract the local relationship between inputs, but they also use the internal correlation information to strengthen the connection between the two features and improve the network accuracy while capturing the internal structural dependencies. The DD module is shown in Figure 2.

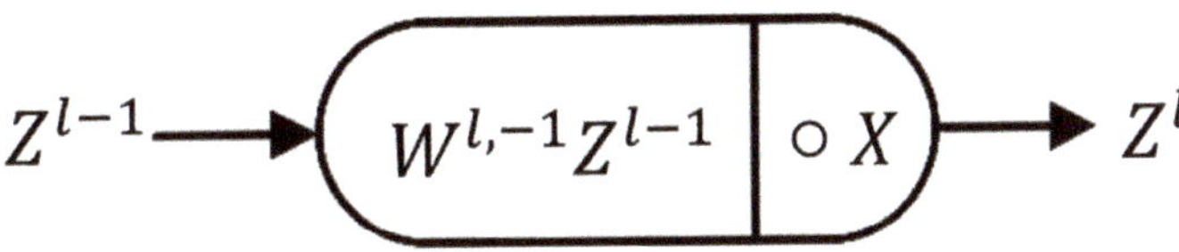

Figure 2. Dendrite module.

The expression of the dendrite module is as follows:

$$Z^l = W^{l,l-1} Z^{l-1} \circ X \tag{9}$$

where Z^{l-1} is the input of the module, Z^l is the output of the module. $W^{l,l-1}$ is the weight matrix from the $(l-1)$th layer to the lth layer. $X = \{x_1, x_2, \cdots, x_n\}$ represents the original input. "$\circ$" represents the Hadamard product, which represents the multiplication of corresponding elements. It is used to establish the logical relationship between inputs.

So, it produces another matrix with the same dimensions as the original product matrix. The features are deeply fused by multiplying corresponding elements, which can better reflect the meaning of feature intersection.

The DD module is essentially an information processing method, similar to a variant of the self-attention mechanism. It pays attention to the input feature variable itself, increases the interaction between information, and strengthens the connection between features. By training its own information to update the parameters, it gives the network stronger information processing capability and better generalization capability.

3.4. Developing a Dendritic Neural Network Model

The artificial neuron structure ignores the capability of dendrites to process information interaction and only uses dendrites to transmit information. However, in actual biological information, each neuron can have one or more dendrites. In this article, multiple DD modules are used to simulate dendritic functions to form a single dendritic neuron [26]. The dendrite extracts logical information from the data transmitted by the upstream neurons. Then, the information is interactively preprocessed and passed to the cell body for linear weighting. The signal is nonlinearly activated and, finally, output by the axon. Figure 3 shows the structure of neurons. As shown in Figure 3a, the biological neurons have dendrites, cell bodies, axons, and other organizations. Figure 3b shows the filter, accumulator, and balancer in dendritic neurons to simulate the organizational function of biological neurons, and the expression is as follows:

$$\begin{cases} Z^1 = W^{1,0}X \circ X \\ \hat{y} = f\left(W^{2,1}Z^1 + b\right) \end{cases} \tag{10}$$

where $X = \{x_1, x_2, \cdots, x_n\}$ is the feature input, n is the number of influencing factors, $W^{1,0}$ and $W^{2,1}$ represent the weight matrix from layer 0 to layer 1 and the weight matrix from layer 1 to layer 2, respectively. b is bias, f is nonlinear activation function, where ReLU is selected as the activation function.

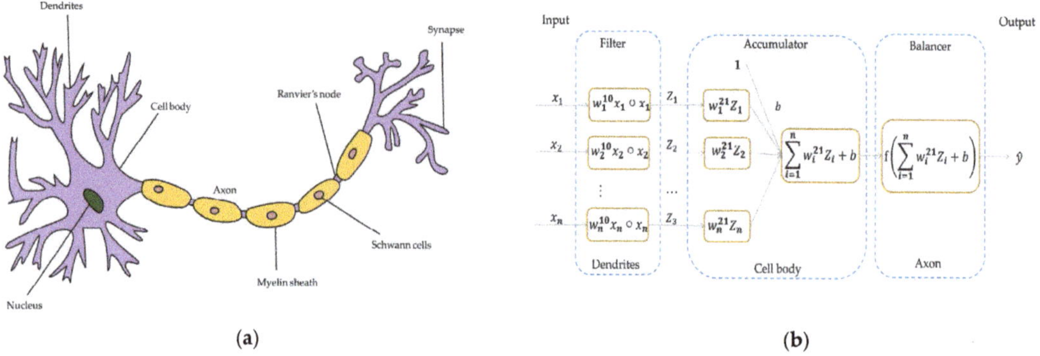

Figure 3. The structure of neurons: (**a**) biological neuron structure; and (**b**) functional structure diagram of dendritic neuron model.

According to the source of the transmission error, five influencing factors are divided into two dimensions. The information of different dimensions was represented by different neurons. Figure 4 shows the fusion of information from different neurons, and the transmission error prediction model was constructed.

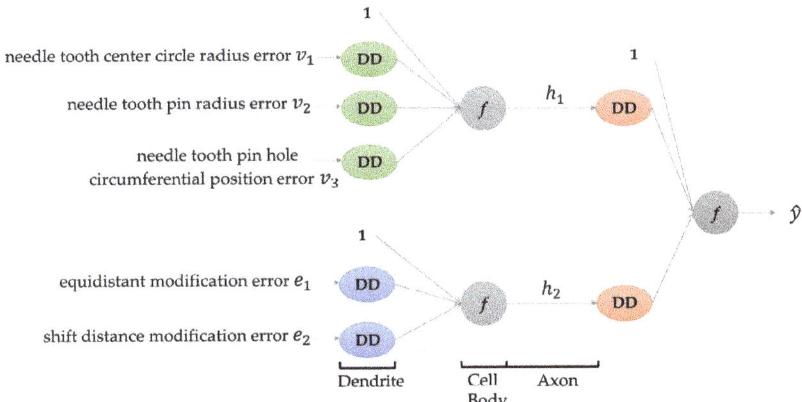

Figure 4. Dendritic neural network (DDNN) model.

The architecture of DDNN can be expressed as follows:

$$\begin{cases} Z_v^1 = W_v^{1,0} X \circ X \\ h_1 = f(W_v^{2,1} Z_v^1 + b) \end{cases} \tag{11}$$

$$\begin{cases} Z_e^1 = W_e^{1,0} X \circ X \\ h_2 = f(W_e^{2,1} Z_e^1 + b) \end{cases} \tag{12}$$

$$\begin{cases} A_h = W_h^{1,0} H \circ H \\ \hat{y} = f(W_h^{2,1} A_h + b) \end{cases} \tag{13}$$

where $V = \{v_1, v_2, v_3\}$, v_1 represents the radius error of needle tooth center circle δr_p, v_2 represents the needle tooth pin radius error δr_{rp}, v_3 represents the needle tooth pin hole circumferential position error δt_Σ, all from the component pin wheel. $E = \{e_1, e_2\}$, e_1 represents the equidistant modification error $\delta \Delta r_p$, e_2 represents the shift distance modification error $\delta \Delta r_{rp}$, all from components cycloid wheel. $H = \{h_1, h_2\}$. Feature vectors are combined into tensor, which is fed into DDNN model for training. The information is interactively preprocessed in the DD module, and then transmitted to the cell body for linear weighting. Finally, the axon performs nonlinear function activation and transmits information h_1 and h_2 to the next layer of DD module. Each DD module selects input units from upper layer without repeating to connect and repeats the above learning rules. Finally, axon output the predicted value. The overall process of DDNN model prediction is shown in Figure 5. Firstly, the sample data set is divided into training set and test set. Secondly, the factors affecting the quality of reducer in the training set and test set are denoted as eigenvalues, and the transmission errors are denoted as labels. The training set is imported into DDNN model for training. Equations (11)–(13) aim to obtain the minimum value of the loss function. They use the dynamic Adam optimization algorithm and the error signal back propagation algorithm to update threshold and weight parameters. When the convergence of the loss function of model training reaches the expectation, the training is stopped and the model is saved. Finally, the test set is brought into the trained model to obtain prediction results.

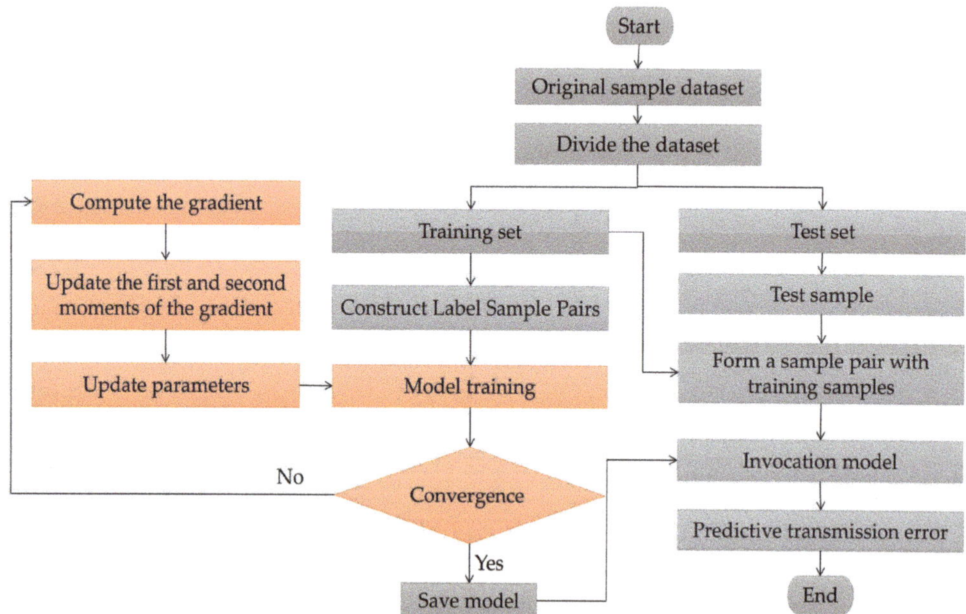

Figure 5. Prediction flow chart of the DDNN model.

4. The Model Solution

4.1. Preprocessing of Test Data

This article takes the RV-40E-121 reducer as an example to analyze. Five influencing factors related to transmission error during assembly were selected: needle tooth center circle radius error δr_p, needle tooth pin radius error δr_{rp}, needle tooth pin hole circumferential position error δt_{Σ}, equidistant modification error $\delta \Delta r_p$, shift distance modification error $\delta \Delta r_{rp}$. The design upper and lower limits of the five influencing factors are shown in Table 1. Dimensions of parts are in millimeters.

Table 1. Design upper and lower limits of parameter error of main parts (mm).

Parameter	δr_{rp}	δt_{Σ}	δr_p	$\delta \Delta r_p$	$\delta \Delta r_{rp}$
Upper limits	0.03	0.001	0.015	0.078	0.048
Lower limits	0.01	-0.003	0.005	0.026	0.016

In order to avoid the large difference in the value range of each feature affecting the efficiency of the gradient descent method. The data set needs to be preprocessed before model training. First, removing the data that deviate too much, then using the Max–Min normalization method to linearly transform the sample eigenvalues. Thus, making the result mapped in the interval [0, 1], and the scaling function is as follows:

$$x_{\text{new}} = \frac{x - \min(x)}{\max(x) - \min(x)} \tag{14}$$

where x is the original data, $\max(x)$ is the maximum, and $\min(x)$ is the minimum values in the data. x_{new} is the standardized data.

Through the error test platform, the size parameters of the above-mentioned main components were collected as characteristic inputs of the influencing factors, and the transmission error of the reducer was used as the output index. A data set was constructed with a total of 300 samples information. Some sample point data are shown in Table 2, where the unit of transmission error is angular minutes.

Table 2. Some sample data example (mm).

Serial Number	δr_{rp}	δt_{Σ}	δr_p	$\delta \Delta r_p$	$\delta \Delta r_{rp}$	Transmission Error/$'$
1	0.025	0.001	0.015	0.038	0.036	1.880
2	0.017	−0.001	0.023	0.040	0.041	2.151
3	0.015	0.001	0.020	0.032	0.036	1.193
4	0.020	0	0.020	0.038	0.068	1.419
5	0.010	−0.001	0.010	0.016	0.026	0.786
6	0.025	−0.002	0.020	0.026	0.052	1.762
7	0.020	−0.003	0.020	0.040	0.063	2.001
8	0.023	0.001	0.017	0.040	0.041	2.508
9	0.027	−0.002	0.027	0.032	0.052	1.952
10	0.020	0.001	0.013	0.032	0.031	1.166

Firstly, 300 groups of samples were randomly shuffled, and then the data set was divided. A total of 90% of the samples were set as the training set, and 10% of the samples were set as the test set.

4.2. Parameter Selection

The framework used in the experiment in this chapter is TensorFlow2.0, and the test was implemented through Python3.6. The specific hardware environment is: the CPU is Intel i7, the GPU model is NVIDIA 2080Ti, the CUDA version is 11.2.1, and the CUDNN version is 8.1.1.33.

The loss function can measure the difference between the output value of the model and the true value. It is a measure to evaluate the fitting capability of the model. The advantage of using Log-Cosh function as loss function is that when the feature errors between samples is small, the Log-Cosh function converges faster. when the feature errors between samples is large, the Log-Cosh function is not susceptible by outliers. Compared with other functions, the characteristic curve is smoother and can be derivable twice. To a certain extent, the robustness of the model can be improved [27]. The expression of the Log-Cosh function is as follows:

$$\text{Loss}(y_i, \hat{y}_i) = \sum_{i=1}^{m} \log(\coth(\hat{y}_i - y_i)) \tag{15}$$

where y_i is the label vector of the ith training sample, $\hat{y}_i$ represents the predicted value of the ith output sample, m represents the total number of training samples. The relevant training parameter conditions: the loss function was the Log-Cosh function, the batch-size was 128, the training epoch was 200, the initial learning rate was 0.001, the optimizer was Adam. After iterative training according to the settings of the above training parameters, the loss function convergence curve of DDNN model, as shown in Figure 6, was obtained.

When the convergence of the loss function of model training reached the expectation, the training was stopped and the model was saved; the weight parameters were extracted from the saved model. Table 3 shows the weight information of each layer of the model parameters of the last training (keep values to three decimal places); here, symbol "—" means no weight value. Finally, the test set was brought into the trained model to get the prediction result and compared with the label.

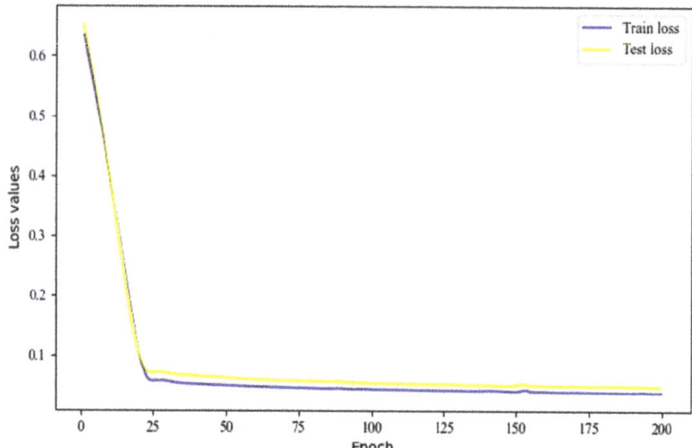

Figure 6. Convergence of loss function of reducer transmission accuracy error.

Table 3. Optimal weights.

The Weight of Each Layer	Weight Value								
The first layer	0.183	−0.070	−0.745	0.478	−0.765	The second layer	0.826	0.465	0.205
	−0.601	0.091	−0.313	−0.006	−0.498		0.656	−0.973	—
	0.146	−0.325	−0.538	−0.181	−0.119		—	—	—
	0.484	1.127	−0.681	−0.747	0.447		—	—	—
	−0.412	0.665	0.513	0.280	0.117		—	—	—
The third layer	0.987	0.756	—	—	—	The fourth layer	1.029	−0.725	—
	−1.273	0.987	—	—	—				

The prediction curve of the DDNN model is shown in Figure 7, where the blue line represents real sample values, the red line represents the predicted values.

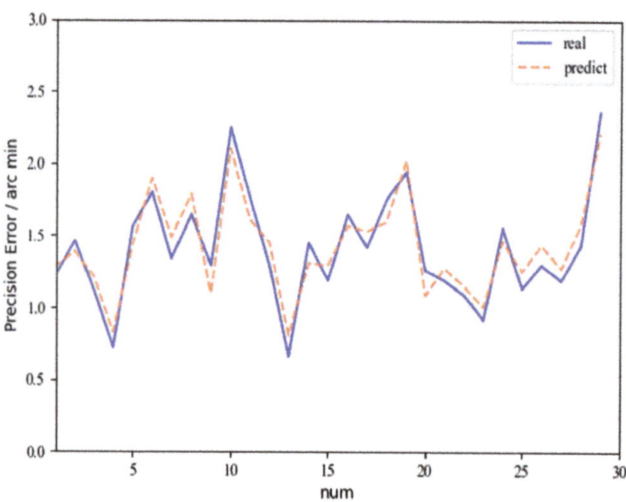

Figure 7. Prediction result of transmission error sample of RV-40E-121 reducer.

4.3. Predictive Performance Analysis

In order to further quantify the prediction accuracy of DDNN model, the mean absolute error and mean square error were introduced as performance evaluation indicators. The effectiveness of the DDNN prediction method was verified by analyzing the performance of model predictions. The specific definitions are as follows:

MSE (Mean squared error)

$$\text{MSE} = \frac{1}{m} \sum_{i=1}^{m} (\hat{y}_i - y_i)^2 \tag{16}$$

MAE (Mean absolute error)

$$\text{MAE} = \frac{1}{m} \sum_{i=1}^{m} |\hat{y}_i - y_i| \tag{17}$$

where y_i is the label vector of the ith training sample, $\hat{y}_i$ represents the predicted value of the ith output sample, and m represents the total number of training samples.

MSE reflects the mean squared prediction error of the model. MAE describes the average value of the absolute error between the predicted value and the observed value [28]. The smaller the value of MSE and MAE, the better. The method proposed in this paper was compared with the other three commonly used prediction models, which are BP neural network, support vector regression with Gaussian kernel (SVR-R) and general regression neural network (GRNN) [29]. The initial parameter settings for all models are shown in Table 4.

Table 4. The initial parameter values of models.

Model	Parameter	Value
BP neural network	Learning rate	0.001
	Optimizer	Stochastic Gradient Descent
GRNN	Smooth factor	0.03
SVR-R	Penalty term	40
	Gamma	4
Dendritic Neural Network	Learning rate	0.001
	Optimizer	Adam

Table 5 shows the comparison results of the prediction performance and computational efficiency of the above four models. From the error results in the table, it can be concluded that for transmission error prediction, the prediction accuracy of the DDNN model is better than the other three models. For computational efficiency, the running time of GRNN and SVR-R outperforms the remaining two models. Considering the prediction performance and calculation efficiency comprehensively, the DDNN model has the highest prediction accuracy and can also meet the time requirements of actual assembly.

Table 5. Comparison of the DDNN with other models.

Model	Mean Squared Error	Mean Absolute Error	Computational Efficiency/s
BP neural network	0.1411	0.2545	13.708
GRNN	0.051	0.229	0.057
SVR-R	0.048	0.216	0.036
Dendritic Neural Network	0.032	0.121	9.751

5. Conclusions and Future Work

In the actual production and assembly process of reducer, the success rate of one-time selection of parts is low, and repeated disassembly and assembly will lead to the generation and accumulation of errors. Through the analysis of the factors affecting the quality of RV reducer, five factors with larger sensitivity indices were selected as sample eigenvalues in the data set, and transmission error was noted as sample label. The prediction results show that DDNN model can capture the logical relationship between features and strengthen the internal information correlation; it can effectively avoid the failure of the loss function to converge due to large fluctuations of parameters when updating; it has better generalization ability and can effectively predict reducer dynamic transmission error. The research in this article has practical guiding significance for the selection of parts and components in the RV reducer assembly, which can improve the assembly qualification rate, avoid repeated disassembly and assembly, and reduce the waste of labor and time costs.

However, there is still room for improvement in this study, such as the deficiencies in selection of quality influencing factors, and the applicability of the method for different examples. In the follow-up, the influence of quality factors on transmission error of reducer will be comprehensively analyzed to obtain a more ideal DDNN prediction model.

Author Contributions: Conceptualization; Methodology; Project administration, S.J.; Writing; Software and Validation, Y.C.; Review, Y.S.; Supervision; Editing, Y.W. All authors have read and agreed to the published version of the manuscript.

Funding: This research was supported by National High-tech R&D Program of China (Grant No.2015AA043002), Natural Science Foundation of Zhejiang Province (Grant No. LQ22E050017), Postdoctoral Science Foundation of China (Grant No. 2021M702894), and Zhejiang Provincial Postdoctoral Science Foundation (Grant No. ZJ2021119).

Institutional Review Board Statement: Not applicable.

Informed Consent Statement: Not applicable.

Data Availability Statement: The data presented in this study are available on request from the corresponding author. The data are not publicly available due to corporate data privacy.

Acknowledgments: This work is supported by Zhejiang Shuanghuan Transmission Machinery Co., Ltd. providing the RV reducer sample and some data.

Conflicts of Interest: The authors declared no potential conflict of interest with respect to the research, authorship, and/or publication of this article.

References

1. Xue, J.; Qiu, Z.R.; Fang, L.; Lu, Y.H.; Hu, W.C. Angular Measurement of High Precision Reducer for Industrial Robot. *IEEE Trans. Instrum. Meas.* **2021**, *70*, 1–10. [CrossRef]
2. Kumar, M.S.; Kannan, S. Optimum manufacturing tolerance to selective assembly technique for different assembly specifications by using genetic algorithm. *Int. J. Adv. Manuf. Technol.* **2007**, *32*, 5–6. [CrossRef]
3. Gentilini, I.; Shimada, K. Predicting and evaluating the post-assembly shape of thin-walled components via 3D laser digitization and FEA simulation of the assembly process. *Comput.-Aided Des.* **2010**, *43*, 316–328. [CrossRef]
4. Khodaygan, S.; Ghaderi, A. Tolerance–reliability analysis of mechanical assemblies for quality control based on Bayesian modeling. *Assem. Autom.* **2019**, *39*, 769–782. [CrossRef]
5. Burggraef, P.; Wagner, J.; Heinbach, B.; Steinberg, F. Machine Learning-Based Prediction of Missing Components for Assembly—A Case Study at an Engineer-to-Order Manufacturer. *IEEE Access* **2021**, *9*, 105926–105938. [CrossRef]
6. Bruennhaeusser, J.; Gogineni, S.; Nickel, J.; Witte, H.; Stark, R. Assembly issue resolution system using machine learning in Aero engine manufacturing. In Proceedings of the International Conference on Advances in Production Management Systems, Novi Sad, Serbia, 30 August–30 September 2020; Volume 591, pp. 149–157.
7. Takahashi, N.; Oertner, T.G.; Hegemann, P.; Larkum, M.E. Active cortical dendrites modulate perception. *Science* **2016**, *354*, 1587–1590. [CrossRef]
8. Trenholm, S.; McLaughlin, A.J.; Schwab, D.J.; Turner, M.H.; Smith, R.G.; Rieke, F.; Awatramani, G.B. Nonlinear dendritic integration of electrical and chemical synaptic inputs drives fine-scale correlations. *Nat. Neurosci.* **2014**, *17*, 1759–1766. [CrossRef]
9. Li, X.Y.; Tang, J.S.; Zhang, Q.T.; Gao, B.; Yang, J.J.; Song, S.; Wu, W.; Zhang, W.Q.; Yao, P.; Deng, N.; et al. Power-efficient neural network with artificial dendrites. *Nat. Nanotechnol.* **2020**, *15*, 776–782. [CrossRef]
10. Antic, S.D.; Zhou, W.L.; Moore, A.R.; Short, S.M.; Ikonomu, K.D. The decade of the dendritic NMDA spike. *J. Neurosci. Res.* **2010**, *88*, 2991–3001. [CrossRef]
11. Todo, Y.; Tamura, H.; Yamashita, K.; Tang, Z. Unsupervised learnable neuron model with nonlinear interaction on dendrites. *Neural Netw.* **2014**, *60*, 96–103. [CrossRef]
12. Tang, C.; Ji, J.; Tang, Y.; Gao, S.; Tang, Z.; Todo, Y. A novel machine learning technique for computer-aided diagnosis. *Eng. Appl. Artif. Intell.* **2020**, *92*, 103627. [CrossRef]

13. Song, S.; Chen, X.; Tang, C.; Song, S.; Tang, Z.; Todo, Y. Training an Approximate Logic Dendritic Neuron Model Using Social Learning Particle Swarm Optimization Algorithm. *IEEE Access* **2019**, *7*, 141947–141959. [CrossRef]

14. Gao, S.C.; Zhou, M.C.; Wang, Y.R.; Cheng, J.J.; Yachi, H.; Wang, J.H. Dendritic Neuron Model with Effective Learning Algorithms for Classification, Approximation, and Prediction. *IEEE Trans. Neural Netw. Learn. Syst.* **2019**, *30*, 601–614. [CrossRef] [PubMed]

15. Liu, G.; Wang, J. Dendrite Net: A White-Box Module for Classification, Regression, and System Identification. *IEEE Trans. Cybern.* **2021**, *in press*. [CrossRef]

16. Liu, G.; Wang, J. A relation spectrum inheriting Taylor series: Muscle synergy and coupling for hand. *Front. Inform. Technol. Elect. Eng.* **2022**, *23*, 145–157. [CrossRef]

17. Xu, A.J.; Deng, X.Z.; Zhang, J.; Xu, K.; Li, J.B. A New Numerical Algorithm for Transmission Error Measurement at Gears Meshing. *Adv. Mater. Res.* **2012**, *472*, 1563–1567. [CrossRef]

18. García-Ayuso, L.E.; Velasco, J.; Dobarganes, M.C.; de Castro, M.D.L. Accelerated extraction of the fat content in cheese using a focused microwave-assisted soxhlet device. *J. Agric. Food Chem.* **1999**, *47*, 2308–2315. [CrossRef]

19. Zhang, Z.H.; Wang, J.X.; Zhou, G.W.; Pei, X. Analysis of mixed lubrication of RV reducer turning arm roller bearing. *Ind. Lubr. Tribol.* **2018**, *70*, 161–171. [CrossRef]

20. Bahubalendruni, M.V.A.R.; Gulivindala, A.; Kumar, M.; Biswal, B.B.; Annepu, L.N. A hybrid conjugated method for assembly sequence generation and explode view generation. *Assem. Autom.* **2019**, *39*, 211–225. [CrossRef]

21. Chen, C.; Yang, Y.H. Structural Characteristics of Rotate Vector Reducer Free Vibration. *Shock Vib.* **2017**, *2017*, 4214370. [CrossRef]

22. Jin, S.S.; Tong, X.T.; Wang, Y.L. Influencing Factors on Rotate Vector Reducer Dynamic Transmission Error. *Int. J. Autom. Technol.* **2019**, *13*, 545–556. [CrossRef]

23. Hu, Y.H.; Li, G.; Zhu, W.D.; Cui, J.K. An Elastic Transmission Error Compensation Method for Rotary Vector Speed Reducers Based on Error Sensitivity Analysis. *Appl. Sci.* **2020**, *10*, 481. [CrossRef]

24. Kingma, D.P.; Ba, J.L. Adam: A method for stochastic optimization. In Proceedings of the 3rd International Conference on Learning Representations (ICLR 2015), San Diego, CA, USA, 7–9 May 2015.

25. Gidon, A.; Zolnik, T.A.; Fidzinski, P.; Bolduan, F.; Papoutsi, A.; Poirazi, P.; Holtkamp, M.; Vida, I.; Larkum, M.E. Dendritic action potentials and computation in human layer 2/3 cortical neurons. *Science* **2020**, *367*, 83–87. [CrossRef] [PubMed]

26. Liu, G. It may be time to improve the neuron of artificial neural network. *TechRxiv*, 2020, *preprint*.

27. Mishra, B.P.; Panigrahi, T.; Wilson, A.M.; Sabat, S.L. A robust diffusion algorithm using logarithmic hyperbolic cosine cost function for channel estimation in wireless sensor network under impulsive noise environment. *Digit. Signal Process. Rev. J.* **2022**, *123*, 103384. [CrossRef]

28. Zulfiqar, A.K.; Tanveer, H.; Ijaz, U.H.; Fath, U.M.U.; Sung, W.B. Towards efficient and effective renewable energy prediction via deep learning. *Energy Rep.* **2022**, *8*, 10230–10243.

29. Song, Z.Y.; Tang, C.; Ji, J.K.; Todo, Y.; Tang, Z. A Simple Dendritic Neural Network Model-Based Approach for Daily $PM_{2.5}$ Concentration Prediction. *Electronics* **2021**, *10*, 373. [CrossRef]

Review

Operation and Maintenance Optimization for Manufacturing Systems with Energy Management

Xiangxin An [1], Guojin Si [1], Tangbin Xia [1,*], Qinming Liu [2], Yaping Li [3] and Rui Miao [4]

1   State Key Laboratory of Mechanical System and Vibration, Department of Industrial Engineering, School of Mechanical Engineering, Shanghai Jiao Tong University, Shanghai 200240, China
2   Department of Industrial Engineering, Business School, University of Shanghai for Science & Technology, Shanghai 200093, China
3   Department of Management Science and Engineering, College of Economics and Management, Nanjing Forestry University, Nanjing 210037, China
4   School of Naval Architecture, Ocean and Civil Engineering, Shanghai Jiao Tong University, Shanghai 200240, China
*   Correspondence: xtbxtb@sjtu.edu.cn

Abstract: With the increasing attention paid to sustainable development around the world, improving energy efficiency and applying effective means of energy saving have gradually received worldwide attention. As the largest energy consumers, manufacturing industries are also inevitably facing pressures on energy optimization evolution from both governments and competitors. The rational optimization of energy consumption in industrial operation activities can significantly improve the sustainability level of the company. Among these enterprise activities, operation and maintenance (O&M) of manufacturing systems are considered to have the most prospects for energy optimization. The diversity of O&M activities and system structures also expands the research space for it. However, the energy consumption optimization of manufacturing systems faces several challenges: the dynamics of manufacturing activities, the complexity of system structures, and the diverse interpretation of energy-optimization definitions. To address these issues, we review the existing O&M optimization approaches with energy management and divide them into several operation levels. This paper addresses current research development on O&M optimization with energy-management considerations from single-machine, production-line, factory, and supply-chain levels. Finally, it discusses recent research trends in O&M optimization with energy-management considerations in manufacturing systems.

Keywords: sustainable; energy optimization; operation and maintenance; manufacturing systems

Citation: An, X.; Si, G.; Xia, T.; Liu, Q.; Li, Y.; Miao, R. Operation and Maintenance Optimization for Manufacturing Systems with Energy Management. *Energies* **2022**, *15*, 7338. https://doi.org/10.3390/en15197338

Academic Editor: David Borge-Diez

Received: 23 August 2022
Accepted: 30 September 2022
Published: 6 October 2022

Publisher's Note: MDPI stays neutral with regard to jurisdictional claims in published maps and institutional affiliations.

1. Introduction

With the rising concern about the environmental pollution of fossil energy, the concept of sustainable development has gained worldwide attention. A huge number of measures have been proposed to achieve the global sustainable goal from various perspectives of energy policy, technology, etc. [1–3]. Renewable energy is recommended to gradually replace fossil energy, and the calls for applications of energy efficiency increases and effective energy saving have become much more urgent [4]. As the important pillar of development as well as the main sector of energy consumption, industry consumed 28.9% of the world's total final energy, 38% of total electricity consumption, and especially 81.9% of coal consumption [5]. Therefore, reasonably improving the sustainable development level of the industry, especially the optimization of energy consumption, plays an important role in promoting the transformation to a sustainable manufacturing mode [6].

Under the background of sustainable development, industrial enterprises not only need to improve their energy efficiency and reduce waste to obtain greater competitiveness but also have to face a more challenging energy transformation environment such as energy source changes and emission punishment, which brings more considerations and constraints for the decision-making process. Effective energy-saving measures can help enterprises save costs, resist the risk of policy changes, and can also reflect the social responsibility of enterprises, driving long-term sustainable development [7].

For a classical industrial enterprise, the energy consumption sources can be mainly divided into two categories according to the usage purpose of the consumption terminal [8]. One is used to support the normal functioning of enterprises, such as lighting and heating of all buildings, which is more stable and easier to predict. This kind of energy consumption can be reduced through the systematic application of green building and management [9,10]. The other is mainly consumed in operation and maintenance (O&M) activities, which are directly related to the manufacturing process and have a strong correlation with the energy consumption of the system [11]. Compared with the former energy consumption source, O&M-related energy optimization faces more severe challenges. Since the production activities involve both products and machines, the randomness of production orders, the diversity of production-line structures, and the complexity of the machine working environment all increase the difficulty of O&M decision-making.

In recent years, O&M decision-making with energy-management considerations has been noticeable and studied from various perspectives. Many valuable reviews have been published to summarize different ways of energy management in manufacturing systems. Park et al. [12] reviewed energy consumption reduction strategies and energy-saving technologies in different countries. Jasiulewicz [13] concluded recent maintenance technologies for sustainable manufacturing to avoid sudden breakdown and decrease energy and material consumption. In addition, Gham et al. [14] constructed a study framework for energy-efficient scheduling and made a sufficient review. These studies on energy management in the O&M field usually focus on totally different objects and are scattered among different processing stages. However, they have not carried out more specific and systematic generalizations for the concrete optimization problems and solving methodologies.

This paper classifies various energy management approaches in the field of O&M, focusing on the energy optimization of manufacturing systems, and reviews from mainly four processing levels: machine level, production-line level, factory level, and supply-chain level. The remainder of this paper is arranged as follows: In Section 2, the O&M methods for manufacturing systems with an energy-saving consideration are reviewed, and the challenge factors are elaborated. Sections 3 and 4 summarize the new research on energy optimization in the field of O&M from the machine level and production-line level, respectively. Section 5 briefly introduces current developments of O&M optimization with energy management in the factory level and supply-chain level. Section 6 further discusses the future development trend in this research field, and, finally, Section 7 summarizes the work of the full text.

2. O&M Methods for Manufacturing Systems with Energy-Saving Consideration

O&M are the main business activities within manufacturing enterprises. In a broad sense, they refer to all related activities needed to preserve the functioning and productivity of a system. For a manufacturing system, O&M are the collection of a series of system management activities, including product scheduling, machine management, product quality control, inventory control, and supply-chain management, as well as after-sale service. These are used to analyze the characteristics of different manufacturing system structures, apply systematic optimization methods, and make improvements to the performance of the whole system, including production cost, system efficiency, machine reliability, and product quality.

Since O&M activities are carried out in multiple forms at different levels of manufacturing systems, the optimization can be categorized according to the organization of manufacturing systems [15], which can also be defined as the manufacturing process. In this context, the manufacturing optimization activities can be decomposed into several levels: the machine level where unit optimization is carried out, the production-line level with the consideration of interactions between machines and products, and the factory level, which is defined as monitoring the whole performance of the factory through various aspects, including building energy monitoring and manufacturing system energy optimization, and the multi-factory and supply-chain level, where energy management is applied from suppliers to customers. The main activities and elements concerning in different decision levels are presented in Figure 1.

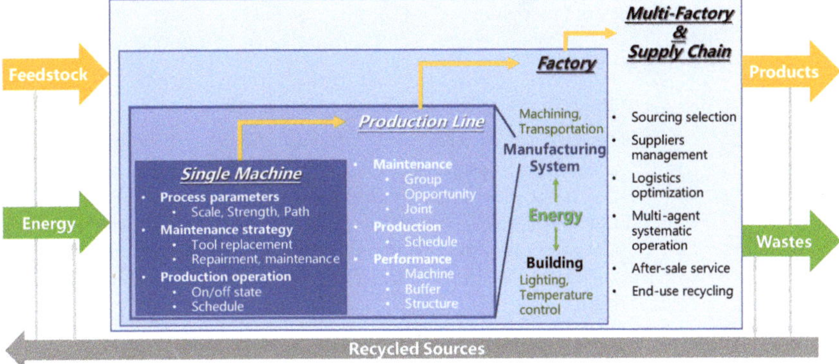

Figure 1. Decision levels of O&M in a manufacturing process.

The O&M research on the machine level mainly studies the degradation trend and machine status through in-depth analysis of the processing details, so as to improve the processing performance. Inputs for a machine processing procedure are the processing parameters and operation schemes, and some noticeable performance outputs include machine degradation status, product quality, and processing costs. Therefore, O&M on a single-machine system usually focuses on the process parameters of machines, degradation of machines [16], production schedules, and single-machine maintenance strategies [17,18], which can be further divided into sub-tool replacement, whole-machine maintenance, etc. Many studies also take product quality into consideration and perform joint modeling and optimization [19,20]. The analysis of the single machine is also the foundation for system-level optimization and provides a more specific optimization angle to realize energy savings.

The system research on the production-line level pays more attention to the structure of the production line and the processing sequence. In addition to the machine status and scheduling problems, which are considered at the machine level, production-line-level optimization starts to concentrate on the interaction among procedures and between machines and products.

At the same time, the application of the Manufacturing Execution System (MES) broadens the meaning of O&M to the factory level [21]. Through additional management systems, the synchronous optimization can be accomplished by real-time monitoring and precise simulation, combined with systematic analysis. In addition, building energy optimization can be included in the operational activities of enterprises. Moreover, energy optimization between factories within the supply chain is also considered to be a promising research field, which includes supplier management, logistics optimization, and so on.

This review focuses on the O&M optimization strategies in the manufacturing system, that is, the related activities at the machine level and production-line level. Since the system profit and energy consumption are directly related to the product production schemes and machine maintenance behaviors, O&M optimization in the manufacturing system pays more attention to the production scheduling and system maintenance strategies. As two key activities in the manufacturing system management, these two kinds of activities are coupled with each other through the reliability and availability of the machine [22]. Maintenance aims to ensure the health of the machine and thus ensure production efficiency and product quality [23], whereas production scheduling can in turn affect the formulation of machine degradation. Therefore, the difficulties of joint decision-making of maintenance and production scheduling are mainly reflected in three aspects:

- The independence of the machine degradation

The function and working environment differ between machines in different procedures among the whole system, and the fact that the degradation distributions of any two machines are independent of each other due to the variation among machine components further proves the necessity of full consideration of independent machines' degradation modeling [24].

- Multiple couplings brought by production operation

The production of workpieces often needs to go through multiple processes; as a result, the machine in each process is connected through the processing sequence of the workpieces. It also unexpectedly influences the reliability of machines and increases the difficulty of overall decision-making at the system level in the formulation of production and maintenance strategies. Moreover, maintenance actions are performed in machines, which occupies the feasible time for product processing and influences the final productivity of the system.

- Interactions between energy consumption and O&M operation

Maintenance and production are related to machine status and thus related to the energy consumption levels of machines. The processing machine usually consumes more energy when it is on for production, and is shut down or kept idle during maintenance. Therefore, reasonable and effective maintenance and production scheduling can reasonably utilize the status of each machine in the system, so as to find the opportunity for energy balance, thereby reducing energy consumption.

Meanwhile, more and more attention has been paid to energy consumption and energy efficiency during processing in the context of sustainable development [25]. The consideration of energy savings in O&M optimization of the manufacturing system also brings changes to the traditional joint decision-making of production and maintenance:

1. Additional targets or constraints brought from energy consideration

The consideration of energy consumption can usually be added to the model in two ways, one as an optimization objective and the other as a model constraint. However, no matter what form the addition takes, it increases the solution complexity of the original model. In particular, when energy consumption is taken as the optimization objective, it may greatly affect the original solution space and the optimization direction of the solution.

2. Changes in modeling methods under a new energy policy background

Simple additional goals or constraints will not change the original model too much. However, with the gradual deepening of energy-saving considerations, the updated energy consumption policies will also become a new research point for system optimization. Under the influence of these circumstances, in addition to the revision of the original model, it is more likely to completely subvert the original model, and force researchers to build a new system optimization model and seek new energy-saving opportunities.

Based on the difficulties and changes mentioned above, the following sections will give a detailed overview of O&M optimization methods with the consideration of energy management from the perspective of single-machine and production-line levels.

3. New Research Developments on Machine Level

The manufacturing system is formed by series and parallel connections of machines with different functions in different processes. Therefore, the machine can be regarded as the most important part, constituting the production line. At the same time, the machine itself is a complex system. From the structural aspect, typical machining equipment is composed of tools, spindles, motors, and other components. In addition, from the perspective of the machining process, it can be divided into rotation, positioning, feeding, and other processes, which have different energy consumption characteristics. Therefore, machine-level operations can be divided into two categories. One is to optimize the processing parameters for different processes to reduce energy consumption and the other is to achieve energy consumption minimization through whole-machine actions such as on/off control and single-machine production scheduling. Similarly, the maintenance of machines can also be summarized into two types. On the one hand, the machine can be taken as a whole, and data-driven methods can be used to optimize the overall maintenance action. On the other hand, each component can be taken as the focus of optimization to study the maintenance and replacement strategies of each part. In addition, some researchers also additionally introduce reliability assessment or optimization with product quality control consideration [26]. Therefore, the O&M optimization of energy management on the machine level can usually be summarized into three main aspects:

3.1. Process Parameters Optimization

The most common type of processing in manufacturing systems is machining, and nearly 99% of the environmental impact is caused by the energy consumption of CNC machines [27]. Typically, the machining process can be divided into a set of steps including positioning, spindling, tool feeding, cutting, and leaving [28,29]. And a typical machine tool power usage in a complete machining process is calculated and shown in Figure 2 [30]. Of these, inherent energy consumption plays a dominant role in total consumption [31].

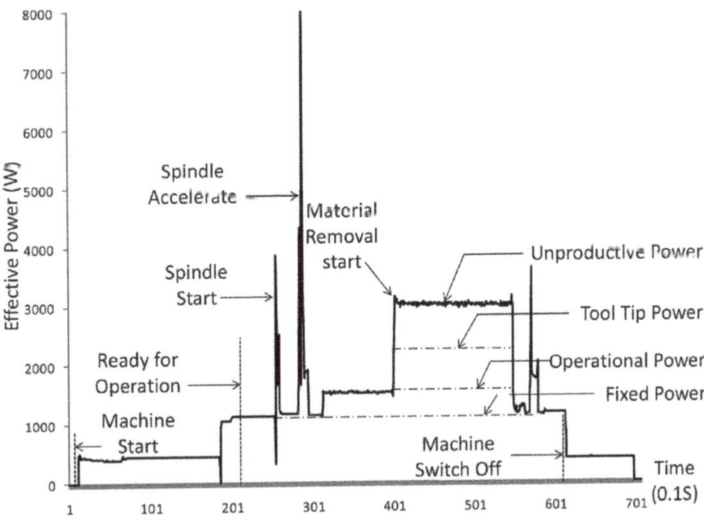

Figure 2. Machine tool power usage in a complete machining process.

Optimization in terms of non-value-added behavior, therefore, has greater potential for energy savings, such as tool path optimization and position location optimization. Research in this area can not only improve machining efficiency but also achieve energy savings. However, it usually requires a balance between the goals of faster movement speeds and more accurate positioning accuracy. Xu et al. [32] developed a model of the machine-dependent energy potential field and modeled specific power requirements at any contact point with any feed direction, thus finding a balance between the minimum energy consumption and tool path planning.

At the same time, the machining process refers to the process of chipping material from the surface of an object and the energy consumption in this process is related to numerous factors [27]. From the perspective of process parameter optimization, many papers focus on the optimal design of cutting parameters [27,33–35]. This kind of research is mainly carried out in three ways:

1. Real experimental-design-based optimization

The first is based on the experimental design of machining, such as the Taguchi method, in order to explain the influence of cutting parameters on energy consumption and to obtain the optimal combination of parameters for energy consumption. Bilga et al. [36] investigated the influence of process parameters on the turning process with the aid of Taguchi's experimental design and signal-to-noise ratio analysis. From the experiment, it was pointed out that the cutting depth variable was the largest influencing factor in the energy consumption of the turning process.

2. Optimization based on fitting experimental data

The second method is based on fitting experimental data to obtain a regression model between parameters and energy consumption. Classical methods include the response surface method. Campatelli et al. [37] established a regression equation between milling parameters and energy consumption by the response surface method and obtained the optimal combination through experimental analysis. They also demonstrated that energy consumption decreases with increases in the material removal rate. Zhong [27] identified energy consumption as the optimization objective, denoted as a function of spindle speed, cutting speed, feed rate, and cutting depth. A decision rule for selecting the best cutting parameters to achieve the lowest energy consumption was also proposed based on the candidate options of the experimental design.

3. Impact-analysis-based optimization

The last kind of approach to construct an energy consumption optimization is by analyzing the impact characteristics of cutting parameters on energy consumption. For example, He et al. [38] established a multi-objective optimization model for process parameters with energy consumption, cutting force, and machining time as three objectives, and solved the problem based on a matching genetic algorithm. Bi [35] developed an energy consumption evaluation model based on the kinematic and dynamic behavior of the machine tool to dynamically evaluate the machine's posture and dynamically optimize the process parameters to reduce energy consumption. Simulation experiments were conducted to verify the accuracy of the method concerning energy consumption and machine posture.

In addition, there is a complex mapping of machining energy consumption to tool selection and tool usage conditions in the machining process. It has been demonstrated that both cutting parameters and energy consumption also vary significantly depending on the tool wear conditions when using worn tools [39]. Therefore, tool selection [29,40], tool wearing conditions [41–43], and machining quality control [26,44,45] are usually also taken into account in new studies. Chen et al. [40] developed an integrated approach for cutting tool and cutting parameter optimization to analyze the energy footprint by considering the flexibility of multiple tool selection and cutting parameters, and the multi-objective cuckoo search algorithm was used to find the optimal solution. The effectiveness was proved through an actual milling process with a step feature. They also proved that the optimization result of minimum production time does not necessarily satisfy minimum energy consumption.

In contrast, Xie et al. [43] proposed a co-optimization multi-objective model to select the optimal cutting parameters in order to consider the effects of tool wear and cutting parameter combinations at a certain material removal rate for energy saving and quality assurance and used the NSGA-III algorithm to solve the problem.

The state of the processing machine in turn also greatly affects the quality of the product, and, for machined products, the main embodiment of quality is surface roughness.

Therefore, Wang et al. [33] classified the energy in the machining process into direct energy and embodied energy, and considered the cost items such as operation cost, consumables cost, tool cost, etc. They established a multi-objective optimization model with surface roughness as the embodiment of machining quality and solved it by using a non-dominated classification genetic algorithm to obtain the optimal set of cutting parameters. Ma et al. [34] developed an energy prediction and optimization model for three-axis milling machines, characterizing the material removal rate as an energy consumption function, and controlling the cutting width and cutting depth parameters by optimizing the spindle speed.

3.2. Energy-Efficient Maintenance Optimization

The most common maintenance actions used in maintenance optimization decisions for a single machine can be divided into two categories: corrective maintenance (CM) and preventive maintenance (PM). CM, which is also known as minimal maintenance [46], refers to actions of taking the necessary repairs to restore the machine to function when a failure has already occurred. This type of maintenance action is not effective enough in ensuring machine reliability because failures are unpredictable and happen randomly. In contrast, PM is a pre-act maintenance strategy, in which the machine is overhauled and repaired according to a pre-defined strategy before a failure occurs, thereby eliminating the failure in advance and avoiding major economic losses caused by sudden breakdown. In the construction of a single-machine maintenance strategy, the above two actions are usually considered at the same time, and the strategy is optimized mainly for preventive maintenance.

For the construction of the maintenance optimization model, machine reliability, productivity, maintenance cost, etc., are usually considered as objectives. The impact of energy consumption is also integrated into the decision model in order to fully consider the impact of sustainability objectives on the decision-making process. Yan et al. [47] modeled the relationship between reliability and energy consumption based on historical data to calculate the energy consumption of machine tools, thus accurately analyzing the relationship between energy consumption and machine reliability. Hoang et al. [48] proposed a state maintenance model based on energy-efficiency thresholds, thus directly considering energy impact in maintenance optimization and combining the degradation of the machine with energy conversion efficiency.

Focusing on the machine itself, it is also composed of multiple sub-components. Therefore, some constructions of the model also address the machine maintenance optimization integrated preventive maintenance and component replacement strategies to achieve a system improvement. For example, Xia et al. [49] integrated energy consumption mechanisms of tool wear and the degradation of machine systems to optimize the combination of machine maintenance and tool replacement strategies. The application of this approach effectively reduced non-value-added energy consumption in sustainable manufacturing systems compared with traditional maintenance strategies.

In terms of the solution methods, the models mentioned above usually use mathematical modeling methods, which construct objective functions, add constraints, and perform model solving to obtain optimal maintenance policies. However, some studies choose model stochastic process simulation methods. An increasing number of studies use Markov methods for detailed modeling and optimization of machine condition transformation. Xu et al. [50] used a partially observable Markov decision process (POMDP) framework to develop a decision model to infer the status of the machine tool through joint observations of machining energy consumption and manufactured workpiece quality. It was proved to be an optimal strategy that maximizes the total expected energy efficiency return of the production process in a limited range. Wu et al. [51] developed a proactive maintenance framework with two dimensions—service age and severity of energy consumption—to guide repairment and spare parts replacement. A semi-Markov chain was constructed to model the energy consumption process and analyze the energy consumption optimization process.

3.3. Energy-Efficient Production Optimization

Production-related energy consumption optimization usually consists of two aspects: one is the production scheduling problem and the other is the machine on/off state switching problem.

1.   Production scheduling

The single-machine scheduling problem is the set of all scheduling problems that perform sequencing on a single machine and is usually used to guide the optimization of the workpiece processing order. The general description is as follows: A number of N and mutually independent workpieces are sorted according to specific sorting rules and scheduled to be processed on a single machine. Each workpiece can only be processed once on that machine and the machine can only process one workpiece at the same time.

In a sustainable manufacturing context, machining energy consumption is usually added to the model as a goal or constraint. Che et al. [52] designed a method to obtain the exact Pareto front based on minimizing the total energy consumption and maximum tardiness. The constructed model considered the differences in workpiece release times of different types and used the production scheduling plan and machine switching operations as decision variables for the systematic optimization of production operations. Some studies also took time-dependent energy costs as research background and carried out the production schedule. Chen [53] solved the single-machine scheduling problem for a set of independent jobs under time-of-use tariffs and designed the corresponding scheduling rules and pseudo-polynomial-time algorithm to solve it. In contrast, Zhou [54] conducted a study on the energy-efficient scheduling problem of a single batch processing machine in the context of time-of-use tariffs, where a model was constructed for different job sizes and release times, taking power cost and productivity into account, and a hybrid multi-objective metaheuristic algorithm based on particle swarm optimization (PSO) was designed to solve it.

2.   On/off control

In addition to considering production scheduling sequence optimization, some studies were also devoted to energy consumption optimization with the help of machine state switching. For machines with high standby power and long production waiting intervals, through timely machine status switching, the standby energy consumption of machines can be saved by a certain amount. Moreover, production scheduling can be further used to optimize the energy consumption of machines systematically. Mouzon [55] considered the fact that a significant amount of energy can be saved when non-bottleneck machines are switched off after a period of inactivity. Then, several scheduling rules based on the machine on/off status switching were proposed, while a set of non-dominated solutions was obtained through multi-objective optimization to determine the most efficient production sequence for the joint optimization of total completion time and energy consumption.

3.   Joint optimization

Single-machine production scheduling does not always bring a significant effect on energy consumption reduction; in this context, some studies have explored a series of joint optimization considering both machine status and production-related elements. Usually, a multi-objective model with energy and productivity objectives is designed in these kinds of studies. Aghelinejad et al. [56] reduced the energy cost of a production system through multi-layer modeling. The model divided the optimization step into two layers: the first layer uses the machine state transition matrix to evaluate the machine state and thus reduce the energy consumption of idle machines; the second layer performs the energy consumption optimization of production scheduling. Wang et al. [57] designed the optimization objective of minimizing the makespan and total energy consumption under the condition that workpieces are machined in batches. In addition, an *ε-constraints* method was adapted to get the Pareto front of the problem.

All O&M optimization topics at the single-machine level mentioned above are summarized in Table 1.

Table 1. Energy-optimization opportunities and challenges in O&M at single-machine level.

Area	Sub-Area	Energy-Optimization Principles	Challenges
Process Parameters Optimization	Feed parameters	Reduce energy consumption by improving machining efficiency and quality	· Processing steps' division
	Tool movement trajectory	Reduce energy consumption during movement through efficient path planning	· Energy consumption calculation · Optimization modeling
Machine Maintenance	Tool replacement	Reduce excess energy consumption due to machine degradation by improving machine reliability	· Machine degradation modeling · Reliability threshold determination
	System maintenance		· Maintenance strategy selection
Machine Production	Machine state control	Reduce idle time by switching on/off	· Settings of switch principles
	Batch scheduling Production scheduling	Reduce idle time through efficient production scheduling	· Construction of efficient scheduling models
	State analysis	Reduce idle time using simulation analysis	· State transition analysis

4. New Research Developments on Production-Line Level

With the widespread use of complex manufacturing systems consisting of different types of multiple machines in series and parallel construction forms according to process requirements, production-line level O&M optimization strategies have become a hot topic for urgent research. O&M optimization at the production-line level is usually planned using traditional operations research methods and solved by intelligent algorithms.

Compared with single-machine O&M planning, production-line-level optimization scheduling is more complex in terms of decision constraint consideration, overall optimization objectives, machine-related analysis, and dynamic optimization processes. If there is no dependency between machines, the single-device optimization model can be directly adopted. However, the multi-device optimization strategy not only needs to consider the characteristics of each machine itself, but also should comprehensively analyze the interdependencies between machines, which include economic dependencies (the productivity of the system is dependent on the bottleneck machine with the lowest production efficiency), fault dependencies (different health decline processes of the machine may interfere with the health status of other devices), and structure dependency (downtime of one machine in the system structure can mean simultaneous downtime of several machines).

In the production-line-level optimization stages, it is necessary to systematically analyze the interdependencies between machines and perform the systematic scheduling of each machine planning to achieve the decision goal of improving the overall system efficiency. At the same time, as the system structure becomes more complex, it also expands the research space on sustainable optimization. The system structure can be fully utilized to find new opportunities for energy savings. Production-line-level O&M optimization with energy management has been studied to some extent, and the most widely used methods among this research can be summarized into three main categories: system maintenance strategy, production scheduling optimization, and system performance optimization. Based on the above three categories of studies, joint optimization in multiple directions is often performed. New angles of energy savings are also explored to add to the traditional model in order to make certain extensions to the current study.

4.1. System Maintenance

The energy-consumption-oriented system maintenance strategy is an extension of the single-machine maintenance strategy, and the rational scheduling of the maintenance plan has a positive effect on the optimization of the system energy consumption [58]. In the process of production-line level maintenance planning, it is necessary to fully consider the maintenance synchronization between machines and the energy-saving opportunities brought by production, effectively use group maintenance and opportunity maintenance strategies to take advantage of system scale, and reduce system energy-consumption costs.

Maintenance planning at the production-line level has been maturely studied, and the research was mainly optimized to ensure system reliability and production efficiency. While the research on maintenance planning considering energy consumption optimization is limited, the related current research methods are mainly carried out in the following areas:

1. Group maintenance

Group maintenance means that in a multi-machine manufacturing system, if one of the same types of machines under the same operating conditions fails or undergoes maintenance, the system maintenance can be performed together on the same type of machines that have not yet failed [59,60]. Group maintenance strategies can be classified into static and dynamic group maintenance strategies according to the different decision modes [61]. In the energy-consumption-oriented maintenance optimization, this group maintenance timing can be used to adjust the system reliability status, so as to reduce the energy cost caused by maintenance operations and the degradation impacts on the system energy utilization efficiency. Zhou [62] utilized the timing of sudden machine failures as a group maintenance opportunity window for serial production systems with finite buffers. In addition, the continuous degradation of the system and the impact of different maintenance thresholds on energy efficiency were analyzed by Monte Carlo simulation.

2. Opportunistic maintenance

The basic idea of the opportunistic maintenance policy is that the existence of inter-machine dependencies means that the decision analysis of a production machine in a manufacturing system should be carried out in a comprehensive way, taking into account the influence of its interaction with related machines [63]. Since the difference in system structures produce various types of inter-machine dependencies, opportunistic maintenance can be used to dynamically schedule the maintenance plan for multi-machine production lines based on different structural configurations of the system [64]. Various system downtime opportunities can be utilized to achieve comprehensive system energy savings. Research on opportunistic maintenance has flourished in recent years for production-line-level optimization.

Among all research, Zou [65] used machine shutdown during production for energy consumption control as an opportunity window for identifying preventive maintenance tasks. Through the application of this strategy, the desired production throughput and energy efficiency were maintained and the maintenance costs during non-production shifts was greatly minimized. Zhou [66] developed a maintenance model with a degradation constraint threshold, energy constraint threshold, and quality constraint threshold for a batch production system, using the transition time between batches as an energy-saving window, and obtained the optimal maintenance combination scheme with maximum energy saving through a Monte Carlo simulation. Xia et al. [67] minimized energy consumption through an energy-oriented selective maintenance policy. The modeling was constructed with the constraints of limited maintenance resources and the objective of minimizing the total energy consumption during the whole processing time. The basic idea was to select the machine and corresponding maintenance actions at each production shift time window.

3. Maintenance in conjunction with production

Both maintenance and production bring changes to the machine's status. In actual industrial practice, machine failures during the operation of a multi-machine production line can interrupt normal production, which requires maintenance operations to ensure system reliability; and the maintenance activities inevitably consume production time and disrupt the original production schedule. Therefore, in recent years, academics have attempted to analyze these two research areas in an integrated model and establish a system-level interactive optimization strategy that combines maintenance planning and production scheduling. At the same time, since the energy consumption related to production accounts for a larger proportion of the total system energy consumption than that of maintenance, maintenance strategies in conjunction with production are more likely to achieve better energy savings at the production-line level.

Sun et al. [68] considered energy control and maintenance implementation jointly to address the problem of energy consumption, smart maintenance, and throughput improvement simultaneously. Multiple measures were evaluated using a single objective (cost minimization). An et al. [69] developed an integrated model for flexible job shop systems to solve a series of problems including flexible job shop scheduling, forklift transportation scheduling, and imperfect cutting tool maintenance. The superiority of the proposed method was demonstrated by a hybrid multi-objective evolutionary algorithm.

4.2. System Production Scheduling

In the optimization of production systems, production scheduling is another cornerstone of system O&M management in addition to maintenance planning. Research in production scheduling is also mature enough. In recent years, with the promotion of sustainable manufacturing, the optimization of production energy consumption has been added to the modeling of production scheduling, which can be reflected in three main aspects:

1. Energy-constraint-based scheduling

One type is to model production scheduling under a finite energy consumption constraint. This type of approach is mainly applied in industries where there may be strict constraints on the peak power demand, and such methods can be further extended to the optimization of other sustainability indicators such as carbon emissions.

Artigues [70] discussed the production scheduling problem involving electrical energy constraints and designed a two-step programming approach to solve the problem. It was demonstrated that this model can accurately characterize the energy demand in production activities and has a strong application value in industrial environments with power-constrained system scheduling. Modos [71] conducted a study on the energy-constrained discrete manufacturing scheduling problem by integrating the peak energy demand constraint, and designed an adaptive local search algorithm for solving the mixed-integer linear programming model. Masmoudi [72] considered a comprehensive energy consumption optimization job shop scheduling problem that not only sets the objective to minimize energy and production costs but also considers the peak limit of processing power. The dual optimization of energy consumption from both constraint and objective perspectives effectively addresses the importance of energy optimization.

2. Energy-target-based scheduling

Often, it is more common to introduce energy consumption as an objective into the original production scheduling model. For example, Zhang et al. [73] introduced the energy-consumption objective into the job shop scheduling model and proposed a multi-objective genetic algorithm with a local improvement strategy. The overall solution quality was improved by locally solving two constrained sub-problems of the original problem.

Wang [74] carried out effective process planning and production scheduling optimization from both process and system stages and applied artificial neural networks to achieve multi-objective optimization, providing an effective sustainable development example for enterprises. Han et al. [75] constructed an energy-saving integrated model by combining production scheduling and ladle scheduling in the steel industry. They fully considered the time dependence of the two scheduling decisions and designed an enhanced migrating birds optimization algorithm for a high-quality solution to achieve effective energy saving.

3. Time-dependent energy-cost-based scheduling

As sustainable energy becomes more widely used, energy guidance policies such as demand-side management (DSM) have gained a fast promotion. Among all DSM strategies, time-of-use (TOU) tariffs are the most widely used worldwide [76]. It is a power charging model that divides the 24 h power supply into several periods and charges the electricity fee according to the average marginal cost of system operation, in order to encourage customers to optimize their ways of electricity consumption [77]. Taking into account the changes in energy prices during a day, high energy-consuming behaviors can be shifted to off-peak periods (with a lower energy charging price) through joint control of machine status switching, maintenance, and production scheduling, thus realizing a power shift from peak to off-peak periods and significantly reducing energy costs and lowering the environmental impact of operations.

In 2014, Shrouf [78] was the first to introduce demand response into a single-machine production scheduling problem. A mathematical model that minimizes energy cost was established to decide the start-up time, idle time, and on/off time of machines, thus reducing energy costs during periods with high energy prices. Fang [79], after that, proved that the single-machine scheduling problems under both uniform and non-uniform processing rates are strongly NP-hard, and studied the structural properties of optimal scheduling for both problems. Optimization under TOU tariffs was also gradually extended to more complex system organizations. Wang [80] considered a two-machine permutation flow shop scheduling problem and designed a heuristic algorithm based on Johnson's rule to achieve the optimal power cost under TOU tariffs. In contrast, Tan [81], constructed a mixed integer linear programming model integrating parallel machine batch load adjustment and production scheduling under TOU tariffs. Production scheduling optimization under TOU tariffs was also studied in the hybrid flow shop system. Schulz [82] integrated three objectives—makespan, total energy cost, and peak load—into the model of a hybrid flow shop, and used a multi-stage iterative local search algorithm to obtain the integrated optimal solutions. The result proved that the energy awareness in the multi-objective model shows great competitiveness in solution quality.

4.3. System Performance Optimization

In addition to maintenance and production decisions, direct control of machine conditions and other energy-saving opportunities have been studied to some extent. To take full advantage of energy-saving opportunities, in addition to taking maintenance and production actions, it is also possible to proactively adjust the machine's working states. Therefore, this type of production-line-level optimization often draws on changes to machine status, using operations such as switching on and off, to actively find and even create energy-saving opportunities and achieve effective reductions in system energy costs.

From the switching management perspective, Huang et al. [83] developed a multi-stage data-driven model of a manufacturing system with data obtained from distributed sensors so as to evaluate the real-time losses or benefits of maintenance on energy savings. The maintenance operation is performed based on real-time maintenance cost rates while switching on and off the machines, thus achieving system energy savings. Gong et al. [84] designed a joint maintenance–production scheduling problem based on switching control, aiming to simultaneously optimize three objectives—makespan, number of switching actions, and energy consumption—to achieve effective system control. The designed rules and algorithms can both effectively solve the problem with robustness.

From the system optimization perspective, researchers have used system simulation and analytical models to find the space for energy consumption optimization. Fernandez [85] proposed a "just-for-peak" buffer inventory management method for serial production lines to minimize inventory and energy costs and achieve productivity balance under TOU tariffs. Li et al. [86] established a method for monitoring the energy efficiency status of a manufacturing system integrating multiple machines and buffers using Markov chains, and performed real-time system management and optimization for the control of energy efficiency to achieve the real-time output of optimization strategies. Wang et al. [87]

used a Petri net with dynamic adaptive fuzzy reasoning to define the system machine state decision reasoning and performed real-time model validation based on production information of discrete stochastic manufacturing systems.

The energy-optimization opportunities and challenges in O&M optimization at the production-line level are summarized below in Table 2.

Table 2. Energy-optimization opportunities and challenges in O&M at production-line level.

Area	Sub-Area	Energy-Optimization Opportunities	Challenges
System Maintenance	Group maintenance	· Machine maintenance shutdown	· Grouping strategies · Maintenance energy calculation
	Opportunistic maintenance	· Machine breakdown	· Opportunity searching · Maintenance energy calculation
	Maintenance and production	· Preventive maintenance schedule · Sudden breakdown · Production changeover	· Multi-objective scheduling · Action interactivities
System Production Scheduling	Energy target and constraints	· Production scheduling (machine selection, processing period selection)	· Adding objectives/constraints to the original model · Innovative modeling and analysis
	Energy usage background		
System Performance Optimization	Machine states	· Machine on/off control	· Effects on production scheduling · Effects on machine reliability
	Buffer states	· Buffer threshold	· System structure analysis
	Dynamic observation	· Integrated methods	· Observation index selection

5. New Research Developments on Factory and Supply-Chain Levels

In addition to the O&M optimization for manufacturing systems considering machines and production lines, in a broader sense, systems at any level in the manufacturing process can be identified as manufacturing-related systems, such as manufacturing factories and supply-chain systems. The O&M optimization of these two levels often faces more difficulties in coordination between system elements, and the scenarios are more complex, bringing more diverse problems into the research field.

The O&M at the factory level is mainly aimed at the management of system components beyond the production lines, such as inventory systems, and building facilities. Such management includes but is not limited to factory facility control, such as lighting and heating control, floor layout optimization, and energy monitoring of the whole factory [15]. With the advancement of Industry 4.0, data-driven management methods have also been identified as the main contributors to factory monitoring and optimization [88,89]. Ebrahimi et al. [90] investigated an energy-aware scheduling–layout optimization problem, both energy consumption from machining and transportation were considered in the model, and a hybrid ant colony and simulated annealing algorithm were proposed to solve the problem. Gourlis [91] explored the combination of building information modelling and building energy modeling methodology in the optimization of energy-efficient industrial buildings, through which 50% of energy savings can be obtained in the overall energy consumption of the industrial facility. It also showed great potential in the application of building energy management and design.

For the O&M optimization at the multi-factory and supply-chain level, the coordination and collaboration between factories are usually considered, and typical problems involve logistics management, warehousing location optimization, etc. These optimization problems focus on objectives such as logistics cost and transportation efficiency. In the context of sustainable manufacturing, energy-related objectives are gradually considered in the modeling of the original problems. For example, Macrina et al. [92] studied a green vehicle routing problem (VRP) considering partial battery re-charging with mixed fleet and time windows, in which a joint energy-related cost objective consisted of battery recharging cost during the route, recharging cost to the depot, traveling cost, and fuel cost. Hooshmand [93] introduced a time-dependent cost function of each arc in the green VRP problem to model the various traffic conditions in real scenarios and used fuel consumption cost as the objective since CO2 emission is proportional to fuel consumption. Then, a two-phase heuristic algorithm was developed to efficiently solve the problem.

6. Summary and Future Research Trends

This paper aims at providing an overview and classification of O&M energy-optimization methods for manufacturing systems under a sustainable background. The sustainable manufacturing paradigms have been promoted with the deepening of the energy revolution to ensure the competitiveness of enterprises. Moreover, a more urgent and specific methodology for O&M optimization with energy management needs to be constructed. This paper mainly focuses on the energy optimization of O&M activities at the single-machine and production-line levels. In such studies, operations research methods have been applied to different levels of O&M strategies.

At the single-machine level, current energy consumption analysis focuses on the machine itself and the machining process, where the energy consumption of each stage is modeled and analyzed according to the actual physical processing stages of machining. The energy-consumption level of a single machine often depends on multiple factors such as process parameters, machine selection, machine status, tool selection, etc. Among them, the optimization of process parameters has been widely and deeply studied, while less research has been devoted to maintenance and production strategies, and there is still much space for improvement.

At the production-line level, the main consideration is the system consumption structure and the interaction between workpieces and machines, often taking the system productivity and energy consumption as objectives, modeling the problem using optimization methods, and solving it using exact or heuristic algorithms. The research at the production-line level focuses more on the scheduling problems of production and maintenance. Through effective scheduling strategies, scheduling optimization is used to reduce the cost of sustainable production and achieve energy savings without over-changing the system structure.

At the system level, energy consumption optimization is often based on the application of management systems and overall policy considerations. Compared with the above two levels, its optimization is more inclined to the upgrading of management methods and innovation of management modes, and there is still a long way for improvement with practical implementation and application.

Based on the above analysis, we believe that future research trends are mainly reflected in the following points:

1. From the methodology perspective:

    - Unification of the analysis and evaluation system

The current model construction is more individualized, and the actual energy consumption flow and energy consumption transformation relationship of the research object have not yet been comprehensively described clearly, either at the single-machine level or at the production-line level. Therefore, the subsequent research can consider unifying the definition of process energy consumption calculation, defining system boundaries, unifying input and output variables, and finally establishing a unified evaluation system to provide systematic guidance for the research of sub-problems.

- Construction of systematic theory

At present, the optimization at each level is basically isolated. Through the construction of systematic energy consumption optimization theory such as the manufacturing system energy flow model, the relationship of each sub-problem can be clarified and further interaction between more decision variables can be discovered.

- Refinement of management theory

To put these energy-optimization methodologies into practice, we need to rely on the efficient management tools of enterprises. At present, the consideration of energy optimization at the system management level is not sufficient; hence, the promotion of technology application is somewhat limited. Innovations in energy management theory can be made in the future to promote energy-saving methods.

2. From the application perspective:
    - Development of control and management software

As a key driving factor of Industry 4.0, the development of industrial software is an important measure of the manufacturing competitiveness of countries. In a sustainable background, good industrial software can not only improve production efficiency but also achieve energy savings. From the machine level, machines are driven by control software. Therefore, in addition to focusing on production efficiency and quality, the development of control software can put more attention on the combining of the optimization of process parameters and energy consumption. Moreover, from the factory level, the combination of energy management in the MES is also gradually becoming an essential part of the industry site management measures.

- Application of new technologies

With the progress of science and technology, the iteration of machine tools and the upgrading of manufacturing paradigms bring challenges to optimizing energy consumption. Industry 5.0, characterized by being human-centric, resilient, and sustainable, further promotes the industrial transformation toward an environment-friendly manufacturing model. Explorations on clean production technology, energy savings, environmental protection technology, and recycling manufacturing all have great prospects.

7. Conclusions

The problem of energy consumption optimization in manufacturing systems has received attention with the rising concern of sustainability. This paper reviews articles in the field of energy optimization of O&M of manufacturing systems, covering several aspects related to the optimization procedures such as model construction, solution methods, and application areas. The main contributions of this paper are as follows:

1. The current research perspectives on energy optimization of O&M are described.
2. A detailed classification and problem overview of energy-optimization methods of the O&M of manufacturing systems are outlined.
3. The current research framework is summarized and future research trends are proposed.

For the research related to energy optimization, we believe that it has high application value and vast research space. Through reasonable O&M scheduling, energy savings can be achieved without changing production system structures, which will pave the way for the progressive promotion of sustainable manufacturing paradigms. Research in this area is expected to achieve greater attention and policy preferences. The systematic review of this paper can provide some references for future studies.

Author Contributions: Conceptualization, methodology, investigation, writing—original draft preparation, X.A.; editing, visualization, G.S.; validation, formal analysis, Q.L.; writing—review, supervision, project administration, funding acquisition, T.X.; data curation, Y.L.; resources, funding acquisition, R.M. All authors have read and agreed to the published version of the manuscript.

Funding: This research was funded by the National Key R&D Program of China (No. 2022YFF0605700), National Natural Science Foundation of China (No. 51875359 and 71971139), Natural Science Foundation of Shanghai (No. 20ZR1428600), Shanghai Science and Technology Innovation Center for System Engineering of Commercial Aircraft (No. FASE-2021-M7), Oceanic Interdisciplinary Program of Shanghai Jiao Tong University (No. SL2021MS008), and CSSC-SJTU Marine Equipment Forward Looking Innovation Foundation (No. 22B010432).

Data Availability Statement: Not applicable.

Acknowledgments: The authors are indebted to the reviewers and editors for their constructive comments, which greatly improved the contents and exposition of this paper.

Conflicts of Interest: The authors declare no conflict of interest.

References

1. Mitlin, D. Sustainable Development: A Guide to the Literature. *Environ. Urban.* **1992**, *4*, 111–124. [CrossRef]
2. Hopwood, B.; Mellor, M.; O'Brien, G. Sustainable Development: Mapping Different Approaches. *Sustain. Dev.* **2005**, *13*, 38–52. [CrossRef]
3. Oikonomou, V.; Becchis, F.; Steg, L.; Russolillo, D. Energy Saving and Energy Efficiency Concepts for Policy Making. *Energy Policy* **2009**, *37*, 4787–4796. [CrossRef]
4. Tanaka, K. Review of Policies and Measures for Energy Efficiency in Industry Sector. *Energy Policy* **2011**, *39*, 6532–6550. [CrossRef]
5. IEA. Energy Statistics Data Browser, IEA, Paris. 2022. Available online: https://www.iea.org/data-and-statistics/data-tools/energy-statistics-data-browser (accessed on 15 August 2022).
6. Yadav, G.; Kumar, A.; Luthra, S.; Garza-Reyes, J.A.; Kumar, V.; Batista, L. A Framework to Achieve Sustainability in Manufacturing Organisations of Developing Economies Using Industry 4.0 Technologies' Enablers. *Comput. Ind.* **2020**, *122*, 103280. [CrossRef]
7. Ke, J.; Price, L.; Ohshita, S.; Fridley, D.; Khanna, N.Z.; Zhou, N.; Levine, M. China's Industrial Energy Consumption Trends and Impacts of the Top-1000 Enterprises Energy-Saving Program and the Ten Key Energy-Saving Projects. *Energy Policy* **2012**, *50*, 562–569. [CrossRef]
8. Hesselbach, J.; Herrmann, C.; Detzer, R.; Thiede, L.M.S.; Hesselbach, J.; Herrmann, C.; Detzer, R.; Martin, L.; Thiede, S.; Lüdemann, B. Energy Efficiency through Optimized Coordination of Production and Technical Building Services. In Proceedings of the 15th Conference on Life Cycle Engineering, Sydney, Australia, 17–19 March 2008; pp. 624–628.
9. Smarra, F.; Jain, A.; de Rubeis, T.; Ambrosini, D.; D'Innocenzo, A.; Mangharam, R. Data-Driven Model Predictive Control Using Random Forests for Building Energy Optimization and Climate Control. *Appl. Energy* **2018**, *226*, 1252–1272. [CrossRef]
10. González-Briones, A.; De La Prieta, F.; Mohamad, M.S.; Omatu, S.; Corchado, J.M. Multi-Agent Systems Applications in Energy Optimization Problems: A State-of-the-Art Review. *Energies* **2018**, *11*, 1928. [CrossRef]
11. Xia, T.; Dong, Y.; Pan, E.; Zheng, M.; Wang, H.; Xi, L. Fleet-Level Opportunistic Maintenance for Large-Scale Wind Farms Integrating Real-Time Prognostic Updating. *Renew. Energy* **2021**, *163*, 1444–1454. [CrossRef]
12. Park, C.-W.; Kwon, K.-S.; Kim, W.-B.; Min, B.-K.; Park, S.-J.; Sung, I.-H.; Yoon, Y.S.; Lee, K.-S.; Lee, J.-H.; Seok, J. Energy Consumption Reduction Technology in Manufacturing—A Selective Review of Policies, Standards, and Research. *Int. J. Precis. Eng. Manuf.* **2009**, *10*, 151–173. [CrossRef]
13. Jasiulewicz-Kaczmarek, M.; Gola, A. Maintenance 4.0 Technologies for Sustainable Manufacturing—An Overview. *IFAC-PapersOnLine* **2019**, *52*, 91–96. [CrossRef]
14. Gahm, C.; Denz, F.; Dirr, M.; Tuma, A. Energy-Efficient Scheduling in Manufacturing Companies: A Review and Research Framework. *Eur. J. Oper. Res.* **2016**, *248*, 744–757. [CrossRef]
15. Reich-Weiser, C.; Vijayaraghavan, A.; Dornfeld, D. Appropriate Use of Green Manufacturing Frameworks. In Proceedings of the 17th CIRP International Conference on Life Cycle Engineering, Hefei, China, 19–21 May 2010.
16. Yang, Z.; Djurdjanovic, D.; Ni, J. Maintenance Scheduling in Manufacturing Systems Based on Predicted Machine Degradation. *J. Intell. Manuf.* **2008**, *19*, 87–98. [CrossRef]
17. Cui, W.-W.; Lu, Z.; Pan, E. Integrated Production Scheduling and Maintenance Policy for Robustness in a Single Machine. *Comput. Oper. Res.* **2014**, *47*, 81–91. [CrossRef]
18. Sortrakul, N.; Nachtmann, H.L.; Cassady, C.R. Genetic Algorithms for Integrated Preventive Maintenance Planning and Production Scheduling for a Single Machine. *Comput. Ind.* **2005**, *56*, 161–168. [CrossRef]
19. Lu, B.; Zhou, X.; Li, Y. Joint Modeling of Preventive Maintenance and Quality Improvement for Deteriorating Single-Machine Manufacturing Systems. *Comput. Ind. Eng.* **2016**, *91*, 188–196. [CrossRef]
20. Yin, H.; Zhang, G.; Zhu, H.; Deng, Y.; He, F. An Integrated Model of Statistical Process Control and Maintenance Based on the Delayed Monitoring. *Reliab. Eng. Syst. Saf.* **2015**, *133*, 323–333. [CrossRef]

21. Shojaeinasab, A.; Charter, T.; Jalayer, M.; Khadivi, M.; Ogunfowora, O.; Raiyani, N.; Yaghoubi, M.; Najjaran, H. Intelligent Manufacturing Execution Systems: A Systematic Review. *J. Manuf. Syst.* **2022**, *62*, 503–522. [CrossRef]
22. Hu, J.; Jiang, Z.; Liao, H. Joint Optimization of Job Scheduling and Maintenance Planning for a Two-Machine Flow Shop Considering Job-Dependent Operating Condition. *J. Manuf. Syst.* **2020**, *57*, 231–241. [CrossRef]
23. Gu, X.; Jin, X.; Guo, W.; Ni, J. Estimation of Active Maintenance Opportunity Windows in Bernoulli Production Lines. *J. Manuf. Syst.* **2017**, *45*, 109–120. [CrossRef]
24. Xia, T.; Dong, Y.; Xiao, L.; Du, S.; Pan, E.; Xi, L. Recent Advances in Prognostics and Health Management for Advanced Manufacturing Paradigms. *Reliab. Eng. Syst. Saf.* **2018**, *178*, 255–268. [CrossRef]
25. Gutowski, T.; Dahmus, J.; Thiriez, A. Electrical Energy Requirements for Manufacturing Processes. In Proceedings of the 13th CIRP International Conference on Life Cycle Engineering, Leuven, Belgium, 31 May–2 June 2006; Volume 31, pp. 623–638.
26. Guo, Y.; Loenders, J.; Duflou, J.; Lauwers, B. Optimization of Energy Consumption and Surface Quality in Finish Turning. *Procedia CIRP* **2012**, *1*, 512–517. [CrossRef]
27. Zhong, Q.; Tang, R.; Peng, T. Decision Rules for Energy Consumption Minimization during Material Removal Process in Turning. *J. Clean. Prod.* **2017**, *140*, 1819–1827. [CrossRef]
28. Tuo, J.; Liu, F.; Liu, P. Key Performance Indicators for Assessing Inherent Energy Performance of Machine Tools in Industries. *Int. J. Prod. Res.* **2019**, *57*, 1811–1824. [CrossRef]
29. Wang, H.; Zhong, R.Y.; Liu, G.; Mu, W.; Tian, X.; Leng, D. An Optimization Model for Energy-Efficient Machining for Sustainable Production. *J. Clean. Prod.* **2019**, *232*, 1121–1133. [CrossRef]
30. Li, W.; Zein, A.; Kara, S.; Herrmann, C. An Investigation into Fixed Energy Consumption of Machine Tools. In Proceedings of the 18th CIRP International Conference on Life Cycle Engineering, Braunschweig, Germany, 2–4 May 2011; Springer: Berlin/Heidelberg, Germany, 2011; pp. 268–273. [CrossRef]
31. Gutowski, T.; Murphy, C.; Allen, D.; Bauer, D.; Bras, B.; Piwonka, T.; Sheng, P.; Sutherland, J.; Thurston, D.; Wolff, E. Environmentally Benign Manufacturing: Observations from Japan, Europe and the United States. *J. Clean. Prod.* **2005**, *13*, 1–17. [CrossRef]
32. Xu, K.; Luo, M.; Tang, K. Machine Based Energy-Saving Tool Path Generation for Five-Axis End Milling of Freeform Surfaces. *J. Clean. Prod.* **2016**, *139*, 1207–1223. [CrossRef]
33. Wang, Q.; Liu, F.; Wang, X. Multi-Objective Optimization of Machining Parameters Considering Energy Consumption. *Int. J. Adv. Manuf. Technol.* **2014**, *71*, 1133–1142. [CrossRef]
34. Ma, F.; Zhang, H.; Cao, H.; Hon, K.K.B. An Energy Consumption Optimization Strategy for CNC Milling. *Int. J. Adv. Manuf. Technol.* **2017**, *90*, 1715–1726. [CrossRef]
35. Bi, Z.M.; Wang, L. Energy Modeling of Machine Tools for Optimization of Machine Setups. *IEEE Trans. Autom. Sci. Eng.* **2012**, *9*, 607–613. [CrossRef]
36. Bilga, P.S.; Singh, S.; Kumar, R. Optimization of Energy Consumption Response Parameters for Turning Operation Using Taguchi Method. *J. Clean. Prod.* **2016**, *137*, 1406–1417. [CrossRef]
37. Campatelli, G.; Lorenzini, L.; Scippa, A. Optimization of Process Parameters Using a Response Surface Method for Minimizing Power Consumption in the Milling of Carbon Steel. *J. Clean. Prod.* **2014**, *66*, 309–316. [CrossRef]
38. He, K.; Tang, R.; Jin, M. Pareto Fronts of Machining Parameters for Trade-off among Energy Consumption, Cutting Force and Processing Time. *Int. J. Prod. Econ.* **2017**, *185*, 113–127. [CrossRef]
39. Zhou, G.; Yuan, S.; Lu, Q.; Xiao, X. A Carbon Emission Quantitation Model and Experimental Evaluation for Machining Process Considering Tool Wear Condition. *Int. J. Adv. Manuf. Technol.* **2018**, *98*, 565–577. [CrossRef]
40. Chen, X.; Li, C.; Tang, Y.; Li, L.; Du, Y.; Li, L. Integrated Optimization of Cutting Tool and Cutting Parameters in Face Milling for Minimizing Energy Footprint and Production Time. *Energy* **2019**, *175*, 1021–1037. [CrossRef]
41. Shi, K.N.; Zhang, D.H.; Liu, N.; Wang, S.B.; Ren, J.X.; Wang, S.L. A Novel Energy Consumption Model for Milling Process Considering Tool Wear Progression. *J. Clean. Prod.* **2018**, *184*, 152–159. [CrossRef]
42. Tian, C.; Zhou, G.; Zhang, J.; Zhang, C. Optimization of Cutting Parameters Considering Tool Wear Conditions in Low-Carbon Manufacturing Environment. *J. Clean. Prod.* **2019**, *226*, 706–719. [CrossRef]
43. Xie, N.; Zhou, J.; Zheng, B. Selection of Optimum Turning Parameters Based on Cooperative Optimization of Minimum Energy Consumption and High Surface Quality. *Procedia CIRP* **2018**, *72*, 1469–1474. [CrossRef]
44. Yan, J.; Li, L. Multi-Objective Optimization of Milling Parameters—The Trade-Offs between Energy, Production Rate and Cutting Quality. *J. Clean. Prod.* **2013**, *52*, 462–471. [CrossRef]
45. Verma, A.; Rai, R. Energy Efficient Modeling and Optimization of Additive Manufacturing Process. In Proceedings of the 2013 International Solid Freeform Fabrication Symposium, University of Texas at Austin, Austin, TX, USA, 12–14 August 2013. [CrossRef]
46. Barlow, R.; Larry, H. Optimum Preventive Maintenance Policies. *Oper. Res.* **1960**, *8*, 90–100. [CrossRef]
47. Yan, J.; Hua, D. Energy Consumption Modeling for Machine Tools after Preventive Maintenance. In Proceedings of the 2010 IEEE International Conference on Industrial Engineering and Engineering Management, Macao, China, 7–10 December 2010; pp. 2201–2205. [CrossRef]
48. Hoang, A.; Do, P.; Iung, B. Investigation on the Use of Energy Efficiency for Condition-Based Maintenance Decision-Making. *IFAC-PapersOnLine* **2016**, *49*, 73–78. [CrossRef]

49. Xia, T.; Shi, G.; Si, G.; Du, S.; Xi, L. Energy-Oriented Joint Optimization of Machine Maintenance and Tool Replacement in Sustainable Manufacturing. *J. Manuf. Syst.* **2021**, *59*, 261–271. [CrossRef]

50. Xu, W.; Cao, L. Optimal Maintenance Control of Machine Tools for Energy Efficient Manufacturing. *Int. J. Adv. Manuf. Technol.* **2019**, *104*, 3303–3311. [CrossRef]

51. Wu, D.; Han, R.; Ma, Y.; Yang, L.; Wei, F.; Peng, R. A Two-Dimensional Maintenance Optimization Framework Balancing Hazard Risk and Energy Consumption Rates. *Comput. Ind. Eng.* **2022**, *169*, 108193. [CrossRef]

52. Che, A.; Wu, X.; Peng, J.; Yan, P. Energy-Efficient Bi-Objective Single-Machine Scheduling with Power-down Mechanism. *Comput. Oper. Res.* **2017**, *85*, 172–183. [CrossRef]

53. Chen, B.; Zhang, X. Scheduling with Time-of-Use Costs. *Eur. J. Oper. Res.* **2019**, *274*, 900–908. [CrossRef]

54. Zhou, S.; Jin, M.; Du, N. Energy-Efficient Scheduling of a Single Batch Processing Machine with Dynamic Job Arrival Times. *Energy* **2020**, *209*, 118420. [CrossRef]

55. Mouzon, G.; Yildirim, M.B.; Twomey, J. Operational Methods for Minimization of Energy Consumption of Manufacturing Equipment. *Int. J. Prod. Res.* **2007**, *45*, 4247–4271. [CrossRef]

56. Aghelinejad, M.; Ouazene, Y.; Yalaoui, A. Production Scheduling Optimisation with Machine State and Time-Dependent Energy Costs. *Int. J. Prod. Res.* **2018**, *56*, 5558–5575. [CrossRef]

57. Wang, S.; Liu, M.; Chu, F.; Chu, C. Bi-Objective Optimization of a Single Machine Batch Scheduling Problem with Energy Cost Consideration. *J. Clean. Prod.* **2016**, *137*, 1205–1215. [CrossRef]

58. Xu, W.; Cao, L. Energy Efficiency Analysis of Machine Tools with Periodic Maintenance. *Int. J. Prod. Res.* **2014**, *52*, 5273–5285. [CrossRef]

59. Bruns, P. Optimal Maintenance Strategies for Systems with Partial Repair Options and without Assuming Bounded Costs. *Eur. J. Oper. Res.* **2002**, *139*, 146–165. [CrossRef]

60. Feng, H.; Xi, L.; Xiao, L.; Xia, T.; Pan, E. Imperfect Preventive Maintenance Optimization for Flexible Flowshop Manufacturing Cells Considering Sequence-Dependent Group Scheduling. *Reliab. Eng. Syst. Saf.* **2018**, *176*, 218–229. [CrossRef]

61. Dekkert, R.; Smit, A.; Losekoot, J. Combining Maintenance Activities in an Operational Planning Phase: A Set-Partitioning Approach. *IMA J. Manag. Math.* **1991**, *3*, 315–331. [CrossRef]

62. Zhou, B.; Qi, Y.; Liu, Y. Proactive Preventive Maintenance Policy for Buffered Serial Production Systems Based on Energy Saving Opportunistic Windows. *J. Clean. Prod.* **2020**, *253*, 119791. [CrossRef]

63. Zhou, X.; Lu, Z.; Xi, L. A Dynamic Opportunistic Preventive Maintenance Policy for Multi-Unit Series Systems with Intermediate Buffers. *Int. J. Ind. Syst. Eng.* **2010**, *6*, 276–288. [CrossRef]

64. Xia, T.; Jin, X.; Xi, L.; Ni, J. Production-Driven Opportunistic Maintenance for Batch Production Based on MAM–APB Scheduling. *Eur. J. Oper. Res.* **2015**, *240*, 781–790. [CrossRef]

65. Zou, J.; Arinez, J.; Chang, Q.; Lei, Y. Opportunity Window for Energy Saving and Maintenance in Stochastic Production Systems. *J. Manuf. Sci. Eng.* **2016**, *138*, 121009. [CrossRef]

66. Zhou, B.; Yi, Q. An Energy-Oriented Maintenance Policy under Energy and Quality Constraints for a Multielement-Dependent Degradation Batch Production System. *J. Manuf. Syst.* **2021**, *59*, 631–645. [CrossRef]

67. Xia, T.; Si, G.; Shi, G.; Zhang, K.; Xi, L. Optimal Selective Maintenance Scheduling for Series–Parallel Systems Based on Energy Efficiency Optimization. *Appl. Energy* **2022**, *314*, 118927. [CrossRef]

68. Sun, Z.; Dababneh, F.; Li, L. Joint Energy, Maintenance, and Throughput Modeling for Sustainable Manufacturing Systems. *IEEE Trans. Syst. Man Cybern.* **2020**, *50*, 2101–2112. [CrossRef]

69. An, Y.; Chen, X.; Zhang, J.; Li, Y. A Hybrid Multi-Objective Evolutionary Algorithm to Integrate Optimization of the Production Scheduling and Imperfect Cutting Tool Maintenance Considering Total Energy Consumption. *J. Clean. Prod.* **2020**, *268*, 121540. [CrossRef]

70. Artigues, C.; Lopez, P.; Haït, A. The Energy Scheduling Problem: Industrial Case-Study and Constraint Propagation Techniques. *Int. J. Prod. Econ.* **2013**, *143*, 13–23. [CrossRef]

71. Módos, I.; Šucha, P.; Hanzálek, Z. On Parallel Dedicated Machines Scheduling under Energy Consumption Limit. *Comput. Ind. Eng.* **2021**, *159*, 107209. [CrossRef]

72. Masmoudi, O.; Delorme, X.; Gianessi, P. Job-Shop Scheduling Problem with Energy Consideration. *Int. J. Prod. Econ.* **2019**, *216*, 12–22. [CrossRef]

73. Zhang, R.; Chiong, R. Solving the Energy-Efficient Job Shop Scheduling Problem: A Multi-Objective Genetic Algorithm with Enhanced Local Search for Minimizing the Total Weighted Tardiness and Total Energy Consumption. *J. Clean. Prod.* **2016**, *112*, 3361–3375. [CrossRef]

74. Wang, S.; Lu, X.; Li, X.X.; Li, W.D. A Systematic Approach of Process Planning and Scheduling Optimization for Sustainable Machining. *J. Clean. Prod.* **2015**, *87*, 914–929. [CrossRef]

75. Han, D.; Tang, Q.; Zhang, Z.; Cao, J. Energy-Efficient Integration Optimization of Production Scheduling and Ladle Dispatching in Steelmaking Plants. *IEEE Access* **2020**, *8*, 176170–176187. [CrossRef]

76. Palensky, P.; Dietrich, D. Demand Side Management: Demand Response, Intelligent Energy Systems, and Smart Loads. *IEEE Trans. Ind. Inform.* **2011**, *7*, 381–388. [CrossRef]

77. Cheng, J.; Chu, F.; Xia, W.; Ding, J.; Ling, X. Bi-Objective Optimization for Single-Machine Batch Scheduling Considering Energy Cost. In *Proceedings of the 2014 International Conference on Control, Decision and Information Technologies, CoDIT 2014, Metz, France, 3–5 November 2014*; Institute of Electrical and Electronics Engineers Inc.: Piscataway, NJ, USA, 2014; pp. 236–241. [CrossRef]
78. Shrouf, F.; Ordieres-Meré, J.; García-Sánchez, A.; Ortega-Mier, M. Optimizing the Production Scheduling of a Single Machine to Minimize Total Energy Consumption Costs. *J. Clean. Prod.* **2014**, *67*, 197–207. [CrossRef]
79. Fang, K.; Uhan, N.A.; Zhao, F.; Sutherland, J.W. Scheduling on a Single Machine under Time-of-Use Electricity Tariffs. *Ann. Oper. Res.* **2016**, *238*, 199–227. [CrossRef]
80. Wang, S.; Zhu, Z.; Fang, K.; Chu, F.; Chu, C. Scheduling on a Two-Machine Permutation Flow Shop under Time-of-Use Electricity Tariffs. *Int. J. Prod. Res.* **2018**, *56*, 3173–3187. [CrossRef]
81. Tan, M.; Duan, B.; Su, Y. Economic Batch Sizing and Scheduling on Parallel Machines under Time-of-Use Electricity Pricing. *Oper. Res.* **2018**, *18*, 105–122. [CrossRef]
82. Schulz, S.; Neufeld, J.S.; Buscher, U. A Multi-Objective Iterated Local Search Algorithm for Comprehensive Energy-Aware Hybrid Flow Shop Scheduling. *J. Clean. Prod.* **2019**, *224*, 421–434. [CrossRef]
83. Huang, J.; Chang, Q.; Arinez, J.; Xiao, G. A Maintenance and Energy Saving Joint Control Scheme for Sustainable Manufacturing Systems. *Procedia CIRP* **2019**, *80*, 263–268. [CrossRef]
84. Gong, G.; Chiong, R.; Deng, Q.; Han, W.; Zhang, L.; Huang, D. Energy-Efficient Production Scheduling through Machine on/off Control during Preventive Maintenance. *Eng. Appl. Artif. Intell.* **2021**, *104*, 104359. [CrossRef]
85. Fernandez, M.; Li, L.; Sun, Z. "Just-for-Peak" Buffer Inventory for Peak Electricity Demand Reduction of Manufacturing Systems. *Int. J. Prod. Econ.* **2013**, *146*, 178–184. [CrossRef]
86. Li, L.; Sun, Z. Dynamic Energy Control for Energy Efficiency Improvement of Sustainable Manufacturing Systems Using Markov Decision Process. *IEEE Trans. Syst. Man Cybern.* **2013**, *43*, 1195–1205. [CrossRef]
87. Wang, J.; Fei, Z.; Chang, Q.; Li, S. Energy Saving Operation of Manufacturing System Based on Dynamic Adaptive Fuzzy Reasoning Petri Net. *Energies* **2019**, *12*, 2216. [CrossRef]
88. Karnouskos, S.; Colombo, A.W.; Martinez Lastra, J.L.; Popescu, C. Towards the Energy Efficient Future Factory. In Proceedings of the 2009 7th IEEE International Conference on Industrial Informatics, Cardiff, Wales, UK, 23–26 June 2009; pp. 367–371. [CrossRef]
89. Kamble, S.S.; Gunasekaran, A.; Gawankar, S.A. Sustainable Industry 4.0 Framework: A Systematic Literature Review Identifying the Current Trends and Future Perspectives. *Process. Saf. Environ.* **2018**, *117*, 408–425. [CrossRef]
90. Ebrahimi, A.; Jeon, H.W.; Lee, S.; Wang, C. Minimizing Total Energy Cost and Tardiness Penalty for a Scheduling-Layout Problem in a Flexible Job Shop System: A Comparison of Four Metaheuristic Algorithms. *Comput. Ind. Eng.* **2020**, *141*, 106295. [CrossRef]
91. Gourlis, G.; Kovacic, I. Building Information Modelling for Analysis of Energy Efficient Industrial Buildings—A Case Study. *Renew. Sustain. Energy Rev.* **2017**, *68*, 953–963. [CrossRef]
92. Macrina, G.; Laporte, G.; Guerriero, F.; Di Puglia Pugliese, L. An Energy-Efficient Green-Vehicle Routing Prob-lem with Mixed Vehicle Fleet, Partial Battery Recharging and Time Windows. *Eur. J. Oper. Res.* **2019**, *276*, 971–982. [CrossRef]
93. Hooshmand, F.; MirHassani, S.A. Time Dependent Green VRP with Alternative Fuel Powered Vehicles. *Energy Syst.* **2019**, *10*, 721–756. [CrossRef]

Article

A Mixed Algorithm for Integrated Scheduling Optimization in AS/RS and Hybrid Flowshop

Jiansha Lu, Lili Xu, Jinghao Jin and Yiping Shao *

College of Mechanical Engineering, Zhejiang University of Technology, Hangzhou 310023, China
* Correspondence: syp123gh@zjut.edu.cn

Abstract: The integrated scheduling problem in automated storage and retrieval systems (AS/RS) and the hybrid flowshop is critical for the realization of lean logistics and just-in-time distribution in manufacturing systems. The bi-objective model that minimizes the operation time in AS/RS and the makespan in the hybrid flowshop is established to optimize the problem. A mixed algorithm, named GA-MBO algorithm, is proposed to solve the model, which combines the advantages of the strong global optimization ability of genetic algorithm (GA) and the strong local search ability of migratory birds optimization (MBO). To avoid useless solutions, different cross operations of storage and retrieval tasks are designed. Compared with three algorithms, including improved genetic algorithm, improved particle swam optimization, and a hybrid algorithm of GA and particle swam optimization, the experimental results showed that the GA-MBO algorithm improves the operation efficiency by 9.48%, 19.54%, and 5.12% and the algorithm robustness by 35.16%, 54.42%, and 39.38%, respectively, which further verified the effectiveness of the proposed algorithm. The comparative analysis of the bi-objective experimental results fully reflects the superiority of integrated scheduling optimization.

Keywords: automated storage and retrieval system; hybrid flowshop; genetic algorithm; migratory birds optimization algorithm; GA-MBO

Citation: Lu, J.; Xu, L.; Jin, J.; Shao, Y. A Mixed Algorithm for Integrated Scheduling Optimization in AS/RS and Hybrid Flowshop. *Energies* **2022**, *15*, 7558. https://doi.org/10.3390/en15207558

Academic Editor: George S. Stavrakakis

Received: 30 August 2022
Accepted: 10 October 2022
Published: 13 October 2022

Publisher's Note: MDPI stays neutral with regard to jurisdictional claims in published maps and institutional affiliations.

1. Introduction

With the development of intelligent logistics equipment and technology, automated storage and retrieval systems (AS/RSs) are widely used in manufacturing enterprise logistics due to the advantages of high efficiency and space utilization [1,2]. The use of AS/RS in the factory warehouse is conducted with the just-in-time (JIT) contribution of various materials in the workshop. To realize the JIT distribution, the AS/RS scheduling must be closely connected with the production scheduling, which affects the efficiency of the integrated scheduling. In addition, the different scheduling optimization objectives in the AS/RS and hybrid flowshop are to maximize the efficiency of storage and retrieval and minimize the makespan. With a high warehousing efficiency, the completion time of the producing material distribution may not meet the required demand or may be unable to greatly increase the line inventory. However, the production scheduling with the minimum completion time may cause the warehouse scheduling tasks to be stacked at a certain time and cannot meet the production demand. Therefore, it is important to cooperate with the scheduling that contains the storage allocation, task sequences, and retrieval task sequences in production, which is a significant practice in manufacturing systems.

At present, the main research in AS/RS concerns storage allocation and task scheduling [3–5]. Roshan et al. [6] formulated a multi-objective model in AS/RS considering energy consumption optimization and energy sustainability. Hachemi et al. [7] determined the integration of the storage allocation and picking paths based on storage and retrieval requests with the objective of travel time. Song and Mu [8] studied the sequence sorting problem with large-scale storage/retrieval requests in AS/RS and proposed a heuristic algorithm based on assignment. Geng et al. [9] proposed a new improved Genetic Algorithm (GA) to solve the scheduling problem in AS/RS, and indicated that it is an effective approach. Wu et al. [10] investigated the scheduling problem of retrieval jobs in double-deep type AS/RS, whose objective is to minimize the working distance.

In the hybrid flowshop, Colak et al. [11] conducted a systematic literature survey for hybrid flowshop scheduling problems, which provided a beneficial road map for the following researchers. Zhang et al. [12] proposed a muti-objective migratory birds optimization (MBO) algorithm based on the decomposition of the multi-objective flowshop rescheduling problem, which has proven to be better than other evolutionary algorithms. Zhang et al. [13] introduced lots of streaming into the hybrid flowshop scheduling problem with consistent sublots to fit the real-world scenarios, which verified the feasibility to solve the integrated scheduling problem with the hybrid flowshop. Li et al. [14] researched the distributed hybrid flowshop scheduling problem with sequenced dependent setup time, which was solved by a discrete artificial bee colony algorithm. Reddy et al. [15] solved a muti-objective problem in a flexible manufacturing system, which considered machine and vehicle scheduling. According to these studies in AS/RS and hybrid flowshop, most researchers have ignored the related effects. However, the connection between AS/RS and hybrid flowshop is important to develop smart manufacturing systems.

Problems with storage allocation, operation scheduling, and workshop scheduling are NP-hard problems that lead to low efficiency and time-consumption by using exact algorithms [16,17]. The intelligent optimization algorithm provides an effective and fast method for solving the above complex problems [18–20]. Li et al. [21] conducted a comprehensive survey of the learning-based intelligent algorithm. Katoch et al. [22] discussed the advances and introduced the pros and cons of GA. Duman et al. [23] proposed a new nature-inspired metaheuristic approach named MBO, which was proven to be an effective formation in energy saving. The algorithms of GA, MBO, and their improvement algorithms are used to solve the scheduling problems and achieve better performance. The scheduling objective in AS/RS is usually the total operation time, but it has a shortcoming by neglecting two situations. The first one is that the number of storage tasks is not equal to the number of retrieval tasks, and the other one is that retrieval arrival time needs to be considered in the connection of AS/RS and hybrid flowshop [24,25]. To obtain a better solution to the integrated scheduling problem in AS/RS and hybrid flowshop, the main contributions in this study include: (1) formulation of a bi-objective model to minimize the operation time in AS/RS and the makespan in the hybrid flowshop; (2) proposal of a two-stage mixed algorithm called GA-MBO that combines the strong global optimization ability of GA and the strong local search ability of MBO.

The structure of this paper is as follows: In Section 2, we described and modeled the integrated scheduling problem in AS/RS and hybrid flowshop. In Section 3, we designed the GA-MBO and presented the operation details. In Section 4, experiments were introduced in AS/RS and hybrid flowshop, and the results were analyzed to verify the effectiveness of the GA-MBO. Finally, Section 5 presents the conclusions.

2. Materials and Methods

2.1. Problem Description

AS/RS is an information technology based on the Internet of Things, which is widely used to store and retrieve materials without any human participation. An AS/RS mainly consists of racks, cranes, input/output (I/O) points, and conveyors. The system has a crane in each aisle and a main I/O point along a conveyor. The top view of the layout schematic diagram is shown in Figure 1. The redesign of the material distribution connection in the AS/RS and hybrid flowshop saves the storage area and lowers the handling cost of the production material. The crane can be used to realize the distribution between different tiers. Under the production conditions of the workshop, intelligent logistics equipment, such as conveyors and automatic guided vehicles (AGVs), can be used to distribute materials with JIT to achieve lean logistics. The material distribution problem with the connection between the AS/RS and the hybrid flowshop is researched. The material distribution diagram is shown in Figure 2.

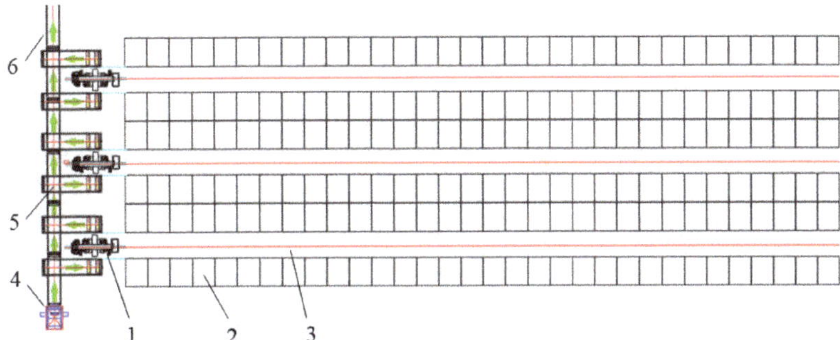

Figure 1. Layout schematic diagram of AS/RS. 1. Crane. 2. Rack. 3. Aisle. 4. Input point. 5. Conveyor. 6. Output point.

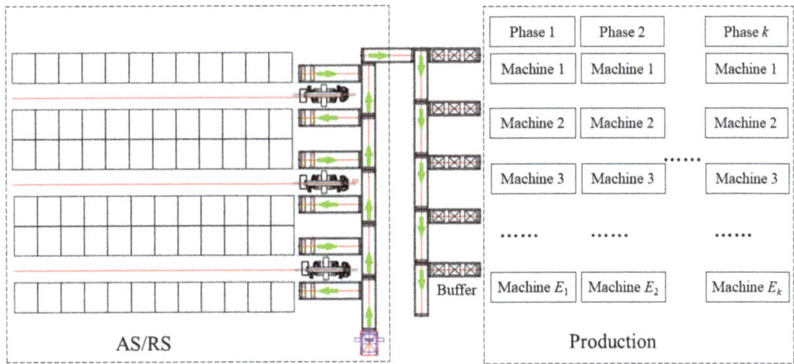

Figure 2. Material distribution diagram in AS/RS and hybrid flowshop.

The integrated scheduling problem in AS/RS and hybrid flowshop is related to the storage allocation, system scheduling, and flowshop scheduling. The objective of the scheduling problem is to minimize the total operation time in AS/RS and the makespan in hybrid flowshop. The research objects are AS/RS and multi-stage hybrid flowshop.

AS/RS can be described as: in a warehouse with a determined status that racks have X rows, Y columns, and Z tiers. The corresponding material locations are the O retrieval task locations, and the number of retrieval tasks is greater than the number of the retrieval material types. The free rack locations are the I retrieval task locations. These storage and retrieval tasks are operated by S cranes, whose operation time is related to task storage allocations (multi-tasks with the same material) and the task operation sequence. Hybrid flowshop can be described as: the O retrieval tasks with the same operation sequence are processed at the K stages. Stage k has $E_k > 1$ independent parallel machines. At this stage, the task processing time is related to both the task material type and the processing machine type. The end operation time of the retrieval task in AS/RS directly affects the starting time of the production task in the hybrid flow workshop scheduling.

In the scheduling of AS/RS, operation modes for storage and retrieval goods are: single command (SC) where the crane completes for storage task or retrieval task and double command (DC) where the crane completes for storage and retrieval tasks. The DC operation should be adopted to improve the efficiency in the system. At the same time, considering the deadline requirements of retrieval tasks, the SC has a shorter retrieval time.

In addition, the mixed command of SC and DC is researched in the operation scheduling problem, which exists under the assumption that the numbers of storage and retrieval tasks are unequal. These situations form a mixed scheduling sequence.

The alternation of storage and retrieval tasks leads to the dwell-point change that affects the task operation time. The location selections of storage and retrieval tasks affect the operation sequence, the system efficiency, and the end operation time of retrieval tasks. The retrieval sequence affects the product sequence and the production efficiency in the hybrid flowshop. This paper integrates and optimizes a batch of storage and retrieval tasks in AS/RS by determining the storage location so that the total operation time in AS/RS and the makespan in workshop production are minimized. Storage selection, task sequences, and production sequences need to be studied at the same time.

2.2. Problem Modelling

2.2.1. Assumption

In the integrated scheduling optimization problem in AS/RS and hybrid flowshop, the main assumptions to simplify the problem under research and the formulated model are:

- An AS/RS has X rows, Y columns, and Z tiers, and the rack coordinates of x row, y column, and z tier can be expressed as (x, y, z). The input point is at $(0, 0, 1)$, and the output point is at $(X + 1, Y + 1, 1)$.
- In the AS/RS storage racks which have the same size. A rack stores one bin or one pallet, and the crane can only load one bin or one pallet at a time.
- The velocity of a crane in the horizontal direction is v_y and in the vertical direction is v_z. Movements in the two directions are independent. The start-up time and braking time of crane can be ignored, while the picking time for any location is fixed.
- The crane stays in position after the task is completed.
- Considering the storage period of the material in the warehouse, it is not allowed to directly check out without checking.
- The equipment failure is not considered during the operation, and the crane is not allowed to interrupt the task.
- The equipment processing in the hybrid flowshop doesn't consider production preparation time.
- The buffer in the equipment room has infinite capacity.

2.2.2. Notation

In order to describe the problem and build a model by a better way, the notations are as follows:

X	Number of rows, $x = 1, 2, \ldots, X$
Y	Number of columns, $y = 1, 2, \ldots, Y$
Z	Number of tiers, $z = 1, 2, \ldots, Z$
J	Number of racks, $j = 1, 2, \ldots, J$
ZJ	Material locations
L	Rack's length
W	Rack's width
H	Rack's high
U	Aisle's width
S	Number of cranes, $s = 1, 2, \ldots, S$
v_y	Velocity of the crane in the horizontal direction
v_z	Velocity of the crane in the vertical direction
C	Load/Unload time of crane
p	Material types, $p = 1, 2, \ldots, p$
N_p	Number of in stock materials p in stock, $n = 1, 2, \ldots, N_p$
I_p	Number of retrieval materials p in stock, $i = 1, 2, \ldots, I_p$
K	Number of production stages, $k = 1, 2, \ldots, K$
E_k	Number of machines at stage k, $e = 1, 2, \ldots, E_k$
T_{pke}	Operation time of material p at stage K in machine e

R_e	Transportation time of output point to machine e
Ω	Large positive number
st_{ns}	Start time of task n in crane s
ct_{ns}	End time of task n in crane s
st_{noke}	Start time of task n at output point o and stage k in machine e
ct_{noke}	End time of task n at output point o and stage k in machine e
α_{nj}	1 if the location of task n is j, otherwise is 0
$\beta_{nn's}$	1 if the task n completes before task n' in crane s, otherwise is 0
$\chi_{nn'ke}$	1 if the task n completes before task n' at stage k in machine e, otherwise is 0

2.2.3. Objective Function

The common operational efficiency in the AS/RS scheduling problem is to measure the operation time, and in the hybrid flowshop scheduling problem, to measure the makespan. The integrated scheduling optimization of these two scheduling problems presents a conflict that needs to be evaluated in model objects.

The crane operates retrieval tasks from the dwell point to the location of the retrieval task and then moves to the output point in the aisle, which becomes a new dwell point. The retrieval operation time of the crane at SC is:

$$ct_{ns} = st_{ns} + 2C + \max\left(\frac{(y_o - y_s) \cdot W}{v_y}, \frac{(z_o - z_s) \cdot H}{v_z}\right) + \max\left(\frac{y_o \cdot W}{v_y}, \frac{(z_o - 1) \cdot H}{v_z}\right) \quad (1)$$

The crane operates storage tasks from the dwell point to the aisle output point to pick up goods and then moves to the storage task location, which becomes a new dwell point. The storage operation time of the crane at SC is:

$$ct_{ns} = st_{ns} + 2C + \max\left(\frac{y_s \cdot W}{v_y}, \frac{(z_s - 1) \cdot H}{v_z}\right) + \max\left(\frac{y_o \cdot W}{v_y}, \frac{(z_o - 1) \cdot H}{v_z}\right) \quad (2)$$

In the DC operation of the crane, the dwell point of input is the output point, which reduces the operation time. The objective function of operation time in AS/RS is:

$$f_1 = \sum_{s=1}^{S} \max(ct_{ns}) \quad (3)$$

The end operation time is the task completion time at the last production stage in hybrid flowshop. The objective function of the makespan is:

$$f_2 = \max(ct_{noke}) \quad (4)$$

To eliminate the influence of different objective dimensions, the above two evaluation objective functions need to be normalized as:

$$f(x) = \frac{f(x) - \min f(x) + 0.001}{\max f(x) - \min f(x) + 0.001} \quad (5)$$

Normalization needs to determine the extreme value of the objective function. The researched problem is NP-hard and it is difficult to obtain the exact solution for which optimization can be obtained by the single-objective function. The weight coefficient method converts the multi-objective optimization into a single-objective description as:

$$F = w_1 \cdot f_1 + w_2 \cdot f_2 \quad (6)$$

The total operating time of the warehouse and the makespan in production are the two study targets of the study. The primary goal in enterprise management is to increase production efficiency, which is also called the makespan. Increasing the operational efficiency in AS/RS cannot directly improve production efficiency but can help to reduce operating costs. Therefore, the weights of the two evaluated objectives are taken as $w_1 = 0.3$ and $w_2 = 0.7$.

2.2.4. Modelling

Based on the above descriptions, the integrated scheduling model of AS/RS and hybrid flowshop can be given as follows:

$$\min(F) = \min(w_1 \cdot f_1 + w_2 \cdot f_2) \tag{7}$$

which is subject to:

$$\sum_{j=1}^{X \cdot Y \cdot Z} \alpha_{nj} = 1 \tag{8}$$

$$ct_{noke} = st_{noke} + T_{pke}; (p = p_{no}) \tag{9}$$

$$st_{n's} \geq ct_{ns} - (1 - \beta_{nn's}) \cdot \Omega \tag{10}$$

$$st_{n'o'ke} \geq ct_{n'o'ke} - (1 - \chi_{nn'ke}) \cdot \Omega \tag{11}$$

$$st_{no'e} \geq ct_{ns} + R_s + \frac{(2S - 1) \cdot \left(L + \frac{u}{2}\right)}{v_x} \tag{12}$$

$$st_{no(k+1)e'} \geq ct_{noke}; (k \neq K) \tag{13}$$

Equation (7) indicates the bi-objective that contains the total operation time in AS/RS and the maximum makespan in production. Equation (8) represents that each task can only correspond to one position in racks. Equation (9) indicates the relationship between the start operation time and the end operation time of the retrieval tasks. Equations (10) and (11) each represent the constraints in the crane and production equipment at the start of the operation time of the next task and at the end of the operation time of the previous task. Equation (12) represents that the start operation time of the production task is larger than the arrival time of the production material. Equation (13) indicates that the production phase in which the start time of the next stage of the task is greater than or equal to the end time of the previous stage of the operation.

3. GA-MBO Design

Intelligent optimization algorithms are more suitable than accurate algorithms to solve NP-hard problems and are conducted to deliver fast solutions. GA is a classic intelligent optimization algorithm which has been widely used in production scheduling, combinatorial optimization, etc. To fix the poor local search ability of GA, this paper adopts the MBO algorithm which has high local search efficiency and outstanding convergence performance. GA-MBO algorithm is proposed to solve the integrated scheduling optimization problem in AS/RS and workshop.

The GA-MBO algorithm consists of three modules, including coding rules, GA rules, and MBO rules. The flow chart of the GA-MBO is shown in Figure 3. The following parameters are defined as: *NG* represents the number of populations in the GA phase; *Mgen* represents the number of iterations in the GA phase; *Pc* represents the probability of crossover; *Pm* represents the probability of mutation; *NM* is the number of flocks in the MBO phase; *a* represents the number of neighborhood solutions generated by an individual; *b* indicates the number of neighborhood solutions that each individual passes to the next individual; and *G* represents the number of tours.

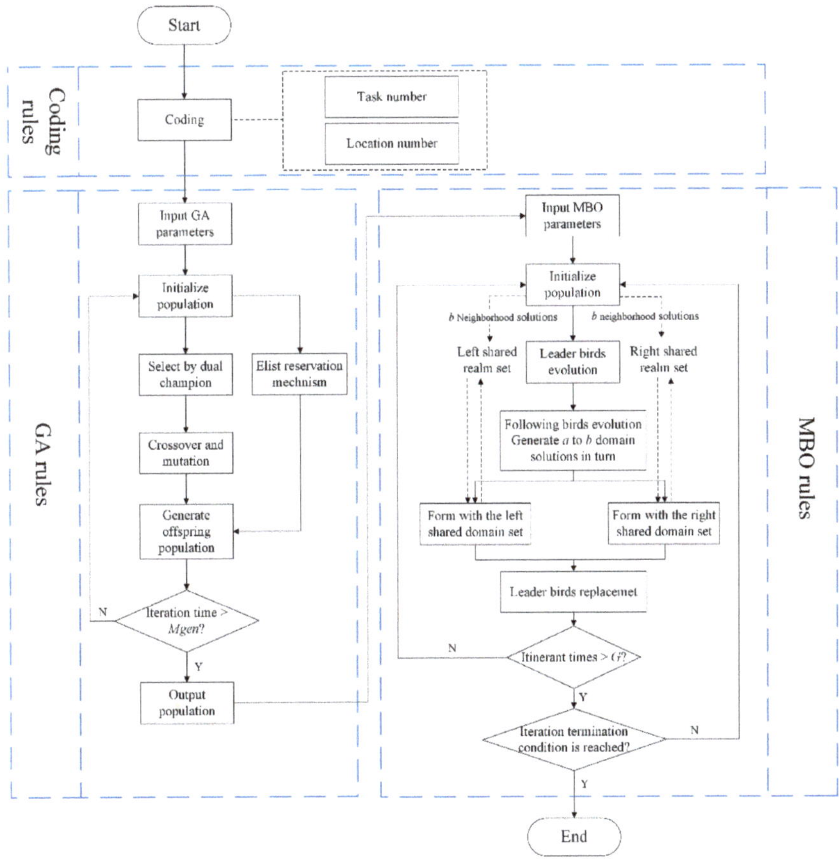

Figure 3. Flow chart of GA-MBO.

3.1. Coding Rules

The real coding method is used to solve the problems, such as multiple storage locations in the AS/RS, various material types of storage and retrieval tasks, mixed operation of storage and retrieval tasks, and complex operation sequences of production tasks. The coding and the decoding mechanism are illustrated with the warehousing information in the AS/RS with 6 rows, 6 columns, and 5 tiers, which are shown in Table 1. The storage and retrieval tasks are numbered as shown in Figure 4. The storage racks are successively numbered with rows, columns, and tiers, starting from 1. The total number of racks is:

$$J = X \cdot Y \cdot Z \tag{14}$$

Table 1. The warehousing information.

Number of Retrieval Types	Number of Bins	Number of Storage Types	Number of Kins
1	2	3	4
2	3	4	2
3	2	5	3

	Retrieval tasks							Storage tasks								
Task number	1	2	3	4	5	6	7	8	9	10	11	12	13	14	15	16
Material number	1	1	2	2	2	3	3	3	3	3	3	4	4	5	5	5

Figure 4. Mixed storage and retrieval tasks and material numbers.

3.1.1. Coding

Considering that coding of rack locations, which is much longer than storage and retrieval tasks, increases the complexity of the algorithm and reduces the search efficiency, coding is performed mainly for the storage and retrieval tasks, as shown in Figure 5. The number of storage and retrieval tasks is used as the coding length, and the tasks are arranged in the order of execution. In the retrieval tasks, it is randomly selected from the corresponding material locations in AS/RS. In the storage tasks, it is randomly selected from the free locations in AS/RS.

Task number	8	4	6	11	10	3	15	5	2	14	1	16	13	12	9	7
Location number	78	67	18	86	133	5	46	8	61	24	2	104	113	25	138	134

Figure 5. Coding diagram.

3.1.2. Decoding

Each code represents the storage locations of the storage and retrieval tasks and the operation sequence in AS/RS. The code needs to be decoded to find the operation sequence of the crane and the production stage of the machines. The crane number is determined by the location number in multi-crane operations. The racks of a crane can store two rows, which are numbered as:

$$S = \left[\frac{j}{2Y \cdot Z}\right] + 1 \tag{15}$$

The decoding progress is:

Step 1: Determine the crane number by the location number in AS/RS and determine the task sequence by the task number to obtain the operation sequence in cranes.

Step 2: The total operation time of cranes is obtained by the end operation time of the task on each crane.

Step 3: The travel times of shelves in different rows to the production equipment are calculated based on the end operation time of the crane retrieval tasks in step 1 and obtain the arrival time from retrieval tasks to the production stage.

Step 4: According to the arrival time of the retrieval task to the production stage, the operation time sequence of tasks can be obtained by the rules of "First-Come-First-Service" of tasks, "First Idle", and "Capacity Priority" (workpiece processing time) of machines.

Step 5: The makespan is obtained by the end operation time of each task at the final production stage.

3.2. GA Rules

GA includes the operations of the population initialization, selection, crossover, and mutation. The population initialization method is randomly generated according to the encoding method. The selection mechanism is based on the elite reservation and the binary tournament selection mechanism. The crossover and mutation operations are needed because the coding method is two-layer coding. The operation tasks are allocated. The retrieval tasks are constrained by the corresponding material storage locations in AS/RS, and the storage tasks are constrained by the free locations in AS/RS.

3.2.1. Crossover

The main purpose of the GA phase is to obtain a better solution, which is to make full use of the global optimization performance of GA and to avoid the generation of useless solutions. Crossover is designed as follows: In retrieval tasks, an improved crossover operation that extracts the corresponding code in process coding is designed based on the type of material. In storage tasks, a uniform crossover operation that extracts partial code of different storage locations is adopted, which considers the same retrieval location under different solutions. After that, the extraction code is put back to the original solution's extraction position accordingly, and the specific process is shown in Figure 6.

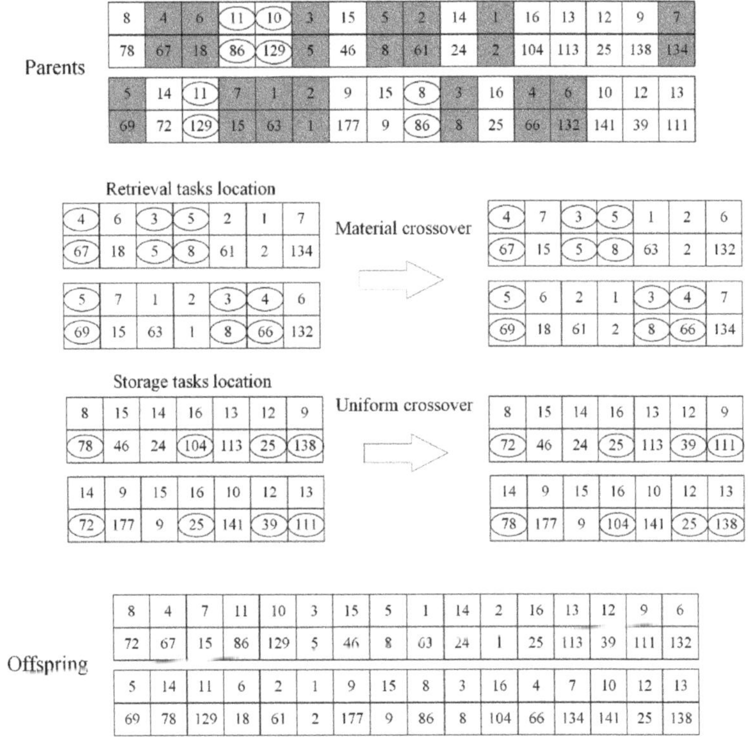

Figure 6. Schematic diagram of the intersection of storage and retrieval tasks.

Useless solutions can be avoided by the above operations under the constraints of task locations. It is beneficial for the global search ability of the algorithm by fully adjusting the old solutions according to different cross operations of storage and retrieval tasks.

3.2.2. Mutation

The sequence of storage and retrieval tasks is generated by four mutation operations, namely random exchange, pre-insertion, post-insertion, and sequential pair exchange. The allocation of storage and retrieval tasks is designed by the single-point mutation and the exchange mutation. The single-point mutation randomly makes a location in a feasible location set of a task to mutate a new location that is not in the solution. The exchange mutation randomly exchanges the locations of two tasks in the set of retrieval tasks or the set of storage tasks of the same material in the solution.

3.3. MBO Rules

The MBO is a neighborhood search-based algorithm that performs a sufficient local search on the neighborhood of each solution, and it compensates for the deficiency of the GA's local search ability. In addition, the MBO algorithm can deeply explore the optimal solution at the late population convergence stage of the algorithm.

3.3.1. Neighborhood Structure Design

The neighborhood structure directly affects the MBO solution quality and the convergence speed of the algorithm. An efficient neighborhood structure needs to be identified. For the sequence of tasks, the formation of the sequence neighborhood structure considers random exchange, pre-insertion, post-insertion, sequence pair exchange, optimal insertion, and optimal exchange.

For the product allocation, the scale of the feasible storage location is larger than the scale of the feasible retrieval locations. Therefore, different operations are designed to construct different neighborhood structures for the storage and retrieval tasks.

- Neighborhood structure for retrieval tasks. A neighborhood structure based on an improved optimal exchange operation is designed. The process is to randomly select a retrieval task for its feasible location set, then delete the original location of the task from that set, and the number of operations is the number that feasible storage locations minus 1. If a storage location of other tasks in the code overlaps with the new feasible storage location of the indicated task, this storage location of other tasks needs to be replaced by the original location of the indicated task. The values of objective function values for these new individuals are calculated respectively and the most optimal one is chosen.
- Neighborhood structure for storage tasks. A neighborhood structure based on an improved optimal exchange operation is designed. In a retrieval task, a new location that is not in individual code is randomly selected from the free locations to replace the original location and the number of operations is the number of retrieval tasks. The values of objective function values for these new individuals are calculated respectively and the most optimal one is chosen.

3.3.2. Adaptive Adjustment of Neighborhood Structure

The total of eight neighborhood structures above present different search effects in the solution at different stages of the algorithm. At the early stage of the algorithm, the effect of the eight neighborhood structures to get better solutions is not much different, but the search efficiencies of random exchange, pre-insertion, post-insertion, and sequence pair exchange are higher, which improve the application frequencies. At the late stage of the algorithm, the two neighborhood structures based on optimal insertion, optimal exchange, and storage assignment are better than the other four structures. The application frequencies of these four structures should be increased. Therefore, an adaptive adjustment strategy is introduced to control the usage frequency of each structure during the search period to optimize the algorithm's efficiency.

Weight ω_0 is assigned to each neighborhood structure. The roulette method is used to randomly select the neighborhood structure according to the neighborhood structure weight ω_i to generate the neighborhood solution, and the weight is updated after each iteration. The adjust weight is contributed by:

$$\omega_{i,seg+1} = (1 - \eta_i) \cdot \omega_{i,seg} + \frac{\eta_i \cdot \beta_{i,seg}}{\alpha_{i,seg}} \tag{16}$$

where α_i is the number of times of structure i; β_i is the cumulative score of structure i. If the solution is better than the original solution generated by structure i, $\beta_i = \beta_i + 1$, $\eta_i \in [0, 1]$ is the speed of the response to the effect of structure i of weight ω_i.

4. Simulation Experiments

4.1. Test Examples

The algorithms for the integrated scheduling problem are coded in MATLAB2016a and run on the Intel i7-7700 k CPU with 16 GB memory. The operation process in an AS/RS of a manufacturing enterprise is used as an example to research the scheduling problem and is modified to obtain the experimental data. The production process consisted of an AS/RS and a hybrid flowshop. The AS/RS has 6 rows, 60 columns, and 15 layers, and the number of racks is 5400. The coordinates of the input point and the output point is (0, 0, 1) and (7, 61, 1). The parameters of each index are shown in Table 2. There are 30 kinds of materials in the system, and the quantity of each material is $N \in U(30, 50)$; the types of storage and retrieval materials are $P \in \{5, 10, 15, 20\}$; the quantity of each material is $O_P \in U(1, 5)$; the number of scheduling stages is $K \in \{3, 5, 7\}$; the number of machines was $E_k \in U(2, 6)$; the material processing time is $T \in U(10, 70)$; material weight is $M \in U(10, 30)$, and the frequency of storage and retrieval is $f \in U(1, 10)$. There are 12 groups of experiments set by the types of storage and retrieval materials and the number of scheduling stages. The data format $U[x, y]$ represents a discrete uniform distribution between x and y.

Table 2. Each index parameter of the AS/RS.

Describe	Value
Rack length/width/high (m/m/m)	1/0.8/0.6
Velocity of the crane in the horizontal direction (m/s)	3
Velocity of the crane in the vertical direction (m/s)	1
Pick-up and set-down times of the crane (s)	5
Velocity of the conveyor (m/s)	0.5
Output point to production machine time (s)	60

4.2. Parameter Settings

The relevant parameters of the GA-MBO are *NG, Mgen, Pc, Pm, NM, a, b, G*. Some parameters of reference were set: $Pc = 0.8$, $Pm = 0.05$, $a = 3$, $b = 1$ [25,26]. It was found that this setting has a good effect on the solution of this paper by experiments. *NG, Mgen, NM*, and *G* are related to the problem scale. The corresponding Taguchi experiment was designed for the factor levels [27]. The parameter level table is shown in Table 3. Experiments are carried out with an example of $p = 10$, $K = 3$, and the algorithm is run 10 times independently under each combination of parameters. The maximum running time is $10(K + 1) \cdot \sum_{p=1}^{P} (O_p + I_p) \cdot s$ [28]. The average values of 10 experimental results are taken as the response values, as shown in Table 4.

Table 3. Parameter level of GA-MBO.

Parameters	Level Value			
	1	2	3	4
NG	150	200	250	300
Mgen	100	150	200	250
NM	25	51	81	101
G	5	8	10	15

Table 4. Orthogonal matrix and response value.

Test Numbers	Parameters				Response Value
	NG	Mgen	NM	G	
1	150	100	25	5	0.09346
2	150	150	51	8	0.09180
3	150	200	81	10	0.08936
4	150	250	101	15	0.09304
5	200	100	51	10	0.09175
6	200	150	25	15	0.09232
7	200	200	101	5	0.09175
8	200	250	81	8	0.08790
9	250	100	81	15	0.09397
10	250	150	101	10	0.08947
11	250	200	25	8	0.08876
12	250	250	51	5	0.08636
13	300	100	101	8	0.09258
14	300	150	81	5	0.09072
15	300	200	51	15	0.08941
16	300	250	25	10	0.08974

The results under the combinations of parameters are analyzed by Minitab 17 in Figure 7, the parameter change trend diagram, and Table 5, the average response value. It can be seen the performance of the algorithm is the best when $NG = 250$, $Mgen = 250$, $NM = 51$, and $G = 10$. This parameter scheme is adopted in subsequent experiments.

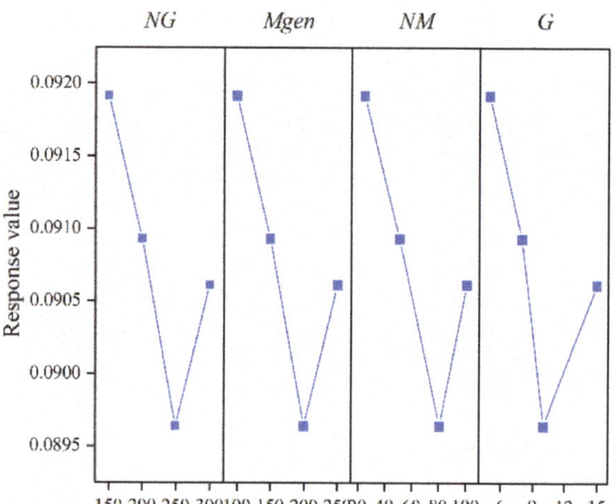

Figure 7. The graph of parameter change trend.

Table 5. Response value of different parameters.

Level	NG	Mgen	NM	G
1	0.09191	0.09294	0.09107	0.09057
2	0.09093	0.09108	0.08983	0.09026
3	0.08964	0.08982	0.09049	0.09008
4	0.09061	0.08926	0.09171	0.09218
Delta	0.00227	0.00368	0.00188	0.0021
Rank	2	1	4	3

4.3. Algorithms Comparison

In the integrated scheduling optimization problem of AS/RS and hybrid flowshop, a comparison with the improved GA, improved particle swarm optimization (PSO) algorithm, and hybrid algorithm of GA and PSO (GA-PSO), the algorithm performance of GA-MBO is verified. With the NP-hard characteristics of the problem, the evaluation indexes, including average (Avg.) and the standard deviation (Std.), are solved by the repeated experiments of 10 times of the four algorithms. The Avg. is used to measure the efficiency and the Std. is used to measure the robustness of the algorithm. The 12 groups for the comparative analysis are shown in Table 6, which set the types of storage and retrieval materials $P \in \{5, 10, 15, 20\}$ and the number of scheduling stages in AS/RS $K \in \{3, 5, 7\}$. From Table 6, compared with three algorithms, including IGA, IPSO, and GA-PSO, it is obvious that GA-MBO has the most optimal solutions of Avg. and Std.

Table 6. Comparison results of four algorithms.

Group	p	K	GA-MBO		IGA		IPSO		GA-PSO	
			Avg.	Std.	Avg.	Std.	Avg.	Std.	Avg.	Std.
1		3	0.10159	0.00192	0.11131	0.00253	0.12152	0.00454	0.10707	0.00254
2	5	5	0.10440	0.00133	0.11919	0.00328	0.12981	0.00446	0.11451	0.00283
3		7	0.10805	0.00136	0.12076	0.00209	0.13058	0.00271	0.11423	0.00269
4		3	0.08701	0.00167	0.09689	0.00209	0.10896	0.00255	0.08997	0.00190
5	10	5	0.08678	0.00148	0.09920	0.00359	0.11416	0.00337	0.09318	0.00353
6		7	0.09021	0.00134	0.10475	0.00321	0.11856	0.00510	0.09533	0.00212
7		3	0.09872	0.00142	0.10968	0.00232	0.12371	0.00688	0.10662	0.00196
8	15	5	0.10328	0.00140	0.11177	0.00211	0.12789	0.00466	0.10536	0.00263
9		7	0.10548	0.00141	0.11464	0.00202	0.12960	0.00244	0.11135	0.00312
10		3	0.09766	0.00148	0.10865	0.00243	0.12666	0.00330	0.10449	0.00205
11	20	5	0.10741	0.00136	0.11451	0.00212	0.12855	0.00512	0.11194	0.00341
12		7	0.11153	0.00286	0.11568	0.00257	0.13227	0.00262	0.11306	0.00365

To display the promotion of the GA-MBO more clearly, the optimization results by comparing with other algorithms are shown in Table 7. In Table 7, the experimental results of IGA, IPSO, and GA-PSO show that: (1) compared with IGA, IPSO, and GA-PSO, GA-MBO, the optimization efficiencies are achieved at 13.88% in Group 6, 23.98% in Group 5, and 8.83% in Group 2, and the average promotions are 9.48%, 19.53%, and 5.12%, respectively; (2) in Group 12, the GA-MBO is not as stable as IGA and IPSO; (3) compared with IGA, IPSO, and GA-PSO, GA-MBO, the optimization robustness are achieved at 59.45% in Group 2, 79.36% in Group 7, and 60.12% in Group 11, and the average promotions are 35.16%, 54.42%, and 39.38%, correspondingly. Although the robustness of GA-MBO is poor in Group 12, it is much better in other groups, and the average value is much higher. Therefore, it is still considered that the GA-MBO has the best robustness.

The T-test uses t-distribution theory to infer the probability of difference and compare whether the difference between two averages is significant. In the test examples, the normally distributed data are assumed, and the operation time is 10 in each group. The t-test is studied to test the statistical difference of the GA-MBO with other three algorithms, and the confidence is 0.95. The results are shown in Table 8. In Table 8, the values of upper confidence and lower confidence between GA-MBO and other algorithms are negative and within the confidence interval, which verifies the effectiveness of the GA-MBO. The advantage of GA-MBO is further proved by indicating that the Avg. of the optimal solution of GA-MBO is stably better than other algorithms again.

Table 7. The optimization efficiency and robustness of algorithms.

Group	OE_1 [1]	OE_2 [2]	OE_3 [3]	OR_1 [4]	OR_2 [5]	OR_3 [6]
1	8.73%	16.40%	5.12%	24.11%	57.71%	24.41%
2	12.41%	19.57%	8.83%	59.45%	70.18%	53.00%
3	10.53%	17.25%	5.41%	34.93%	49.82%	49.44%
4	10.20%	20.15%	3.29%	20.10%	34.51%	12.11%
5	12.52%	23.98%	6.87%	58.77%	56.08%	58.07%
6	13.88%	23.91%	5.37%	58.26%	73.73%	36.79%
7	9.99%	20.20%	7.41%	38.79%	79.36%	27.55%
8	7.60%	19.24%	1.97%	33.65%	69.96%	46.77%
9	7.99%	18.61%	5.27%	30.20%	42.21%	54.81%
10	10.12%	22.90%	6.54%	39.09%	55.15%	27.80%
11	6.20%	16.44%	4.05%	35.85%	73.44%	60.12%
12	3.59%	15.68%	1.35%	−11.28%	−9.16%	21.64%
Average promotion	9.48%	19.53%	5.12%	35.16%	54.42%	39.38%

[1] OE_1 is the optimization efficiency promotion of the GA-MBO relative to IGA. It is calculated as OE_1 = (IGA Avg. − GA-MBO Avg.)/IGA Avg. [2] OE_2 is the optimization efficiency promotion of the GA-MBO relative to IPSO. [3] OE_3 is the optimization efficiency promotion of the GA-MBO relative to GA-PSO. [4] OR_1 is the optimization robustness promotion of the GA-MBO relative to IGA. It is calculated as OR_1 = (IGA Std. − GA-MBO Std.)/IGA Std. [5] OR_2 is the optimization robustness promotion of the GA-MBO relative to IPSO. [6] OR_3 is the optimization robustness promotion of the GA-MBO relative to GA-PSO.

Table 8. T-test of optimal solution between GA-MBO and other algorithms with confidence of 0.95.

Group	GA-MBO~IGA		GA-MBO~IPSO		GA-MBO~GA-PSO	
	Upper Confidence	Lower Confidence	Upper Confidence	Lower Confidence	Upper Confidence	Lower Confidence
1	−0.01212	−0.00765	−0.02377	−0.02014	−0.00496	−0.00097
2	−0.01305	−0.00893	−0.03178	−0.02622	−0.00868	−0.00498
3	−0.01298	−0.00894	−0.03000	−0.01999	−0.00968	−0.00614
4	−0.01186	−0.00758	−0.02290	−0.01697	−0.00799	−0.00299
5	−0.01522	−0.00962	−0.02981	−0.02494	−0.00933	−0.00347
6	−0.00903	−0.00517	−0.02496	−0.01732	−0.00748	−0.00159
7	−0.01038	−0.00662	−0.02766	−0.02157	−0.00425	0.00008
8	−0.01740	−0.01218	−0.02887	−0.02195	−0.01250	−0.00772
9	−0.01724	−0.01184	−0.03249	−0.02421	−0.00703	−0.00321
10	−0.00752	−0.00077	−0.02246	−0.01902	−0.00511	0.00204
11	−0.01102	−0.00730	−0.02622	−0.02202	−0.00814	−0.00361
12	−0.01476	−0.01067	−0.02468	−0.02039	−0.00819	−0.00417

To find the iteration situation of these algorithms, the convergence comparisons between the GA-MBO and other three algorithms are carried out with the calculation example of $p = 10$ and $K = 3$ as shown in Figure 8. In Figure 8, the abscissa is the iteration times of the four algorithms, and the ordinate is the value of objective function which calculated by the algorithms. The optimal solution of the GA-MBO is the best among the other three algorithms when the iteration is greater than 276, and the iteration tends to be flat when the iteration is greater than 350, which means the GA-MBO can find a best value of objective function. Based on the above analysis, the GA-MBO is superior to IGA, IPSO, and GA-PSO in terms of the efficiency and robustness of the solution.

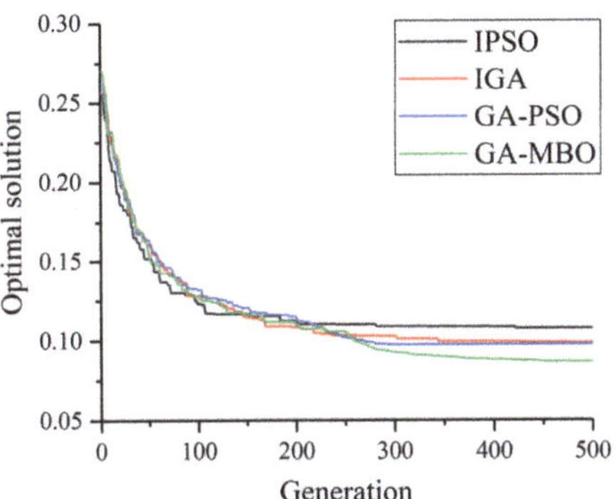

Figure 8. Iteration diagrams of four algorithms.

4.4. Bi-Objective Comparison

To verify the superiority of the integrated scheduling optimization of AS/RS and hybrid flowshop, three experiments with $p = 10$ and $K = 3$ are tested and compared. Objectives of these experiments are operation time in AS/RS, makespan in hybrid flowshop, and the bi-objective. The results are shown in Table 9, which shows that while f_2 only changes 4.43%, f_1 has a 26.34% improvement. This condition is more suitable for the actual production which concerns the total profit and reflects the superiority of the integrated scheduling optimization.

Table 9. Comparison of results between bi-objective optimization and single objective optimization.

Objective	Value	Variation	Value	Variation
f_1	609.8	22.77%	595.4	−38.72%
f_2	997.6	−26.34%	410.2	4.43%
f_1 and f_2	789.6	None	429.2	None

5. Conclusions

In this paper, considering the combination of distribution and production, the study of the integrated scheduling problem of AS/RS and hybrid flowshop is conducted to establish an integrated scheduling optimization model with the bi-objective of minimizing the total operation time and makespan. A mixed optimization algorithm of GA-MBO combining the global optimization performance and search capability of GA with a strong local search ability of the MBO algorithm is proposed to improve the efficiency and robustness of the algorithm solution.

In a simulation case of an AS/RS with 5400 storage locations, the model and algorithm of repeat experiments were verified by a comparison of commonly intelligent algorithms with GA-MBO. The experimental results of IGA, IPSO, and GA-PSO show that the average promotions of efficiencies are 9.48%, 19.53%, and 5.12% and the average promotions of robustness are 35.16%, 54.42%, and 39.38%, correspondingly. Comparative analysis of the bi-objective verified the superiority of integrated scheduling optimization. This verification helps to coordinate the scheduling in warehouse distribution and workshop production to reduce distribution costs. The gap in the integrated scheduling optimization of AS/RS and the hybrid flowshop is filled.

At present, the rapid development of intelligent warehousing systems makes the AS/RS and hybrid flowshop closely connected. The integrated scheduling optimization problem of the AS/RS and hybrid flowshop is one of the problems in the integrated AR/RS and hybrid flowshop, which should be solved to build a complete intelligent factory, and it is a critical problem facing the intelligent factory. The collaboration between machines and between the AS/RS and hybrid flowshop are hard to realize, and the system cannot be actually used in the research [29].

The devices have the effect factors of energy consumption, delivery waste, machine handling time, and maintenance when operated in the AS/RS and hybrid flowshop. In the future, a corresponding optimization model in the scheduling problem should be established by introducing factors that further conform to the real production situation. In addition, different structures of AS/RS and hybrid flowshop as well as different types and numbers of AGVs and cranes may also be researched to adapt the development of the flexible manufacturing workshop and enhance the efficiency.

Author Contributions: Conceptualization, J.L.; Data curation, L.X.; Formal analysis, L.X.; Funding acquisition, Y.S.; Investigation, Y.S.; Methodology, L.X.; Project administration, J.L.; Resources, J.L.; Software, J.J.; Supervision, J.L.; Validation, J.L.; Visualization, J.J.; Writing—original draft, L.X.; Writing—review & editing, Y.S. All authors have read and agreed to the published version of the manuscript.

Funding: This research was funded by the Zhejiang Science and Technology Plan Project, grant number 2018C01003, the Zhejiang Provincial Natural Science Foundation, grant number LQ22E050017, the China Postdoctoral Science Foundation, grant number 2021M702894, and the Zhejiang Provincial Postdoctoral Science Foundation, grant number ZJ2021119.

Data Availability Statement: Not applicable.

Conflicts of Interest: The authors declare no conflict of interest.

References

1. Hu, K.Y.; Chang, T.S. An innovative automated storage and retrieval system for B2C e-commerce logistics. *Int. J. Adv. Manuf. Technol.* **2010**, *48*, 297–305. [CrossRef]
2. Sawicki, P.; Sawicka, H. Optimisation of the Two-Tier Distribution System in Omni-Channel Environment. *Energies* **2021**, *14*, 7700. [CrossRef]
3. Kazemi, M.; Asef-Vaziri, A.; Shojaei, T. Concurrent Optimization of Shared Location Assignment and Storage/Retrieval Scheduling in Multi-Shuttle Automated Storage and Retrieval Systems. *IFAC Pap.* **2019**, *52*, 2531–2536. [CrossRef]
4. Ma, H.P.; Su, S.F.; Simon, D.; Fei, M.R. Ensemble multi-objective biogeography-based optimization with application to automated warehouse scheduling. *Eng. Appl. Artif. Intel.* **2015**, *44*, 79–90. [CrossRef]
5. Kung, Y.H.; Kobayashi, Y.; Higashi, T.; Sugi, M.; Ota, J. Order scheduling of multiple stacker cranes on common rails in an automated storage/retrieval system. *Int. J. Prod. Res.* **2014**, *52*, 1171–1187. [CrossRef]
6. Roshan, K.; Shojaie, A.; Javadi, M. Advanced allocation policy in class-based storage to improve AS/RS efficiency toward green manufacturing. *Int. J. Environ. Sci. Technol.* **2019**, *16*, 5695–5706. [CrossRef]
7. Hachemi, K.; Sari, Z.; Ghouali, N. A step-by-step dual cycle sequencing method for unit-load automated storage and retrieval systems. *Comput. Ind. Eng.* **2012**, *63*, 980–984. [CrossRef]
8. Song, Y.B.; Mu, H.B. Integrated Optimization of Input/Output Point Assignment and Twin Stackers Scheduling in Multi-Input/Output Points Automated Storage and Retrieval System by Ant Colony Algorithm. *Math. Probl. Eng.* **2022**, *2022*. [CrossRef]
9. Geng, S.; Wang, L.; Li, D.D.; Jiang, B.C.; Su, X.M. Research on scheduling strategy for automated storage and retrieval system. *CAAI Trans. Intell. Technol.* **2021**, *7*, 522–536. [CrossRef]
10. Wu, K.Y.; Xu, S.S.D.; Wu, T.C. Optimal Scheduling for Retrieval Jobs in Double-Deep AS/RS by Evolutionary Algorithms. *Abstr. Appl. Anal.* **2013**. [CrossRef]
11. Colak, M.; Keskin, G.A. An extensive and systematic literature review for hybrid flowshop scheduling problems. *Int. J. Ind. Eng. Comp.* **2022**, *13*, 185–222.
12. Zhang, B.A.; Pan, Q.K.; Meng, L.L.; Zhang, X.L.; Jiang, X.C. A decomposition-based multi-objective evolutionary algorithm for hybrid flowshop rescheduling problem with consistent sublots. *Int. J. Prod. Res.* **2022**. [CrossRef]
13. Zhang, B.; Pan, Q.K.; Meng, L.L.; Zhang, X.L.; Ren, Y.P.; Li, J.Q.; Jiang, X.C. A collaborative variable neighborhood descent algorithm for the hybrid flowshop scheduling problem with consistent sublots. *Appl. Soft. Comput.* **2021**, *106*, 107305. [CrossRef]
14. Li, Y.L.; Li, X.Y.; Gao, L.; Zhang, B.; Pan, Q.K.; Tasgetiren, M.F.; Meng, L.L. A discrete artificial bee colony algorithm for distributed hybrid flowshop scheduling problem with sequence-dependent setup times. *Int. J. Prod. Res.* **2021**, *59*, 3880–3899. [CrossRef]
15. Reddy, B.S.P.; Rao, C.S.P. A hybrid multi-objective GA for simultaneous scheduling of machines and AGVs in FMS. *Int. J. Adv. Manuf. Technol.* **2006**, *31*, 602–613. [CrossRef]
16. Gao, K.Z.; Cao, Z.G.; Zhang, L.; Chen, Z.H.; Han, Y.Y.; Pan, Q.K. A Review on Swarm Intelligence and Evolutionary Algorithms for Solving Flexible Job Shop Scheduling Problems. *IEEE-CAA J. Autom.* **2019**, *6*, 904–916. [CrossRef]

17. Vicedo, P.; Gil-Gomez, H.; Oltra-Badenes, R.; Guerola-Navarro, V. A bibliometric overview of how critical success factors influence on enterprise resource planning implementations. *J. Intell. Fuzzy Syst.* **2020**, *38*, 5475–5487. [CrossRef]

18. Tongur, V.; Ulker, E. PSO-based improved multi-flocks migrating birds optimization (IMFMBO) algorithm for solution of discrete problems. *Soft Comput.* **2019**, *23*, 5469–5484. [CrossRef]

19. Fandi, W.; Kouloughli, S.; Ghomri, L. Multi-shuttle AS/RS dimensions optimization using a genetic algorithm-case of the multi-aisle configuration. *Int. J. Adv. Manuf. Technol.* **2022**, *120*, 1219–1236. [CrossRef]

20. Allali, K.; Aqil, S.; Belabid, J. Distributed no-wait flow shop problem with sequence dependent setup time: Optimization of makespan and maximum tardiness. *Simul. Model. Pract. Theory* **2022**, *116*, 102455. [CrossRef]

21. Li, W.; Wang, G.G.; Gandomi, A.H. A Survey of Learning-Based Intelligent Optimization Algorithms. *Arch. Comput. Method Eng.* **2021**, *28*, 3781–3799. [CrossRef]

22. Katoch, S.; Chauhan, S.S.; Kumar, V. A review on genetic algorithm: Past, present, and future. *Multimed. Tools Appl.* **2021**, *80*, 8091–8126. [CrossRef]

23. Duman, E.; Uysal, M.; Alkaya, A.F. Migrating Birds Optimization: A new metaheuristic approach and its performance on quadratic assignment problem. *Inform. Sci.* **2012**, *217*, 65–77. [CrossRef]

24. Kouloughli, S.; Feciane, M.K.; Sari, Z. Mobile rack AS/RS dimensions optimization for single cycle time minimization. *Int. J. Adv. Manuf. Technol.* **2022**, *121*, 1815–1836. [CrossRef]

25. Defersha, F.M.; Chen, M.Y. Mathematical model and parallel genetic algorithm for hybrid flexible flowshop lot streaming problem. *Int. J. Adv. Manuf. Technol.* **2012**, *62*, 249–265. [CrossRef]

26. Xu, W.J.; He, L.J.; Zhu, G.Y. Many-objective flow shop scheduling optimisation with genetic algorithm based on fuzzy sets. *Int. J. Prod. Res.* **2021**, *59*, 702–726. [CrossRef]

27. Naderi, B.; Gohari, S.; Yazdani, M. Hybrid flexible flowshop problems: Models and solution methods. *Appl. Math. Model.* **2014**, *38*, 5767–5780. [CrossRef]

28. Naderi, B.; Ruiz, R.; Zandieh, M. Algorithms for a realistic variant of flowshop scheduling. *Comput. Oper. Res.* **2010**, *37*, 236–246. [CrossRef]

29. Munoz, E.R.; Jabbari, F. An Octopus Charger-Based Smart Protocol for Battery Electric Vehicle Charging at a Workplace Parking Structure. *Energies* **2022**, *15*, 6459. [CrossRef]

Article

Evaluation of the Operational Efficiency and Energy Efficiency of Rail Transit in China's Megacities Using a DEA Model

Hao Zhang [1], Xinyue Wang [1], Letao Chen [1], Yujia Luo [2] and Sujie Peng [3,*]

[1]   Business School, University of Shanghai for Science and Technology, Shanghai 200093, China
[2]   School for Business and Society, University of York, York YO10 5GD, UK
[3]   Faculty of Business and Management, Beijing Normal University-Hong Kong Baptist University United International College, Zhuhai 519087, China
*   Correspondence: sujiepeng@uic.edu.cn

Abstract: To date, along with the rapid development of urban rail transit (URT) in China, the evaluation of operational efficiency and energy efficiency has become one of the most important topics. However, the extant literature regarding the efficiency of URT at the line level and considering carbon emissions is limited. To fill the gap, an evaluation model based on slacks-based measure (SBM) data envelopment analysis (DEA) is proposed to measure the efficiencies, which is applied to 61 URT lines in China's four megacities. The findings are summarized as follows: (1) The average operational efficiency and energy efficiency of URT lines are low, and both have great room for improvement. (2) There are significant disparities in the efficiency of URT lines in the case cities. For instance, the average operational efficiency of URT lines in Guangzhou is higher than that of other cities, while the average energy efficiency of URT lines in Shanghai is higher than that of other cities. (3) The URT lines operated by state-owned enterprises have higher average operational efficiency, while the lines operated by joint ventures have higher average energy efficiency. Finally, some suggestions are provided to improve the efficiencies.

Keywords: urban rail transit; operational efficiency; energy efficiency; carbon emission; data envelopment analysis

Citation: Zhang, H.; Wang, X.; Chen, L.; Luo, Y.; Peng, S. Evaluation of the Operational Efficiency and Energy Efficiency of Rail Transit in China's Megacities Using a DEA Model. *Energies* **2022**, *15*, 7758. https://doi.org/10.3390/en15207758

Academic Editors: Xi Gu, Tangbin Xia, Ershun Pan, Rongxi Wang and Yupeng Li

Received: 27 August 2022
Accepted: 7 October 2022
Published: 20 October 2022

Publisher's Note: MDPI stays neutral with regard to jurisdictional claims in published maps and institutional affiliations.

1. Introduction

Over the past two decades, urban rail transit (URT) has rapidly developed to mitigate traffic congestion in China's megacities [1]. According to statistics, by the end of 2021, 50 cities on the Chinese mainland operated 283 URT lines with a total length of 9206.8 km [2]. Compared with other means of public transportation, URT is faster, more frequent, and punctual, which is an important part of urban public transportation. Due to the rapid increase in modernization and the advance of rail transit planning in urban agglomerations, URT has a larger potential development space in China. Improving the operational efficiency of URT makes a great impact on economic and social activities. Operational efficiency evaluation can identify sources of inefficiency and improve URT's operation, which has become one of the most important investigation topics [3,4].

In the literature, URT is usually considered a complex system with multiple inputs (e.g., train, line, station, and energy) to provide transit services and thereby produce multiple outputs (e.g., passenger kilometers, passenger volume, and train kilometers). The efficiency evaluation of public transport is always investigated by comparing multiple inputs and outputs comprehensively [5–8]. In this study, the operational efficiency of URT can be defined as the conversion efficiency between the input system and the output system. Multi-criteria decision analysis (MCDA) methods can be used to comprehensively evaluate alternatives [9–11]. However, different MCDA methods often produce contradictory results when comparing, and decisionmakers may obtain different decisions even using the same criteria weights and criterial evaluations of variants [11]. As one of the non-parametric approaches, data envelopment analysis (DEA) has the advantage of having no pre-determined weights, which is applicable in estimating the relative efficiency of decision-making units (DMUs) with multiple inputs and outputs. Since first proposed by Charnes et al. [12], DEA has been successfully and widely applied to measure efficiency in the public transport sector, such as railways (e.g., [13–15]), highway bus transit (e.g., [16–18]), shipping and ports (e.g., [19–21]), and airlines and airports (e.g., [22–24]).

In terms of the efficiency of URT, it can be measured at different levels, such as the city level and the company level. In this sense, Karlaftis [19] used the DEA model to measure the efficiency and effectiveness of 256 US URT systems, and the results showed that efficiency is positively correlated with effectiveness. Jain et al. [25] applied DEA to explore the relationship between technical efficiency and ownership structure for 15 global URT systems and found that privatization directly and positively impacts efficiency. Qin et al. [26] adopted a slacks-based multi-stage network DEA to assess the efficiency of 17 URT systems in China in 2012 and found that lower average overall efficiency is more related to inefficiencies in the earning stage and construction stage. Tsai et al. [27] used DEA to measure the efficiency of 20 international URT systems from 2009 to 2011 and suggested that the number of stations and population density impact efficiency significantly. Costa et al. [28] utilized DEA to compute the efficiency of four URT systems in Portugal from 2009 to 2018 and explored the impact of the ownership model on efficiency. The findings indicated that privately managed firms were more efficient than public firms. Although the above studies made great progress, estimation at the city or company level cannot identify the efficiency of specific lines or provide deeper insight into the improvement of efficiency at the line level.

To the best of our knowledge, studies on the efficiency of URT at the line level are scarce. Kang et al. [29] developed a mixed network DEA model and a hybrid two-stage network DEA model to explore the efficiency of two metro systems, including six lines in Taipei, and found that the efficiency results between the two models differed significantly. Le et al. [30] used the DEA model to measure the operational efficiency, cost efficiency, and revenue efficiency of 18 URT lines in the Tokyo Metropolitan Area in 2017. The results indicated that the in-vehicle congestion rate can be a reflection of the service quality in the operational efficiency measurement. Unfortunately, these two studies did not consider carbon emissions in the efficiency evaluation process. Due to growing environmental concerns, carbon emissions are considered an undesirable output in efficiency estimations in the transportation sector [31–33]. An efficiency measurement without considering carbon emissions may lead to imprecise operational efficiency results, which leaves a research gap.

In addition, with the increase in URT mileage, the corresponding energy consumption is also rising. The measurement of URT's energy efficiency can help operators save electricity and reduce operating costs and carbon emissions. However, while there are many studies on energy efficiency in the transportation sector [7,33–35], few works focus on the URT field. To the best of our knowledge, two studies are closely related to this topic. Xiao et al. [36] applied the DEA model to evaluate the energy efficiency of URT in Beijing Metro Lines 5 and 15 and the Batong Line without considering carbon emissions in the evaluation. To et al. [37] used the dimensional indicator to discuss the energy efficiency of Hong Kong's mass transit railway over the period from 2008–2017 and found that the energy efficiency was between 0.076 and 0.093 kWh per passenger–km and CO_2 emissions were between 0.055–0.071 kg per passenger–km. Notably, the energy efficiency in this study was similar to the energy intensity. The efficiency evaluation did not consider other inputs and outputs and may not provide significant implications. Hence, there exists another gap related to energy efficiency in URT lines, which needs to be explored.

To fill the gaps, this study aims to estimate operational efficiency and energy efficiency considering CO_2 emissions for URT at the line level, which is the novelty of this paper. To achieve this, an evaluation model based on the slacks-based measure (SBM) is developed to assess operational efficiency and energy efficiency synchronously. Furthermore, a method of detecting the improvement potentials of inputs and outputs is proposed. Then, this study applies the proposed model to the URT lines in China's four megacities (Beijing, Shanghai, Guangzhou, and Shenzhen).

In summary, the contributions of this study are listed as follows. First, this study measures the operational efficiency and energy efficiency of the URT in consideration of CO_2 emissions at the line level, which is a step further than previous studies have taken on undesirable outputs. Second, the proposed model can evaluate operational efficiency and energy efficiency simultaneously and provide more precise results. Third, an empirical study of China's 61 URT lines in four major cities verifies the effectiveness of the proposed model. This micro-level research may enrich the theoretical literature and provide new management enlightenment for efficiency improvement in URT operation.

The remainder of this paper is structured as follows. The methodology is presented in Section 2. Section 3 presents the results, and Section 4 provides discussions. Finally, Section 5 illustrates the conclusions and limitations.

2. Methodology

To clearly describe the evaluation method, the input and output variables and the operation process of the URT system are introduced first. Then, the SBM model is developed to measure the operational efficiency of URT lines. Furthermore, a measurement for energy efficiency is proposed.

2.1. Input and Output Variables and Operation Process

Generally, a URT system is invested in by enterprises to provide travel services for citizens. Its operation process is shown in Figure 1. According to previous studies, line mileage, station, train, and energy are indispensable resources for transportation services [19,26,29,38,39]. Hence, these four resources are considered input variables in the operation process. Passenger transport volume and revenue passenger kilometers are taken as the two desirable output variables, while energy-related CO_2 emission is considered one undesirable output variable.

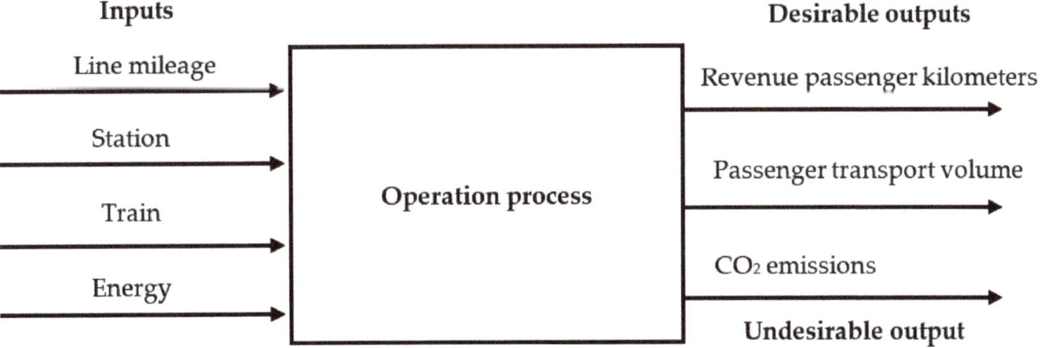

Figure 1. The operation process of a URT system.

2.2. Efficiency Evaluation Model Based on SBM-DEA

This study aims to measure the operational efficiency and energy efficiency of Chinese URT lines with the SBM model. As a non-radial DEA approach, the SBM model directly captures each "input excess" and "output shortfall" to identify the inefficiency of DMUs from an overall perspective [40]. Therefore, the SBM model has been widely used to evaluate the efficiency of public transportation systems, such as by Zhang et al. [41], Chu et al. [42], and Tavassoli et al. [43].

Suppose that there are n DMUs, which represent the URT lines, denoted by DMU_j ($j = 1, 2, \ldots , n$). Each DMU utilizes line mileage (XL), station (XD), train (XT), and energy

(XE) and then produces passenger transport volume (YP), revenue passenger kilometers (YR), and CO_2 emissions (YC). The evaluation model for the operational efficiency of the URT line based on SBM can be expressed as follows:

$$\theta_i = \min \frac{1 - \frac{1}{4}\left(\frac{s_l^-}{XL_i} + \frac{s_d^-}{XD_i} + \frac{s_t^-}{XT_i} + \frac{s_e^-}{XE_i}\right)}{1 + \frac{1}{3}\left(\frac{s_p^+}{YP_i} + \frac{s_r^+}{YR_i} + \frac{s_c^-}{YC_i}\right)}$$

$$\text{s.t.} \quad \sum_{j=1}^{n} \lambda_j XL_j + s_l^- = XL_i,$$

$$\sum_{j=1}^{n} \lambda_j XD_j + s_d^- = XD_i,$$

$$\sum_{j=1}^{n} \lambda_j XT_j + s_t^- = XT_i,$$

$$\sum_{j=1}^{n} \lambda_j XE_j + s_e^- = XE_i, \tag{1}$$

$$\sum_{j=1}^{n} \lambda_j YP_j - s_p^+ = YP_i,$$

$$\sum_{j=1}^{n} \lambda_j YR_j - s_r^+ = YR_i,$$

$$\sum_{j=1}^{n} \lambda_j YC_j + s_c^- = YC_i,$$

$$\sum_{j=1}^{n} \lambda_j = 1,$$

$$\lambda_j, s_l^-, s_d^-, s_t^-, s_e^-, s_p^+, s_r^+, s_c^- \geq 0, \ j = 1, 2, \ldots, n.$$

In Model (1), θ_i represents the operational performance score; s_l^-, s_d^-, s_t^-, s_e^-, s_p^+, s_r^+, and s_c^- are slacks of line mileage, station, train, energy, passenger transport volume, revenue passenger kilometers, and CO_2 emission, respectively, representing either the excess of the input or the shortfall of the output. λ_j expresses the participation degree of each DMU in constructing the production frontier. Note that Model (1) is non-linear. To simplify the calculation, a linear form is transformed following the proposed method by Tone [40] as follows:

$$\theta_i = \min\left(t - \frac{1}{4}\left(\frac{S_l^-}{XL_i} + \frac{S_d^-}{XD_i} + \frac{S_t^-}{XT_i} + \frac{S_e^-}{XE_i}\right)\right)$$

$$\text{s.t.} \quad t + \frac{1}{3}\left(\frac{S_p^+}{YP_i} + \frac{S_r^+}{YR_i} + \frac{S_c^-}{YC_i}\right) = 1$$

$$\sum_{j=1}^{n} \eta_j XL_j + S_l^- = tXL_i,$$

$$\sum_{j=1}^{n} \eta_j XD_j + S_d^- = tXD_i,$$

$$\sum_{j=1}^{n} \eta_j XT_j + S_t^- = tXT_i,$$

$$\sum_{j=1}^{n} \eta_j XE_j + S_e^- = tXE_i, \tag{2}$$

$$\sum_{j=1}^{n} \eta_j YP_j - S_p^+ = tYP_i,$$

$$\sum_{j=1}^{n} \eta_j YR_j - S_r^+ = tYR_i,$$

$$\sum_{j=1}^{n} \eta_j YC_j + S_c^- = tYC_i,$$

$$\sum_{j=1}^{n} \eta_j = t,$$

$$\eta_j, S_l^-, S_d^-, S_t^-, S_e^-, S_p^+, S_r^+, S_c^- \geq 0, j = 1, 2, \ldots, n.$$

The variables in Model (1) undergo the following transformations in Model (2): $\lambda t = \eta$, $ts_l^- = S_l^-, ts_d^- = S_d^-, ts_t^- = S_t^-, ts_e^- = S_e^-, ts_p^+ = S_p^+, ts_r^+ = S_r^+, ts_c^- = S_c^-$. The optimal η_j^*, $S_l^{-*}, S_d^{-*}, S_t^{-*}, S_e^{-*}, S_p^{+*}, S_r^{+*}, S_c^{-*}$, and t^* are measured for operational performance, θ_i^*. If $\theta_i^* = 1$ and all optimal slacks are equivalent to 0, the performance is efficient; otherwise, it is inefficient. Moreover, if a larger performance score of a DMU is obtained, it indicates that this DMU operates better than other DMUs.

In DEA theory, the projected point on the production frontier is the optimal target for each inefficient DMU to pursue. Hence, the DEA method can be used to set the optimization targets of inputs and outputs to improve performance. The target energy expresses a minimum level of energy input to achieve optimal operational performance. Naturally, the target energy input can be obtained with the following equation:

$$TE_i = \sum_{j=1}^{n} \lambda_j XE_j \tag{3}$$

Hence, energy efficiency, ρ_i, is defined as the ratio of target energy to its actual consumed energy in this study. It is can be expressed as follows:

$$\rho_i = \frac{TE_i}{XE_i} \tag{4}$$

For ease of reading, the formulas for calculating the improvement potentials of variables are provided in Appendix A.

3. Empirical Study

3.1. Data Source

As for the empirical analysis, the datasets from the URT lines were collected from the yearbook of the China Urban Rail Transit Almanac 2021, which is an annual report released by the China Association of Urban Rail Transit. In total, 61 URT lines from Beijing, Shanghai, Guangzhou, and Shenzhen were considered for analysis. As shown in Figure 2, Beijing, Shanghai, Guangzhou, and Shenzhen are the top four cities in terms of economic strength on the Chinese mainland. Each city has a population of more than 10 million and an urban rail network of hundreds of kilometers. A large number of people take urban rail transit for their daily travel. Overall, data on line mileage, station, train, energy, passenger transport volume, and revenue passenger kilometers were collected from the aforementioned yearbook. While there are no official statistics on CO_2 emissions, we calculated the carbon emission based on energy consumption and the regional grid carbon emission factor in 2019 following the approach of Yu et al. [44]. Descriptive statistics are shown in Table 1.

Table 1. Descriptive Statistics [1].

Variable	Line Mileage (km)	Station	Train	Energy (10^4 kwh)	Passenger Transport Volume (10^4 PT)	Revenue Passenger Kilometers (10^4 PK)	CO_2 (10^4 tons)
Max	81.40	45.00	116.00	29,997.00	56,139.00	499,058.00	24.49
Min	3.90	2.00	4.00	1046.00	180.10	1891.00	0.84
Mean	37.85	22.39	50.38	12,243.28	14,769.75	129,951.56	10.26
SD	15.63	9.96	26.70	6813.24	11,363.46	100,773.41	5.69

[1] PT and PK are short for person-time and passenger kilometers, respectively.

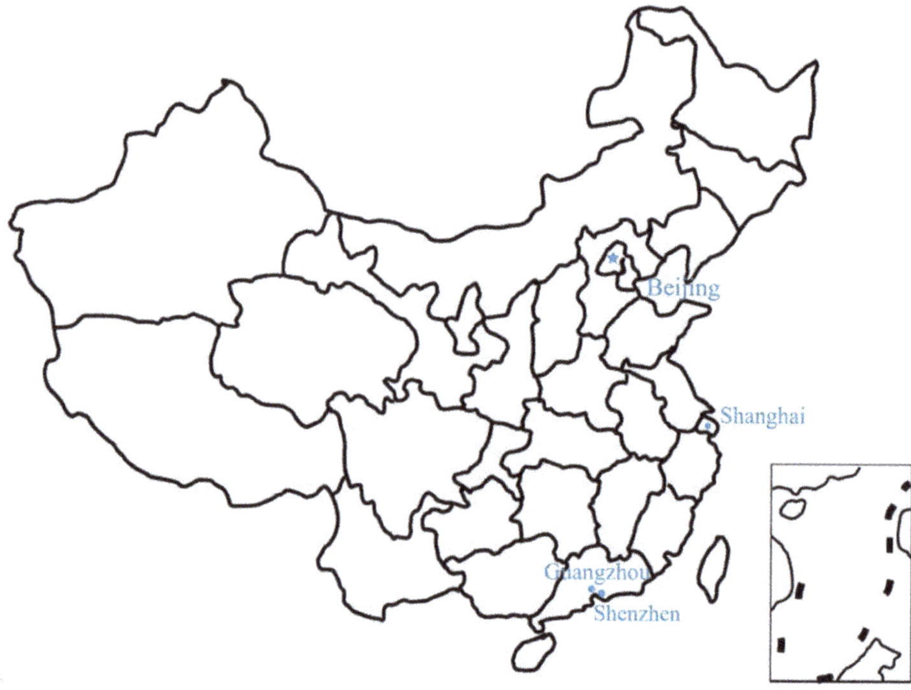

Figure 2. Four megacities in mainland China.

3.2. Efficiency Results

Table 2 and Figure 3 show the efficiency results at the line level and the city level, respectively. As can be seen from Table 2, the average operational efficiency is 0.5634. Overall, the average room for URT lines to improve operational efficiencies is 43.66%. From a line angle, it can be seen that of the operational efficiencies of the 61 observed URT lines, 10 of which are evaluated as being an efficient level, another 15 lines are over the average level, and 36 lines are under the average level. There is a significant difference between URT lines in efficiency. From a city angle, Figure 3 suggests that the average operational efficiency of the URT lines in Guangzhou (0.6453) tops the list. The average operational efficiency of URT lines in Shanghai (0.5921) is higher than the average level, while those of the URT lines in Beijing (0.5054) and Shenzhen (0.5157) are slightly lower than the average level. That is to say, in terms of operational efficiency, there is a slight difference between URT lines at the city level. The reason might be that these megacities are similar in terms of their large population and high economic development level.

Table 2. The efficiency of the URT systems in case cities.

City	Line Name	Operational Efficiency	Energy Efficiency
Beijing	BJ-Line 1	0.4978	0.6264
	BJ-Line 2	0.6100	0.9338
	BJ-Line 4	0.5534	0.7648
	BJ-Line 5	0.6363	0.7953
	BJ-Line 6	0.4514	0.5669
	BJ-Line 7	0.2978	0.3621
	BJ-Line 8	0.2583	0.3937
	BJ-Line 9	0.6827	0.8589
	BJ-Line 10	0.5011	0.6816

Table 2. *Cont.*

City	Line Name	Operational Efficiency	Energy Efficiency
	BJ-Line 13	1.0000	1.0000
	BJ-Line 15	0.4755	0.7136
	BJ-Line 16	0.2337	1.0000
	BJ-Ba Tong Line	0.5034	0.7695
	BJ-Changping Line	0.4839	0.7098
	BJ-Fangshan Line	0.3897	0.7360
	BJ-Capital Airport Express	1.0000	1.0000
	BJ-Yizhuang Line	0.5210	0.7726
	BJ-Line S1	0.4582	0.8313
	BJ-Yanfang Line	0.1703	1.0000
	Daxing Airport Express	0.3835	0.8522
Shanghai	SH-Line 1	0.7284	0.8102
	SH-Line 2	0.6271	0.6593
	SH-Line 3	0.4579	0.6813
	SH-Line 4	1.0000	1.0000
	SH-Line 5	0.3441	0.5530
	SH-Line 6	0.4503	0.8257
	SH-Line 7	0.4973	0.7020
	SH-Line 8	0.6560	0.8782
	SH-Line 9	0.6037	0.9204
	SH-Line 10	0.5341	0.7536
	SH-Line 11	0.5230	0.8840
	SH-Line 12	0.4086	0.6036
	SH-Line 13	0.4006	0.5204
	SH-Line 16	0.4028	0.8354
	SH-Line 17	0.4325	0.6074
	SH-Pujiang Line	1.0000	1.0000
	SH-Maglev Line	1.0000	1.0000
Guangzhou	GZ-Line 1	1.0000	1.0000
	GZ-Line 2	1.0000	1.0000
	GZ-Line 3	1.0000	1.0000
	GZ-Line 4	0.3907	0.5833
	GZ-Line 5	0.7386	0.7233
	GZ-Line 6	0.5449	0.7320
	GZ-Line 7	0.6451	0.6479
	GZ-Line 8	0.6409	1.0000
	GZ-Line 9	0.4179	0.7522
	GZ-Line 13	0.4449	0.6373
	GZ-Line 14	0.3138	0.4875
	GZ-Line 21	0.3319	0.4218
	GZ-APM Line	1.0000	1.0000
	GZ-Guangfo Line	0.5657	0.7845
Shenzhen	SZ-Line 1	0.5881	0.6987
	SZ-Line 2	0.3099	0.5008
	SZ-Line 3	0.5907	0.7982
	SZ-Line 4	0.6102	0.8385
	SZ-Line 5	0.6361	0.7252
	SZ-Line 6	0.3345	0.7195
	SZ-Line 7	0.4164	0.5595
	SZ-Line 9	0.3310	0.3948
	SZ-Line 10	0.3398	1.0000
	SZ-Line 11	1.0000	1.0000
	Average	0.5634	0.7641

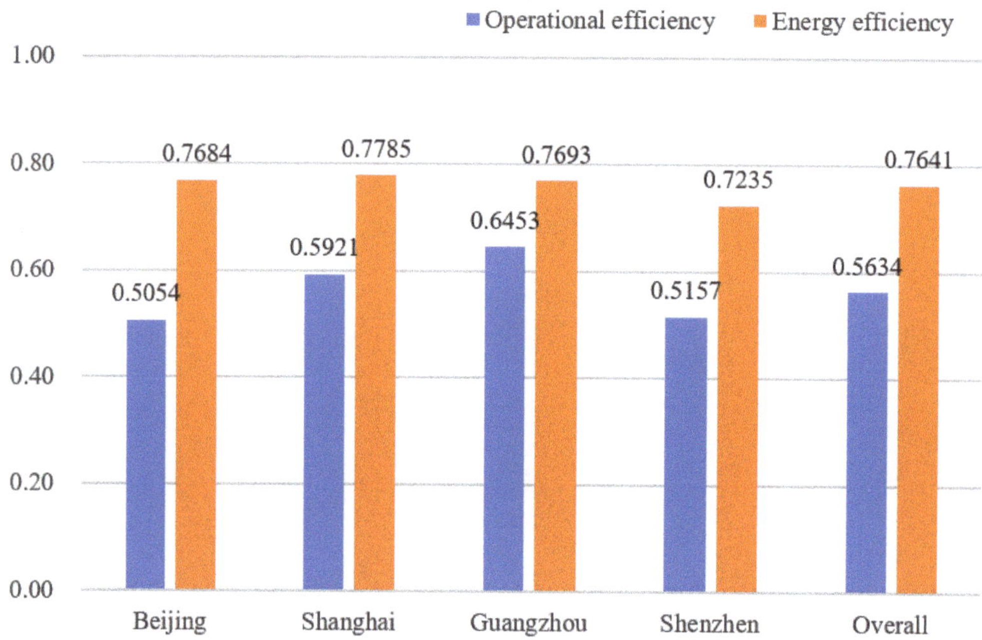

Figure 3. The average efficiency of the URT systems in case cities.

In particular, it can be seen that around five-sixths of the URT lines are inefficient. In Beijing, the operational efficiencies of 2 out of 20 observed URT lines are efficient, another 3 lines are over the overall average level, and 15 lines are under the overall average level. In Shanghai, the operational efficiencies of 3 out of 17 observed URT lines are efficient, another 4 lines are over the overall average level, and 10 lines are below the overall average level. In Guangzhou, the operational efficiencies of 4 out of 14 observed URT lines are efficient, another 4 lines are over the overall average level, and 6 lines are below the overall average level. In Shenzhen, the operational efficiencies of 1 out of 10 observed URT lines are efficient, another 4 lines are over the overall average level, and 5 lines are below the overall average level. Obviously, the operational efficiencies of most URT lines need to be improved further, as they are underperforming. For instance, the operational efficiency of Beijing Line 8 is 0.2583, suggesting that the operational efficiency can be improved by 30.51% and 76.17% to reach the overall average and optimal level, respectively. In a similar vein, in other case cities, the operational efficiencies of SH-Line 5 (0.3441), GZ-Line 14 (0.3138), and SZ-Line 2 (0.3099) can be improved by 65.59%, 68.62%, and 69.01%, respectively, to reach the optimal level. These lines with poor performance should make great efforts to improve operational efficiency to reach the overall average level first and then pursue a higher efficiency.

Similar results are also observed in energy efficiency. Overall, the average energy efficiency of the URT lines is 0.7641. That is to say, the URT lines are recommended to improve their energy efficiency by 23.59% on average to reach the optimal energy utilization level. From a line perspective, it can be found that of the energy efficiencies of the 61 observed URT lines, 14 of which are evaluated as an efficient level, another 17 lines are over the average level, and 30 lines are under the average level. There is a great disparity among URT lines in energy efficiency. From a city perspective, Figure 3 suggests that the average energy efficiency of the URT lines in Shanghai (0.7785) tops the list. The average operational efficiencies of URT lines in Guangzhou (0.7693) and Beijing (0.7684) are higher than the average level, while those of the URT lines in Shenzhen (0.7235) are lower than the average performance level. That being said, there is no significant difference in energy efficiency between URT lines at the city level. It might be that these cities have developed URT in similar periods, with a mixture of new and old facilities and equipment in the lines.

Additionally, the results illustrate that the energy efficiency of most URT lines is inefficient. In Beijing, the operational efficiencies of 2 out of 20 observed URT lines are efficient, another 10 lines are over the overall average level, and 8 lines are below the overall average level. In Shanghai, the operational efficiencies of 3 out of 17 observed URT lines are efficient, another 6 lines are over the overall average level, and 8 lines are below the overall average level. In Guangzhou, the operational efficiencies of 4 out of 14 observed URT lines are efficient, another 2 lines are over the overall average level, and 6 lines are below the overall average level. In Shenzhen, the operational efficiencies of 2 out of 10 observed URT lines are efficient, another 2 lines are over the overall average level, and 6 lines are below the overall average level. Obviously, the energy efficiencies of most URT lines need to be improved further, as they are underperforming. For instance, the energy efficiency of some of the cases is much lower than the average level (e.g., the energy efficiency of BJ-Line 7 is 0.3621), suggesting that the operational efficiencies can be improved by 40.2% and 63.79% to reach the overall average and optimal level respectively. In a similar vein, in other case cities, the operational efficiencies of SH-Line 7 (0.3092), GZ-Line 21 (0.4218), and SZ-Line 9 (0.3974) can be improved by 69.08%, 57.82%, and 59.36%, respectively, to reach the optimal level. These lines with worse performance should make more efforts to improve operational efficiency to reach the overall average level first and then pursue a higher efficiency.

In other words, the efficiency of the energy consumption of these URT systems is optimized. Furthermore, of the 61 observed URT systems, 31 of them are above the average level; the energy efficiency of 14 observed URT systems is optimized. For those higher than the average level, the energy efficiency of 3 out of 20 URT systems in Beijing is optimized; the energy efficiency in 11 URT systems is above the average level). Likewise, 3 out of 17 URT systems in Shanghai are optimized in terms of energy efficiency; nine URT systems in Shanghai perform better than the average level in terms of energy efficiency. Meanwhile, in Guangzhou, 5 out of 14 URT systems reach the ideal level of energy consumption efficiency; the energy efficiency of nine URT systems in Guangzhou is higher than the average level. In Shenzhen, 2 out of 10 URT systems are fully optimized; the energy utilization level of four URT systems in Shenzhen is higher than the average level. In these cases, some of them are close to the optimal level. For example, the energy efficiency of BJ-Line 2 is 0.9338, which demonstrates a significant potential to reach the ideal energy consumption efficiency. In other cases, some of them are under the average level of energy consumption efficiency. For instance, the energy efficiency of the BJ-Fangshan Line is 0.736, which is close to the average value. In other words, there is a potential to further improve performance beyond the average level. Furthermore, the energy efficiency of some of the cases is much lower than the average level (e.g., the energy efficiency of SZ-Line 9 is 0.3948).

In addition to the efficiencies across cities, Table 3 reports a comparison of the efficiencies of URT lines operated by joint ventures and state-owned enterprises. The average operational efficiency of the state-owned lines (0.5684) is higher than that of the joint lines (0.4658). Specifically, there are three lines operated by joint ventures (i.e., BJ-Line 4, BJ-Yanfang Line, and SZ-Line 4). Only the operational efficiency of SZ-Line 4 (0.6102) is higher than the average level.

Table 3. The average efficiency of the URT systems in case cities.

Type	Operational Efficiency	Energy Efficiency
Joint venture	0.4658	0.8678
State-owned enterprise	0.5684	0.7587
Overall	0.5634	0.7641

Regarding energy efficiency, the average energy efficiency of URT lines operated by joint ventures is 0.8678, which is higher than the overall energy efficiency (0.7641), while the average energy efficiency of URT lines operated by state-owned enterprises (0.7587) is slightly lower than the overall value. The reason may be that the joint-owned lines were built in a more recent period, with more new energy-saving technologies. To sum up, state-owned enterprises are better at improving operational efficiency, while joint ventures are more concentrated on energy efficiency. This may be due to the difference between the two ownership models. In this sense, operators are encouraged to learn from each other's management and technology advantages so as to maximize their efficiencies.

3.3. Improvement Analysis

As shown in Table 4 and Figure 4, the improvement potentials of inputs and outputs for the URT lines and case cities are presented. As mentioned in the previous methodology section, line mileage and station are not discussed in the adjustment analysis, as they cannot be easily changed after they are built.

Table 4. The improvement values of the URT systems in case cities.

City	Line Name	Train	Energy (10^4 kwh)	Passenger Transport Volume (10^4 PT)	Revenue Passenger Kilometers (10^4 PK)	CO_2 (10^4 tons)
Beijing	BJ-Line 1	−56.63% (−39.64)	−37.36% (−5503.57)	0.00% (0.00)	0.00% (0.00)	−46.57% (−6.46)
	BJ-Line 2	−51.20% (−25.60)	−6.62% (−593.65)	0.00% (0.00)	67.99% (55475.85)	−20.27% (−1.71)
	BJ-Line 4	−57.68% (−49.60)	−23.52% (−4126.99)	10.42% (2489.35)	0.00% (0.00)	−34.70% (−5.73)
	BJ-Line 5	−52.85% (−32.24)	−20.47% (−2571.82)	0.00% (0.00)	0.00% (0.00)	−32.09% (−3.80)
	BJ-Line 6	−50.63% (−42.53)	−43.31% (−11258.96)	35.91% (7660.91)	0.00% (0.00)	−51.62% (−12.64)
	BJ-Line 7	−71.66% (−48.73)	−63.79% (−10938.69)	0.00% (0.00)	5.20% (4802.56)	−69.08% (−11.16)
	BJ-Line 8	−80.87% (−89.77)	−60.63% (−8717.98)	5.76% (566.00)	0.00% (0.00)	−66.46% (−9.00)
	BJ-Line 9	−50.70% (−19.26)	−14.11% (−982.93)	0.00% (0.00)	30.79% (21903.56)	−26.66% (−1.75)
	BJ-Line 10	−60.80% (−70.53)	−31.84% (−8025.87)	0.00% (0.00)	0.00% (0.00)	−41.80% (−9.93)
	BJ-Line 15	−32.19% (−10.94)	−28.64% (−3132.13)	67.37% (6002.66)	0.00% (0.00)	−39.07% (−4.02)
	BJ-Line 16	−43.35% (−16.47)	0.00% (−0.00)	322.69% (8138.21)	259.66% (65798.21)	−14.82% (−0.81)
	BJ-Ba Tong Line	−53.53% (−19.81)	−23.05% (−1153.24)	21.12% (1163.62)	0.00% (0.00)	−34.55% (−1.63)
	BJ-Changping Line	−37.77% (−12.09)	−29.02% (−2072.49)	64.82% (3608.44)	0.00% (0.00)	−39.56% (−2.66)
	BJ-Fangshan Line	−59.57% (−26.21)	−26.40% (−1476.97)	78.65% (3180.23)	0.00% (0.00)	−37.38% (−1.97)
	BJ-Yizhuang Line	−41.39% (−9.52)	−22.74% (−1110.60)	40.54% (1926.50)	0.00% (0.00)	−34.03% (−1.57)
	BJ-Line S1	0.00% (−0.00)	−16.87% (−455.47)	119.00% (1161.69)	178.52% (9366.26)	−29.63% (−0.75)
	BJ-Yanfang Line	−30.64% (−4.90)	0.00% (−0.00)	579.51% (3000.11)	520.04% (20791.87)	−14.81% (−0.31)
	Daxing airport express	−46.73% (−5.61)	−14.78% (−1055.11)	273.57% (1595.43)	8.57% (1805.91)	−28.15% (−1.89)
Shanghai	SH-Line 1	−44.50% (−36.94)	−18.98% (−3803.09)	5.19% (1592.55)	0.00% (0.00)	−17.74% (−2.82)
	SH-Line 2	−40.11% (−35.29)	−34.07% (−10220.53)	4.29% (1616.63)	0.00% (0.00)	−33.08% (−7.86)
	SH-Line 3	−56.07% (−27.48)	−31.87% (−3348.62)	5.14% (664.59)	0.00% (0.00)	−30.83% (−2.57)
	SH-Line 5	−65.84% (−32.92)	−44.70% (−3071.10)	31.34% (1567.34)	0.00% (0.00)	−44.07% (−2.40)
	SH-Line 6	−69.64% (−43.87)	−17.43% (−1297.95)	0.00% (0.00)	30.46% (22440.93)	−16.17% (−0.95)
	SH-Line 7	−62.17% (−49.11)	−29.80% (−4531.66)	0.93% (190.95)	0.00% (0.00)	−28.72% (−3.46)
	SH-Line 8	−55.80% (−47.99)	−12.18% (−1843.22)	0.00% (0.00)	0.00% (0.00)	−10.84% (−1.30)
	SH-Line 9	−51.77% (−53.84)	−7.96% (−1611.06)	34.10% (9404.77)	0.00% (0.00)	−6.58% (−1.06)
	SH-Line 10	−42.56% (−23.41)	−24.64% (−3725.59)	0.00% (0.00)	15.99% (26597.23)	−23.49% (−2.81)
	SH-Line 11	−39.92% (−32.73)	−11.60% (−2392.19)	60.66% (13652.48)	0.00% (0.00)	−10.28% (−1.68)
	SH-Line 12	−63.97% (−46.70)	−39.64% (−6021.89)	0.00% (0.00)	32.01% (36809.27)	−38.72% (−4.66)
	SH-Line 13	−59.88% (−37.12)	−47.96% (−7897.27)	0.00% (0.00)	21.57% (24977.69)	−47.17% (−6.15)
	SH-Line 16	−55.89% (−34.09)	−16.46% (−1616.47)	170.52% (9836.12)	0.00% (0.00)	−15.30% (−1.19)
	SH-Line 17	−34.49% (−9.66)	−39.26% (−2822.11)	68.25% (3136.13)	0.00% (0.00)	−38.53% (−2.19)
Guangzhou	GZ-Line 4	−55.07% (−31.39)	−41.67% (−5447.91)	23.78% (2768.70)	0.00% (0.00)	−41.77% (−4.39)
	GZ-Line 5	−27.88% (−17.28)	−27.67% (−5869.81)	0.00% (0.00)	0.00% (0.00)	−27.67% (−4.72)
	GZ-Line 6	−40.13% (−22.07)	−26.80% (−4378.77)	0.00% (0.00)	30.83% (47904.33)	−26.80% (−3.52)
	GZ-Line 7	−26.64% (−6.13)	−35.21% (−2014.83)	0.00% (0.00)	9.73% (4676.85)	−35.47% (−1.63)
	GZ-Line 8	−26.27% (−10.25)	0.00% (−0.00)	0.00% (0.00)	61.87% (61432.57)	−0.08% (−0.01)
	GZ-Line 9	0.00% (−0.00)	−24.78% (−1471.49)	157.10% (4923.21)	123.30% (36153.98)	−24.84% (−1.19)
	GZ-Line 13	0.00% (−0.00)	−36.27% (−2710.67)	144.62% (5111.42)	46.95% (22703.99)	−36.34% (−2.18)
	GZ-Line 14	−35.21% (−11.27)	−51.25% (−7174.10)	121.65% (7087.50)	0.00% (0.00)	−51.25% (−5.77)
	GZ-Line 21	−37.15% (−11.14)	−57.82% (−8272.38)	75.70% (4869.10)	0.00% (0.00)	−57.82% (−6.65)

Table 4. *Cont.*

City	Line Name	Train	Energy (10⁴ kwh)	Passenger Transport Volume (10⁴ PT)	Revenue Passenger Kilometers (10⁴ PK)	CO$_2$ (10⁴ tons)
GZ-Guangfo Line		−39.83% (−15.14)	−21.55% (−2120.81)	0.00% (0.00)	4.42% (5293.87)	−21.55% (−1.71)
Shenzhen	SZ-Line 1	−52.26% (−44.42)	−30.13% (−6193.01)	0.00% (0.00)	0.00% (0.00)	−30.13% (−4.98)
	SZ-Line 2	−68.87% (−52.34)	−49.92% (−8033.73)	0.00% (0.00)	33.50% (32952.38)	−49.92% (−6.46)
	SZ-Line 3	−50.58% (−37.94)	−20.18% (−3226.14)	9.10% (2079.68)	0.00% (0.00)	−20.24% (−2.60)
	SZ-Line 4	−50.38% (−24.18)	−16.15% (−1564.35)	0.00% (0.00)	6.23% (7771.33)	−16.15% (−1.26)
	SZ-Line 5	−30.23% (−17.53)	−27.48% (−5734.49)	1.75% (513.67)	0.00% (0.00)	−27.48% (−4.61)
	SZ-Line 6	−53.24% (−14.91)	−28.05% (−1407.52)	74.48% (2708.38)	0.00% (0.00)	−28.05% (−1.13)
	SZ-Line 7	−41.95% (−17.20)	−44.05% (−6390.59)	0.00% (0.00)	80.58% (59081.94)	−44.05% (−5.14)
	SZ-Line 9	−54.02% (−25.93)	−60.52% (−11326.23)	0.00% (0.00)	49.71% (39498.59)	−60.52% (−9.11)
	SZ-Line 10	−39.21% (−10.20)	0.00% (−0.00)	120.56% (4742.59)	143.44% (41412.43)	0.00% (−0.00)

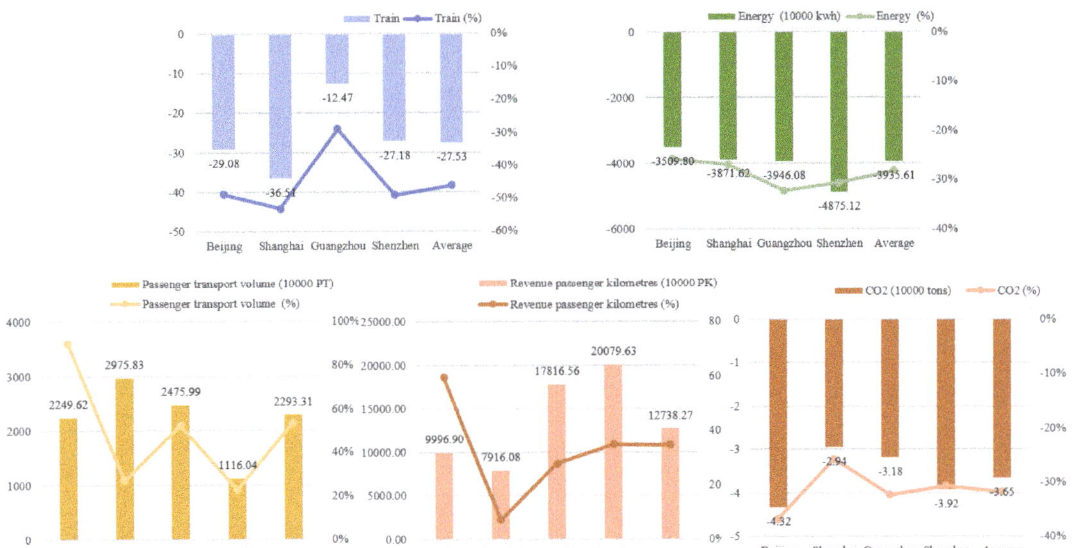

Figure 4. The average improvement values of the URT systems in case cities.

3.3.1. Input Adjustment Plan

In terms of the number of allocated trains, the average improvement value of 51 inefficient lines is 46.07% (27.53). Only three URT lines (i.e., BJ-Line S1, GZ-Line 9, and GZ-Line 13) reach the optimal level. In total, 20 URT lines are under the average level, while 28 lines are above the average level. From the perspective of operation, there is a need to calculate the optimal number of trains and develop a dynamic scheduling mechanism. Different types of trains (e.g., short trains can be used during the off-peak period) should be used to optimize overall efficiency. For instance, for SZ-Line 10, 39.21% (10.20) of trains can be reduced based on optimal efficiency. Furthermore, some lines, such as BJ-Line 8 (80.87%) and SH-Line 6 (69.64%), show a high improvement potential to reach the maximized resource utilization level. In this sense, attention should be paid to such URT lines to optimize the number of allocated trains. At the city level, the average improvement values of the number of allocated trains for Beijing, Shanghai, Guangzhou, and Shenzhen are −48.79%, −53.04%, −28.82%, and −46.07%, respectively. Namely, Shanghai tops the list, while Guangzhou is closer to the ideal level compared with other case cities.

Regarding energy, the average improvement value of the lines is 28.22% (39.35 million kWh). Only four URT lines (i.e., BJ-Line 16, BJ-Yanfang Line, GZ-Line 8, and SZ-Line 10) reach the optimal level. In total, 24 lines are under the average level, while 23 lines are above the average level. That is to say, for most of the URT lines, there is a lot of room to improve overall efficiency by reducing energy. For instance, based on the benchmark, the energy consumed by BJ-Line 6 can be reduced by around 43.31% (11.26 million kWh) to minimize energy wastage. Particularly, some lines (e.g., BJ-Line 7, BJ-Line 8, SH-Line 5, GZ-Line 14, GZ-Line 21, and SZ-Line 9) should take measures to improve the utilization of energy for their greater potential. At the city level, the average improvement values of the energy of Guangzhou (−32.30%) and Shenzhen (−30.72%) are larger than the average level, while those of Beijing (−25.73%) and Shanghai (−26.90%) are smaller than the average level. This indicates that the inefficient URT lines in Guangzhou and Shenzhen deserve more attention in terms of energy conservation.

3.3.2. Output Adjustment Plan

In addition to the input plan, an improvement plan to maximize outputs is demonstrated. Firstly, in terms of passenger transport volume, the average improvement value of the passenger transport volume of observed lines is 53.50% (22.93 million person-times). In total, 21 URT lines (e.g., BJ-Line 1, SH-Line 6, GZ-Line 5, and SZ-Line 1) reach the optimal level. However, 16 lines are under the average level, while 14 lines are above the average level. Some lines (e.g., BJ-Yanfang Line, Daxing Airport Express, and SH-Line 16,) should improve passenger transport volume as much as possible for the lower output. At the city level, the average improvement value of the passenger transport volume of Shenzhen's URT lines is the closest to the optimal level among the case cities (i.e., 22.88%). By contrast, based on the results, the improvement values of Beijing (i.e., 89.96%) and Guangzhou (i.e., 52.28%) are lower than the average level. The lines with great improvement potential should be encouraged to expand passenger transport volume.

As for revenue passenger kilometers, the average improvement value of the URT lines is 34.54% (127.38 million passenger kilometers). In total, 29 URT lines (e.g., BJ-Line 1 and SZ-Line 1) are optimized, while 3 lines are above the average level and 19 lines are lower than the average value. It can be seen that most of the URT lines have produced sufficient passenger turnover output, while some lines have great improvement potential in passenger turnover, such as BJ-Line 16 (i.e., 259.66%) and BJ-Yanfang Line (i.e., 520.04%). From the city perspective, the average improvement value of URT lines in Shanghai is 7.15%, which is closer to the optimal level. At another extreme, the average improvement value of the URT lines in Beijing is 59.49%, which is much lower than the optimal level. The situations for Guangzhou and Shenzhen are between them.

Concerning CO_2 emissions, the average improvement value of URT lines is 31.82% (36.5 kilotons). Only SZ-Line 10 reached the optimal level, while another 25 lines are above the average value and 26 are less than the average value. In particular, some lines are significantly lower than the optimal level, such as BJ-Line 7 (69.08%) and SZ-Line 9 (60.52%). There is a lot of room for these lines to decline CO_2 emissions to maximize environmental sustainability. At the city level, compared with other cities, the average improvement value of CO_2 emissions for the URT lines in Shanghai (25.82%) is closer to the ideal level. On the contrary, the largest gap between the actual CO_2 emissions and the ideal emissions can be found in Beijing's URT lines (36.74%).

4. Discussion

First, the improvement values reveal that the efficiency of the URT systems can be improved by reducing unessential wastage on the input side. In this sense, the number of the same type of trains can be appropriately reduced, and redundant trains can be sent to other lines or other cities to improve utilization. In terms of energy, for one thing, the application of new energy-saving technologies and the dynamic marshaling of trains according to real-time passenger flow can reduce the energy consumption of train traction. For another, new technology in heating and air conditioning equipment can be used to reduce the operation energy consumption of station facilities for heating and cooling. Reducing energy consumption reduces the corresponding undesirable carbon emissions, which is conducive to improving efficiency. In this sense, the infrastructure and facilities can be updated by adopting new technologies or management techniques. In response to this, for the URT lines built in the early period (e.g., BJ-Line 1 and SH-Line 2), the local authorities should encourage operators to upgrade the trains and station facilities by adopting new technologies to improve energy efficiency and reduce carbon emissions. Therefore, in addition, there is also a need for operators to collaborate with other stakeholders (e.g., the local government and research institutions) to develop a multi-dimensional method to improve passenger turnover efficiency for stations in different locations (e.g., a preference policy can be developed for those using other means of transportation during rush hours). Moreover, the efficiency of the URT systems can be enhanced by increasing desirable outputs. Based on the results, it can be seen that the operational efficiency of new lines is relatively lower than those built in the earlier period. Taking Shanghai as an example, the average operational efficiency of SH-Line 1 and SH-Line 2 is higher than that of SH-Line 16 and SH-Line 17. One reason might be that operational efficiency is associated with passenger volume. The operational efficiency of lines close to the city center is relatively high compared with the operational efficiency of those close to suburban areas. This provides a management implication, in that increasing passenger volume can help improve operational efficiency. On the one hand, the government should encourage URT operators to strengthen cooperation with other transportation service providers (e.g., bus companies, taxi companies, bike-sharing companies) and promote their joint operation to provide convenient transfer conditions to attract passenger flow. On the other hand, operators can develop a preference policy and adjust ticket prices, such as discount sales for inefficient lines at certain fixed times, in order to entice citizens to take rail transit. This may be an effective way to improve operational efficiency in the short term.

In addition, more investment should be made in advanced technologies, such as 5G communication technology, big data, artificial intelligence, and industrial Internet, to build smart URT systems to enhance efficiency. In terms of stations, existing stations can be upgraded to smart stations, which can provide passengers with intelligent security checks, intelligent customer service centers, intelligent guidance, and other services. A series of intelligent systems, such as intelligent passenger guide screens, multimedia platform screens, intelligent ticket machines, and intelligent customer service centers, can be installed to provide passengers with refined and intelligent travel services through the real-time perception, acquisition, and transmission of operation information. In terms of lines, on the one hand, new intelligent technologies should be applied to the operation and maintenance of lines to reduce relevant costs. On the other hand, new lines should be fully automated, which can save labor costs and improve efficiency. From this angle, the construction of smart URT systems is an important way to improve operational efficiency and energy efficiency and achieve better development in the URT sector.

5. Conclusions

With the unprecedented development of the URT in China, a certain number of studies have explored the evaluation of URT efficiencies. However, carbon emissions are rarely taken into account in the estimation process in existing studies. Considering the importance of emission reduction and URT line heterogeneity, this paper considers CO_2 as undesirable output and constructs an efficiency evaluation model based on the SBM, which can estimate the operational efficiency and energy efficiency for URT lines.

The proposed model was applied to evaluate the efficiency of 61 URT lines in four megacities in China. The empirical findings show that the URT lines in Guangzhou perform better in terms of operational efficiency, while the average energy efficiency of URT systems in Shanghai is higher than in other case cities. In addition, the average overall operational efficiency of URT lines in case cities is relatively low compared with energy efficiency, and there is a lot of room for improvement. A comparison of the efficiency of URT systems operated by state-owned enterprises and joint ventures indicates that state-owned enterprises are better at improving operational efficiency, while joint ventures are better at improving energy efficiency.

The limitations of this current paper should also be clarified, and some further research can be extended in the future. First, we only adopted the 2020 data of 61 URT lines in China to evaluate operational efficiency and energy efficiency in this paper. A study with more URT lines and multi-year panel data may explore the long-term dynamic changes in efficiency and obtain new management implications. Second, this paper does not consider service quality indicators from the passenger's perspective. URT systems aim to provide comfortable, convenient, and fast transport services for citizens. In future research, service quality factors such as transport congestion and service satisfaction degree can be adopted as outputs to comprehensively evaluate performance. Third, energy efficiency at the station level may provide a new perspective on energy saving and emission reduction for URT operations. In other words, more investigations can be conducted to provide deeper insights regarding energy efficiency at the station level. Last but not least, the convenience of transfer and joint operations between URT and bus systems may be important ways to improve operational efficiency and energy efficiency, which are also two important research directions that need to be further investigated.

Author Contributions: Conceptualization, H.Z. and X.W.; methodology, H.Z. and L.C.; software, Y.L.; data curation, L.C. and X.W.; writing—original draft preparation, H.Z. and S.P.; writing—review and editing, H.Z., X.W. and S.P.; visualization, Y.L.; funding acquisition, H.Z. All authors have read and agreed to the published version of the manuscript.

Funding: This research was funded by the Launching Scientific Research Fund from the University of Shanghai for Science and Technology, grant number BSQD202110.

Data Availability Statement: The data can be found in the yearbook of China Urban Rail Transit Almanac 2021 (in Chinese) and also be available from the corresponding author upon reasonable request.

Conflicts of Interest: The authors declare no conflict of interest.

Appendix A

Based on the proposed model, the indicator of energy improvement potential can be defined as the ratio of the difference between the actual value and the target value to the actual value, i.e.:

$$PE_i = \frac{TE_i - XE_i}{XE_i} \tag{A1}$$

Generally speaking, the infrastructures of the URT system are difficult to adjust further in the short term once they have been constructed. Therefore, we aim to investigate the improvement potentials for train, energy, passenger transport volume, revenue passenger kilometers, and CO_2 emissions. Similarly, the targets of train, passenger transport volume, revenue passenger kilometers, and CO_2 emissions are expressed as follows:

$$TT_i = \sum_{j=1}^{n} \lambda_j XT_j \tag{A2}$$

$$TP_i = \sum_{j=1}^{n} \lambda_j YP_j \tag{A3}$$

$$TR_i = \sum_{j=1}^{n} \lambda_j YR_j \tag{A4}$$

$$TC_i = \sum_{j=1}^{n} \lambda_j YC_j \tag{A5}$$

Likewise, the improvement potentials of train, passenger transport volume, revenue passenger kilometers, and CO_2 emissions can be formulated as follows:

$$PT_i = \frac{TT_i - XT_i}{XT_i} \tag{A6}$$

$$PP_i = \frac{TP_i - YP_i}{YP_i} \tag{A7}$$

$$PR_i = \frac{TR_i - YR_i}{YR_i} \tag{A8}$$

$$PC_i = \frac{TC_i - YC_i}{YC_i} \tag{A9}$$

References

1. Lu, K.; Han, B.; Lu, F.; Wang, Z. Urban Rail Transit in China: Progress Report and Analysis (2008–2015). *Urban Rail Transit* **2016**, *2*, 93–105. [CrossRef]
2. China Association of Urban Rail Transit. 2022. Available online: https://www.camet.org.cn/ (accessed on 30 May 2022).
3. Kuang, X. Evaluation of railway transportation efficiency based on super-cross efficiency. *IOP Conf. Series: Earth Environ. Sci.* **2018**, *108*, 032049. [CrossRef]
4. Michali, M.; Emrouznejad, A.; Dehnokhalaji, A.; Clegg, B. Noise-pollution efficiency analysis of European railways: A network DEA model. *Transp. Res. Part D Transp. Environ.* **2021**, *98*, 102980. [CrossRef]
5. Roy, W.; Yvrande-Billon, A. Ownership, contractual practices and technical efficiency: The case of urban public transport in France. *J. Transp. Econ. Policy (JTEP)* **2007**, *41*, 257–282.
6. Ottoz, E.; Fornengo, G.; Di Giacomo, M. The impact of ownership on the cost of bus service provision: An example from Italy. *Appl. Econ.* **2009**, *41*, 337–349. [CrossRef]
7. Wu, J.; Zhu, Q.; Chu, J.; Liu, H.; Liang, L. Measuring energy and environmental efficiency of transportation systems in China based on a parallel DEA approach. *Transp. Res. Part D Transp. Environ.* **2016**, *48*, 460–472. [CrossRef]
8. Li, T.; Yang, W.; Zhang, H.; Cao, X. Evaluating the impact of transport investment on the efficiency of regional integrated transport systems in China. *Transp. Policy* **2016**, *45*, 66–76. [CrossRef]
9. Cinelli, M.; Kadziński, M.; Miebs, G.; Gonzalez, M.; Słowiński, R. Recommending multiple criteria decision analysis methods with a new taxonomy-based decision support system. *Eur. J. Oper. Res.* **2022**, *302*, 633–651. [CrossRef]
10. Sałabun, W.; Wątróbski, J.; Shekhovtsov, A. Are MCDA Methods Benchmarkable? A Comparative Study of Topsis, Vikor, Copras, and Promethee Ii Methods. *Symmetry* **2020**, *12*, 1549. [CrossRef]
11. Wątróbski, J.; Jankowski, J.; Ziemba, P.; Karczmarczyk, A.; Zioło, M. Generalised framework for multi-criteria method selection. *Omega* **2019**, *86*, 107–124. [CrossRef]
12. Charnes, A.; Cooper, W.W.; Rhodes, E. Measuring the efficiency of decision making units. *Eur. J. Oper. Res.* **1978**, *2*, 429–444. [CrossRef]
13. Lan, L.W.; Lin, E.T. Technical efficiency and service effectiveness for railways industry: DEA approaches. *J. East. Asia Soc. Transp. Stud.* **2003**, *5*, 2932–2947.
14. Lan, L.W.; Lin, E.T. Measuring railway performance with adjustment of environmental effects, data noise and slacks. *Transportmetrica* **2005**, *1*, 161–189. [CrossRef]
15. Oum, T.H.; Waters, W.G.; Yu, C. A survey of productivity and efficiency measurement in rail transport. *J. Transp. Econ. Policy* **1999**, 9–42.
16. Barnum, D.T.; Karlaftis, M.G.; Tandon, S. Improving the efficiency of metropolitan area transit by joint analysis of its multiple providers. *Transp. Res. Part E Logist. Transp. Rev.* **2011**, *47*, 1160–1176. [CrossRef]
17. Fielding, G.J.; Babitsky, T.T.; Brenner, M.E. Performance evaluation for bus transit. *Transp. Res. Part A: Gen.* **1985**, *19*, 73–82. [CrossRef]
18. Georgiadis, G.; Politis, I.; Papaioannou, P. Measuring and improving the efficiency and effectiveness of bus public transport systems. *Res. Transp. Econ.* **2014**, *48*, 84–91. [CrossRef]
19. Karlaftis, M.G. A DEA approach for evaluating the efficiency and effectiveness of urban transit systems. *Eur. J. Oper. Res.* **2004**, *152*, 354–364. [CrossRef]
20. Chen, C.; Lam, J.S.L. Sustainability and interactivity between cities and ports: A two-stage data envelopment analysis (DEA) approach. *Marit. Policy Manag.* **2018**, *45*, 1–18. [CrossRef]
21. Kuo, K.-C.; Lu, W.-M.; Le, M.-H. Exploring the performance and competitiveness of Vietnam port industry using DEA. *Asian J. Shipp. Logist.* **2020**, *36*, 136–144. [CrossRef]
22. Lozano, S.; Gutiérrez, E.; Moreno, P. Network DEA approach to airports performance assessment considering undesirable outputs. *Appl. Math. Model.* **2013**, *37*, 1665–1676. [CrossRef]

23. Huang, Y.K. The Effect of Airline Service Quality on Passengers' Behavioural Intentions Using SERVQUAL Scores: A TAIWAN Case Study. *J. East. Asia Soc. Transp. Stud.* **2010**, *8*, 2330–2343.

24. Lozano, S.; Gutiérrez, E. A slacks-based network DEA efficiency analysis of European airlines. *Transp. Plan. Technol.* **2014**, *37*, 623–637. [CrossRef]

25. Jain, P.; Cullinane, S.; Cullinane, K. The impact of governance development models on urban rail efficiency. *Transp. Res. Part A Policy Pr.* **2008**, *42*, 1238–1250. [CrossRef]

26. Qin, F.; Zhang, X.; Zhou, Q. Evaluating the impact of organizational patterns on the efficiency of urban rail transit systems in China. *J. Transp. Geogr.* **2014**, *40*, 89–99. [CrossRef]

27. Tsai, C.H.P.; Mulley, C.; Merkert, R. Measuring the cost efficiency of urban rail systems an international comparison using DEA and tobit models. *J. Transp. Econ. Policy (JTEP)* **2015**, *49*, 17–34.

28. Costa, Á.; Cruz, C.O.; Sarmento, J.M.; Sousa, V.F. Empirical Analysis of the Effects of Ownership Model (Public vs. Private) on the Efficiency of Urban Rail Firms. *Sustainability* **2021**, *13*, 13346. [CrossRef]

29. Kang, C.-C.; Feng, C.-M.; Chou, P.-F.; Wey, W.-M.; Khan, H.A. Mixed network DEA models with shared resources for measuring and decomposing performance of public transportation systems. *Res. Transp. Bus. Manag.* **2022**, 100828. [CrossRef]

30. Le, Y.; Oka, M.; Kato, H. Efficiencies of the urban railway lines incorporating financial performance and in-vehicle congestion in the Tokyo Metropolitan Area. *Transp. Policy* **2021**, *116*, 343–354. [CrossRef]

31. Song, X.; Hao, Y.; Zhu, X. Analysis of the Environmental Efficiency of the Chinese Transportation Sector Using an Undesirable Output Slacks-Based Measure Data Envelopment Analysis Model. *Sustainability* **2015**, *7*, 9187–9206. [CrossRef]

32. Park, Y.S.; Lim, S.H.; Egilmez, G.; Szmerekovsky, J. Environmental efficiency assessment of U.S. transport sector: A slack-based data envelopment analysis approach. *Transp. Res. Part D Transp. Environ.* **2018**, *61*, 152–164. [CrossRef]

33. Liu, H.; Zhang, Y.; Zhu, Q.; Chu, J. Environmental efficiency of land transportation in China: A parallel slack-based measure for regional and temporal analysis. *J. Clean. Prod.* **2017**, *142*, 867–876. [CrossRef]

34. Cui, Q.; Li, Y. The evaluation of transportation energy efficiency: An application of three-stage virtual frontier DEA. *Transp. Res. Part D Transp. Environ.* **2014**, *29*, 1–11. [CrossRef]

35. Djordjević, B.; Krmac, E. Evaluation of Energy-Environment Efficiency of European Transport Sectors: Non-Radial DEA and TOPSIS Approach. *Energies* **2019**, *12*, 2907. [CrossRef]

36. Xiao, X.; Zhong, Z.; Wang, Y.; Zhang, C.; Wu, H. Research on Energy Efficiency Evaluation of Urban Rail Transit Based on DEA-BCC Model. *IOP Conf. Series: Earth Environ. Sci.* **2020**, *435*, 012038. [CrossRef]

37. To, W.; Lee, P.K.; Yu, B.T. Sustainability assessment of an urban rail system—The case of Hong Kong. *J. Clean. Prod.* **2020**, *253*, 119961. [CrossRef]

38. Yu, M.M. Assessing the technical efficiency, service effectiveness, and technical effectiveness of the world's railways through NDEA analysis. *Transp. Res. Part A Policy Pract.* **2008**, *42*, 1283–1294. [CrossRef]

39. Yu, M.-M.; Lin, E.T. Efficiency and effectiveness in railway performance using a multi-activity network DEA model. *Omega* **2008**, *36*, 1005–1017. [CrossRef]

40. Tone, K. A slacks-based measure of efficiency in data envelopment analysis. *Eur. J. Oper. Res.* **2001**, *130*, 498–509. [CrossRef]

41. Zhang, S.; Zhao, X.; Yuan, C.; Wang, X. Technological Bias and Its Influencing Factors in Sustainable Development of China's Transportation. *Sustainability* **2020**, *12*, 5704. [CrossRef]

42. Chu, J.-F.; Wu, J.; Song, M.-L. An SBM-DEA model with parallel computing design for environmental efficiency evaluation in the big data context: A transportation system application. *Ann. Oper. Res.* **2018**, *270*, 105–124. [CrossRef]

43. Tavassoli, M.; Faramarzi, G.R.; Saen, R.F. Efficiency and effectiveness in airline performance using a SBM-NDEA model in the presence of shared input. *J. Air Transp. Manag.* **2014**, *34*, 146–153. [CrossRef]

44. Yu, A.; You, J.; Zhang, H.; Ma, J. Estimation of industrial energy efficiency and corresponding spatial clustering in urban China by a meta-frontier model. *Sustain. Cities Soc.* **2018**, *43*, 290–304. [CrossRef]

Article

Preventive Maintenance Strategy Optimization in Manufacturing System Considering Energy Efficiency and Quality Cost

Liang Yang [1], Qinming Liu [1,*], Tangbin Xia [2], Chunming Ye [1] and Jiaxiang Li [1]

[1] Department of Industrial Engineering, Business School, University of Shanghai for Science and Technology, 516 Jungong Road, Shanghai 200093, China
[2] State Key Laboratory of Mechanical System and Vibration, School of Mechanical Engineering, Shanghai JiaoTong University, SJTU-Fraunhofer Center, Shanghai 200240, China
* Correspondence: qmliu@usst.edu.cn

Abstract: Climate change is a serious challenge facing the world today. Countries are already working together to control carbon emissions and mitigate global warming. Improving energy efficiency is currently one of the main carbon reduction measures proposed by the international community. Within this context, improving energy efficiency in manufacturing systems is crucial to achieving green and low-carbon transformation. The aim of this work is to develop a new preventive maintenance strategy model. The novelty of the model is that it takes into account energy efficiency, maintenance cost, product quality, and the impact of recycling defective products on energy efficiency. Based on the relationship between preventive maintenance cost, operating energy consumption, and failure rate, the correlation coefficient is introduced to obtain the variable preventive maintenance cost and variable operating energy consumption. Then, the cost and energy efficiency models are established, respectively, and finally, the Pareto optimal solution is found by the nondominated sorting genetic algorithm II (NSGAII). The results show that the preventive maintenance strategy proposed in this paper is better than the general maintenance strategy and more relevant to the actual situation of manufacturing systems. The scope of the research in this paper can support the decision of making energy savings and emission reductions in the manufacturing industry, which makes the production, maintenance, quality, and architecture of the manufacturing industry optimized.

Keywords: preventive maintenance; energy efficiency; quality cost; multiobjective optimization; manufacturing system

Citation: Yang, L.; Liu, Q.; Xia, T.; Ye, C.; Li, J. Preventive Maintenance Strategy Optimization in Manufacturing System Considering Energy Efficiency and Quality Cost. *Energies* **2022**, *15*, 8237. https://doi.org/10.3390/en15218237

Academic Editors: Brian D. Fath and Nicu Bizon

Received: 17 August 2022
Accepted: 2 November 2022
Published: 4 November 2022

Publisher's Note: MDPI stays neutral with regard to jurisdictional claims in published maps and institutional affiliations.

1. Introduction

Climate change is a serious challenge facing the world today. Since the industrial revolution, energy consumption has increased year by year. With the increase in carbon dioxide emissions, the global temperature is gradually rising. Carbon dioxide-based greenhouse gas emissions are the main cause of global warming [1,2]. Thus, controlling carbon emissions and mitigating global warming has become an important global issue and is gradually becoming a global consensus. Taking China as an example, to cope with climate change, carbon peaking and carbon neutrality goals have been proposed to promote the construction of an ecological civilization and achieve high-quality development [3].

By comparing the carbon emissions of various industries, it can be found that the manufacturing industry has long accounted for a large proportion, with relevant data showing that over 70% of the carbon dioxide emissions from China come from industrial production or generative emissions [4,5]. As a result, industry, especially the manufacturing sector, has become the main battleground for reducing carbon emissions in China and the key to achieving carbon peak and neutrality targets. As the main body of the national economy, the manufacturing industry needs to carry out green and low-carbon transformation and development to achieve carbon peak and neutrality targets and realize green manufacturing and intelligent manufacturing [6]. Hassan T et al. [7] found that technology to improve energy efficiency is a crucial method to achieve lower carbon emissions and mitigate global warming. Thus, it is critical to improve energy efficiency in the manufacturing system.

Energy consumption in manufacturing is mainly from production equipment. Thus, we need to pay close attention to the energy efficiency of production equipment. Maintenance plays a crucial role in the normal operation of equipment, and maintenance activities affect the reliability of equipment, indirectly affecting the energy efficiency of the equipment. For this reason, it is crucial to take into account energy efficiency in the optimization of maintenance strategies, gradually achieving a transition from condition-centered maintenance to energy-centered maintenance [8]. Many maintenance methods have been proposed in previous studies, such as breakdown maintenance where maintenance is performed after the equipment has failed to return to its normal function. However, this type of maintenance can affect the production schedule, so preventive maintenance is proposed, which predicts the status of equipment and maintains the equipment in advance to keep it in continuous production [9]. As the detailed literature review below shows, there is a wide range of literature that focuses on the cost of preventive maintenance and the quality of the products produced by the equipment. However, only a few focus on the energy consumption and environmental impact of maintenance, and even fewer articles combine cost, quality, and energy consumption. This paper proposes a new preventive maintenance strategy model. The innovation of this paper is that not only the cost is considered in maintenance activities but also the quality loss cost is introduced to constrain the product quality of equipment, the energy consumption is modeled and calculated, and the recovery of defective products is taken into account. The maximization of energy efficiency and the minimization of maintenance costs are taken as the overall optimization objectives to develop the maintenance strategy.

The remainder of the paper is organized as follows. Section 2 presents a short literature review and shows the contributions of this paper. Section 3 describes the problems associated with equipment maintenance and makes some assumptions about the model. Section 4, a multiobjective decision model is constructed in four steps based on identifying decision variables and optimization objectives and then solved according to the NSGAII algorithmic process. Section 5 validates the model using a numerical case. Conclusions, managerial impacts, and future research scopes are discussed in Section 6.

2. Literature Review

Quality control in equipment maintenance has been studied by scholars for a long time. The relationship between maintenance and quality is discussed, and a broad framework is proposed. Two approaches to connecting and modeling this relationship are discussed in the article. The first approach is based on the idea that maintenance affects the failure modes of the equipment and that it should be modeled with the concept of imperfect maintenance. The second approach is based on the quality approach of Taguchi [10]. Subsequent scholars began to link maintenance and quality closely together. On the one hand, excessive maintenance can lead to unnecessary costs. On the other hand, if the equipment is not correctly maintained, this will lead to failures and result in defective products. In an integrated model of maintenance and quality, the literature [11] correlates the failure rate of equipment with the quality of the product to obtain a function of the variation of the product quality. The control of quality is also reflected in costs such as quality loss and maintenance thresholds, and these models can minimize the total cost and ensure high quality products [12–14].

Scholars have researched energy consumption and environmental impact in equipment maintenance. Jiang et al. [15] considered the ecological impact of equipment degradation, the excessive emissions of equipment, and the energy consumption and obtained maintenance thresholds and inspection intervals that were optimal considering energy consumption and CO_2 emissions by minimizing the average expected cost. Tlili et al. [16]

considered the penalties to be incurred when equipment degradation exceeds a critical level and developed two inspection strategies (periodic and nonperiodic), with separate preventive maintenance thresholds and inspection sequences obtained to reduce cost. Chouikhi et al. [17] proposed a condition-based maintenance strategy for production systems to reduce excess greenhouse gas emissions, translated environmental constraints into maintenance thresholds, and determined optimal maintenance inspection cycles by minimizing maintenance costs. Huang et al. [18] developed a data-driven model from the date of distributed sensors to integrate energy conversation and maintenance to determine the optimal level of maintenance. Liu et al. [19] considered the maintenance of wind turbines and correlated energy consumption with the operating costs of equipment to obtain a maintenance strategy by minimizing the expected costs. Horenbeek et al. [20] developed an economic and ecological analysis tool covering a wide range of maintenance policies. The model developed was validated using the example of a turning machine tool. Saez et al. [21] studied the relationship between production environment, quality, reliability, productivity, and energy consumption and proposed a modeling framework for manufacturing systems that integrates systems, machines, and parts.

The above studies are based on the maintenance cost, where the energy consumption and the environmental impact are regarded as the threshold or other influencing factors in the maintenance cost. The modeling and calculation of the specific energy consumption of equipment are not involved. In terms of modeling the energy consumption of equipment, Yan et al. [22] proposed a method for modelling the energy consumption of a machine tool, using the model to obtain the energy consumption of the machine tool during and after maintenance and converting the energy consumption into carbon emissions, thus effectively controlling the impact on the environment. Zhou et al. [23] analyzed the energy consumption of machine tools commonly found in manufacturing, dividing the machine tool energy consumption model into three parts: a linear cutting energy model, a process-oriented machining energy model, and cutting energy consumption for various specific parameters. After summarizing the power consumption characteristics of heavy machine tools, Shang et al. [24] developed a generic power consumption model for heavy machine tools to predict the power consumption and assess the energy consumption state and developed corresponding energy saving strategies, but they did not take into account the variation of energy consumption. Zhou and Yi [25] have linked energy consumption to equipment degradation, elaborated on the variability of energy consumption, and introduced energy quality thresholds to create an energy-oriented decision model. Mawson and Hughes [26] used new technologies such as digital twins to simulate the energy consumption of equipment. Using a digital twin strategy, Bermeo-Ayerbe et al. [27] proposed an online data-driven energy consumption model. Xia et al. [28] modelled the energy consumption of machine tools and tools and proposed an energy-oriented machine tool maintenance and tool replacement strategy to save energy. Aramcharoen and Mativenga [29] carried out a detailed analysis and calculation of the energy consumption of the entire process of machining a machine, including machine start-up, workpiece set-up, machine warm-up, tool change and cutting, and machine shutdown.

In terms of the energy efficiency calculation of equipment, Zhou et al. [30] proposed the concept of effective energy efficiency by considering the energy saving opportunities arising from machine downtime, obtained the optimal maintenance threshold based on the energy saving opportunity window to maximize energy efficiency, and verified the superiority of the model by comparison. Xia et al. [31] modeled the energy attributes to obtain the multiattribute model (MAM), used the energy savings window (ESW) and constructed the MAM-ESW maintenance policy model by considering energy consumption, mass production, and maintenance. Brundage et al. [32] proposed a control scheme where energy opportunity windows were inserted into various machines to reduce the energy consumption and increase profits. Xia et al. [33] proposed a selective maintenance model for energy-oriented series-parallel systems to find a maintenance strategy for each equipment to maximize the energy efficiency. Hoang et al. [34] defined the concepts of the energy efficiency index (EEI) and remaining energy-efficient lifetime (REEL), calculated the various energy consumptions of equipment, and constructed a model to maximize the energy efficiency index. Frigerio and Matta [35] proposed an aggregate control policy framework that determines the optimal control policy by calculating the energy consumption in each machine state and minimizing the expected energy required by the equipment.

The above literature analysis shows that fewer studies integrate equipment maintenance costs, energy efficiency, and product quality, which still need more attention and research. Most of the existing papers examined several of these components. The main contributions of this article include: (1) a preventive maintenance decision optimization model that takes into account energy efficiency, product quality, and maintenance cost with preventive maintenance thresholds and maintenance efficiency as decision variables; (2) a link between preventive maintenance costs, equipment operation energy consumption, and equipment failure rates to obtain more realistic variable preventive maintenance costs and variable operation energy consumption; (3) a recovery model for defective products produced by equipment to reduce energy consumption, which describes the reduction in the number of defective products to be recovered as the equipment degrades by introducing a recovery factor.

3. Problem Description and Hypotheses

3.1. Problem Description

In reality, equipment cannot be restored to a new health state after use; it is in continuous degradation, and the failure probability of equipment is increasing. The failure rate function of equipment can be obtained by simulating the historical data of equipment. The degradation of equipment failure rate is influenced by controllable factors such as maintenance activities and production schedules and uncontrollable factors such as changes in the production environment. The specific impact of maintenance activities will be described in the hypothesis section, and the impact of the production schedule on equipment degradation can be obtained through historical data analysis. As for the impact of environmental changes in the field, only the degradation of equipment under normal environmental conditions is considered in this paper because the environment in which equipment is located varies and is full of randomness.

Generally speaking, the life cycle of production equipment is relatively long. For the convenience of calculation, the time interval between the brand-new condition of equipment and the next replacement is selected as a period to be considered in this paper. During the operational cycle of equipment, only three maintenance actions are adopted, including breakdown maintenance, preventive maintenance, and replacement, as shown in Figure 1. The different conditions of equipment will determine the adopted type of maintenance actions, and the effect of each type of maintenance varies. When equipment reliability reaches the preventive maintenance threshold, preventive maintenance will be executed, and equipment cannot be restored to a new health state but a state between the new state and the state before adopting maintenance. Breakdown maintenance occurs when equipment fails during preventive maintenance intervals. It is impossible that preventive maintenance is always carried out when the number of preventive maintenance reaches a certain amount. The equipment needs to be replaced to reduce the maintenance cost and improve energy efficiency.

Similarly, in actual production, the product quality decreases as the equipment degrades, and as the equipment continues to operate, the number of defective products will increase, resulting in a large portion of the cost of quality loss. Therefore, the problem of recycling defective products is considered in this paper by introducing the recovery coefficient because recycling defective products can save a part of the energy consumption.

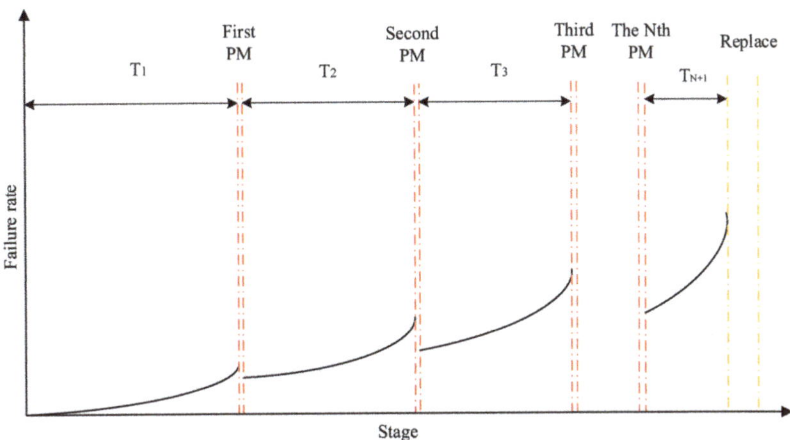

Figure 1. The variation in equipment failure rates and maintenance actions.

3.2. Assumptions

When equipment fails during the preventive maintenance period, it is supposed to be shut down for breakdown maintenance. The time and cost of breakdown maintenance are fixed, and the breakdown maintenance will not change the failure rate of the equipment. By adopting maintenance, equipment will return to the state before failure.

(1) Preventive maintenance is required when equipment reliability reaches the threshold, and the equipment is in a standby state during maintenance. The cost continues to increase with the degradation of equipment, and the health state of equipment after maintenance cannot be repaired to a like-new state.

(2) When the number of preventive maintenance reaches N, equipment will be shut down for replacement at the time of the next preventive maintenance. The time and cost of replacement are fixed, and the health state of the equipment is restored to a brand new condition after replacement.

(3) Equipment-operating energy consumption is variable and increases with degradation. After equipment is shut down and restarted, it needs a warm-up time, expressed as the time required for equipment to run from power to normal operation.

(4) During the production of equipment, the rate of defective products produced increases with degradation.

4. Modeling of the Maintenance Strategy Optimization

The multiobjective maintenance model considering equipment energy efficiency under the variable of cost is described as follows: first, the degradation failure model of equipment is developed by using the Weibull distribution and introducing the failure increasing rate factor and the age reduction factor to simulate the degradation process. Furthermore, based on the relationship between the reliability and failure rate, preventive maintenance intervals are calculated, which lays the foundation for the construction of equipment cost model and energy efficiency model. Then, the variable cost model is developed by considering the cost of different maintenance activities of equipment and considering the quality loss of the products produced by the equipment. Third, by calculating the energy consumption of equipment in each state to obtain the total energy consumption and introducing the recovery coefficient to obtain the effective output of equipment, the energy efficiency model of equipment is constructed. Fourth, the decision variables and optimization goals of the proposed model are determined to build a multiobjective decision-making model. Finally, the NSGAII algorithm is selected, and the model is solved based on the algorithm process. The modeling of the maintenance strategy optimization process is shown in Figure 2.

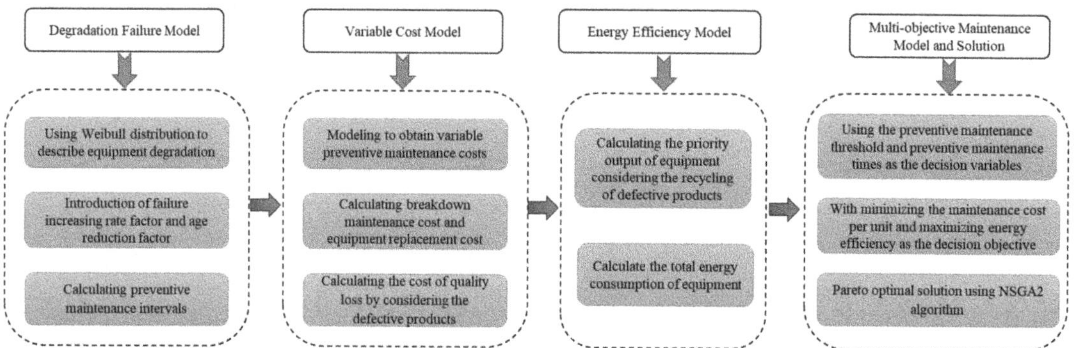

Figure 2. Modeling of the maintenance strategy optimization process.

4.1. Degradation Failure Model

The performance of equipment is continuously degrading as equipment operates. The Weibull distribution is widely used to simulate the cumulative failure analysis of mechanical and electrical equipment. In this paper, the Weibull distribution is used to describe the degradation level of equipment. The failure rate function at running time *t* is expressed as:

$$\lambda(t) = \frac{\beta}{\alpha}\left(\frac{t}{\alpha}\right)^{\beta-1} \tag{1}$$

where α, β are the scale parameters and shape parameters of the Weibull distribution, respectively, which are obtained from the historical maintenance data of equipment. As the health rate of equipment cannot be restored to a new state after preventive maintenance, the failure rate increasing factor $b_i (b_i > 1)$ and service age decreasing factor $a_i (0 < a_i < 1)$ are introduced to express the change in the failure rate. The failure rate increasing factor indicates that the equipment failure rate at each operating moment will increase after preventive maintenance. The service age decreasing factor indicates that the health state of equipment after preventive maintenance will return to a state between the new state and the state before adopting maintenance. The failure rate expression of equipment after the *i*th preventive maintenance can be obtained:

$$\lambda_{i+1} = b_i \lambda_i(t + a_i T_i) \tag{2}$$

There is a certain relationship between the reliability of equipment and the failure rate function. When the equipment reaches the preventive maintenance threshold R^*, preventive maintenance will be carried out. Assuming that *N* times of preventive maintenance are carried out, each preventive maintenance interval can be obtained:

$$e^{-\int_0^{T_i} \lambda_i(t)dt} = R^* \quad i = \{1, 2, \ldots, N\} \tag{3}$$

where T_i represents the operational time of equipment from $i - 1$th preventive maintenance to *i*th preventive maintenance.

4.2. Variable Cost Model

In this section, the maintenance cost in each condition is calculated, and the quality loss of the defective products produced is considered. Then, the total cost is divided by the cycle time to obtain the maintenance cost per unit, which is obtained by dividing the total cost by the cycle time *T*. In this paper, the total maintenance cost is mainly composed of preventive maintenance cost, breakdown maintenance cost, replacement cost, and quality loss cost.

(1) Preventive maintenance cost

According to the assumption of the model, when the reliability of equipment reaches the threshold of preventive maintenance, preventive maintenance will be carried out. In previous research, the cost of preventive maintenance was fixed. However, the cost of preventive maintenance cannot be fixed due to the deteriorating performance of equipment. When equipment performance significantly deteriorates, it will inevitably cost more to maintain. It means that the cost of preventive maintenance will continue to increase with the degradation of equipment, which is fluctuating. Thus, the cost of ith preventive maintenance of equipment is composed of both fixed costs and variable costs, and it is expressed as:

$$C_m^{pmi} = C_s^{pm} + C_{vi}^{pm} = C_s^{pm} + \gamma \lambda_i(T_i) \tag{4}$$

where C_s^{pm} represents the fixed cost of preventive maintenance, C_{vi}^{pm} represents the variable cost of ith preventive maintenance, which is linearly related to the amount of equipment degradation indicated by $\lambda_i(T_i)$, and the correlation coefficient is γ.

Assuming that the maintenance time for each preventive maintenance is t_m^{pm}, the total cost and total time of preventive maintenance in the cycle is therefore expressed as:

$$TC_m^{pm} = \sum_{i=1}^{N} C_m^{pmi} = \sum_{i=1}^{N} C_s^{pm} + \gamma \lambda_i(T_i) \tag{5}$$

$$T_m^{pm} = N \times t_m^{pm} \tag{6}$$

(2) Breakdown maintenance cost

According to assumption 1, the total cost of breakdown maintenance is equal to the number of failures multiplied by each breakdown maintenance cost. The number of failures in a preventive maintenance cycle can be calculated from Equation (3), which is expressed as:

$$N_c^i = \int_0^{T_i} \lambda_i(t)dt = -\ln R^* \tag{7}$$

The time and cost of each breakdown maintenance are t_m^{pm}, C_m^{cm}, respectively. The total time and total cost of breakdown maintenance can be obtained according to Equation (7):

$$TC_m^{cm} = C_m^{cm} \times ((N+1) \times (-\ln R^*)) \tag{8}$$

$$T_m^{cm} = t_m^{cm} \times ((N+1) \times (-\ln R^*)) \tag{9}$$

(3) Quality loss cost

Generally, equipment will produce a certain amount of substandard products during the production process. Most of these substandard products can be caused by equipment designed so that it cannot be reduced, and some are caused by equipment degradation. The cost of quality loss is the loss of inferior products that cannot be sold properly due to quality problems. With the degradation of equipment, the product quality will continue to degrade. At this time, the number of defective products will continue to increase, resulting in a particular cost. According to past sales data, the revenue of each product can be measured, and the cost of quality loss can be measured by the original sales revenue of defective products. Thus, it is necessary to calculate the number of defective products closely related to the defective product rate. According to Assumption 5, the defective rate of equipment varies, and it can be expressed as:

$$p(\lambda_i(t)) = p_0 + \mu[1 - e^{-\sigma \lambda_i(t)^\theta}] \tag{10}$$

where p_0 represents the defective rate in the new state of equipment, μ represents the boundary of quality deterioration, and σ and θ are constants [36].

Assuming that the loss cost of a single product is C_v^d, and the production rate is v, the quality loss cost of equipment in the period T is:

$$TC_d = C_{dv} \times \sum_{i=1}^{N+1} T_i \times v \times \overline{p}_i \tag{11}$$

In Equation (11), $\overline{p}_i$ represents the average defective rate of products produced by equipment during $i - 1$th and ith preventive maintenance intervals. The calculation equation of the average defective rate is as follows:

$$\overline{p}_i = \frac{p(\lambda_i(0)) + p(\lambda_i(T_i))}{2} \tag{12}$$

According to Assumption 3, replacement cost and time are C_r and T_r, respectively.

Currently, the total cost of equipment can be obtained. The next cycle time needs to be calculated. As is shown in Figure 1, the period T from the new state of equipment to the next replacement is composed of preventive maintenance time, equipment normal operation time, breakdown maintenance time, and replacement time. Finally, based on the above analysis, the variable cost model can be developed expressed by the maintenance cost per unit obtained by dividing the total cost during the cycle time T. The expression is as follows:

$$\begin{aligned}
ETC &= \frac{TC_m^{pm} + TC_m^{cm} + C_r + TC_d}{\sum\limits_{i=1}^{N+1} T_i + T_m^{pm} + T_m^{cm} + T_r} \\
&= \frac{\sum\limits_{i=1}^{N} C_s^{pm} + \gamma\lambda_i(T_i) + C_m^{cm} \times ((N+1) \times (-\ln R^*)) + C_r + \sum\limits_{i=1}^{N+1} C_{dv} \times v \times \overline{p}_i \times T_i}{\sum\limits_{i=1}^{N+1} T_i + N \times t_m^{pm} + T_r + t_m^{cm} \times ((N+1) \times (-\ln R^*))}
\end{aligned} \tag{13}$$

4.3. Energy Efficiency Model

The key to achieving carbon neutrality lies in energy conservation and emission reduction. In order to achieve energy conservation and emission reduction, it is necessary to explore the problem of excessive energy consumption in the process of product production and equipment maintenance and to improve energy efficiency. Energy efficiency is an important indicator to measure the input–output ratio of the manufacturing industry. The energy efficiency in the manufacturing process is expressed as the ratio between the total output capacity and the total input energy [37]. Thus, in order to obtain the energy efficiency of equipment, the total energy consumption and the total effective output of equipment need to be calculated. The first step is to find out the energy consumed by equipment. It is common knowledge that the energy consumption of equipment in different states varies. According to the difference, the state of equipment can be divided into normal operation state, standby state, warm-up state, power-on state, and power-off state. Since the time of power-on and power-off is very short, the energy consumption is negligible. The energy consumption of equipment is shown in Figure 3.

The energy consumption considered in this paper includes two parts, the energy consumption of equipment and the energy consumption of maintenance. The energy consumption of equipment includes the energy consumption of the normal operation, the standby energy consumption, and the warm-up energy consumption. The maintenance energy consumption is mainly the energy consumption of three maintenance activities. The various energy consumptions are calculated below.

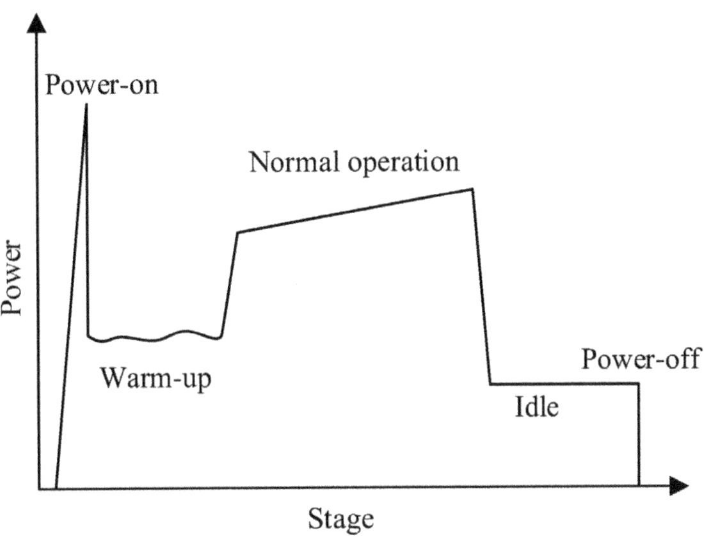

Figure 3. Different energy consumption states of equipment.

(1)   Operation energy consumption of equipment

The energy consumption of equipment during operation increases with degradation, which is linearly related to the failure rate. The energy consumption per unit before *i*th preventive maintenance is expressed as:

$$X_i(t) = X_0 + \omega\lambda_i(t) \tag{14}$$

where X_0 represents the energy consumption of equipment at the initial stage, and ω represents the linear relationship between the variation of equipment energy consumption and equipment failure rate. Thus, the total operation energy consumption of equipment in period T is expressed as:

$$E_o = \sum_{i=1}^{N+1} \int_0^{T_i} X_i(t)dt = \sum_{i=1}^{N+1} \int_0^{T_i} X_0 + \omega\lambda_i(t)dt \tag{15}$$

(2)   Standby energy consumption of equipment

According to Assumption 2, equipment is on standby while preventive maintenance is being performed. The standby time of equipment can be measured by the preventive maintenance time. Assuming that the standby energy consumption per unit of equipment is E_{iv}, the total standby energy consumption can be expressed as:

$$E_i = E_{iv} \times T_m^{pm} = E_{iv} \times N \times t_m^{pm} \tag{16}$$

(3)   Warm-up energy consumption of equipment

According to Assumption 4, the equipment needs to go through a warm-up time after it is turned on. The equipment needs to be shut down for maintenance and replacement. Assuming that the energy required to warm up equipment once is E_{wv}, then the total warm-up energy consumption of equipment is expressed as:

$$E_w = E_{wv} \times \sum_{i=1}^{N+1} N_c^i = E_{wv} \times ((N+1) \times (-\ln R^*)) \tag{17}$$

(4) Equipment maintenance energy consumption

When equipment performs maintenance activities, it needs to consume other energy such as electricity. Assuming that the energy consumption of each breakdown maintenance and preventive maintenance is E_{pv}, and the energy consumption of equipment replacement is E_{pr}, then the maintenance energy consumption of equipment in the period T is:

$$E_p = E_{pv} \times (N + (N+1) \times (-\ln R)) + E_{pr} \tag{18}$$

The above equation has calculated the total energy consumption of equipment, and then the total effective output of equipment needs to be obtained. The total effective output of equipment includes qualified products and defective products that can be recycled. With the deterioration of equipment, the defective rate is increasing, and the recovery factor will change correspondingly. The recovery factor τ^i is introduced to describe the gradual decrease in the amount of recovery. Thus, the final effective output of equipment is obtained by subtracting the number of defective products that cannot be recovered and is expressed as:

$$Y = \sum_{i=1}^{N+1} T_i \times v - \sum_{i=1}^{N+1} v \times \overline{p}_i \times T_i \times (1 - \tau^i) \tag{19}$$

The final energy efficiency model can be obtained and expressed as:

$$
\begin{aligned}
EE \quad &= \frac{Y}{E_o + E_i + E_w + E_p} \\[2mm]
&= \frac{\displaystyle\sum_{i=1}^{N+1} T_i \times v - \sum_{i=1}^{N+1} v \times \overline{p}_i \times T_i \times (1 - \tau^i)}{\displaystyle\sum_{i=1}^{N+1} \int_0^{T_i} X_0 + \omega\lambda_i(t)\,dt + E_{iv} \times N \times t_m^{pm} + E_{wv} \times ((N+1) \times (-\ln R^*)) + E_{pv} \times (N + (N+1) \times (-\ln R)) + E_{pr}}
\end{aligned}
\tag{20}
$$

4.4. Multiobjective Maintenance Model

In order to achieve a balance between the economic benefits and social benefits of the enterprise, not only maintenance costs but also energy efficiency need to be considered. Thus, this paper aims to minimize the maintenance cost per unit and maximize energy efficiency, using the preventive maintenance threshold and preventive maintenance times as the decision variables. In addition, a multiobjective maintenance model can be constructed. The expression is as follows:

$$D = \begin{cases} \min ETC(R^*, N^*) \\ \max EE(R^*, N^*) \end{cases} \tag{21}$$

4.5. Solution of the Maintenance Strategy Optimization

NSGAII is a multiobjective genetic algorithm widely used to analyze and solve multiobjective optimization problems due to its advantages of fast solution speed and good convergence of solutions. In this paper, the relationship between energy efficiency and cost per unit needs to be reconciled to satisfy each objective as far as possible. For this reason, the NSGAII algorithm is used to solve the model; the model solution process is shown in Figure 4. Its processes are as follows.

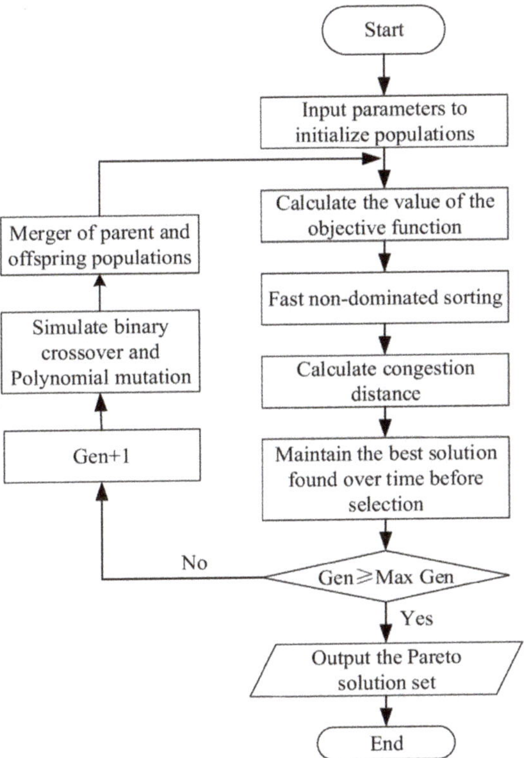

Figure 4. NSGAII algorithm processes.

Step 1: Parameter input. Input relevant parameters of the algorithm such as the number of populations, the maximum number of iterations, upper and lower bounds on the preventive maintenance threshold, crossover rate, and variation rate. Initialize the population and generate a random population P of N individuals.

Step 2: Calculate the maintenance cost and energy efficiency per unit for each individual in the population.

Step 3: Fast nondominated sorting. The individuals in the population P are classified by the fast nondominated sorting algorithm. According to the dominance relationship between the objective function values, the current optimal solution is selected and marked as rank 1. Then after excluding the solutions in the dominant rank 1, the optimal solution is selected from the remaining population and marked as rank 2, and so on, until the whole population is graded. The nondominated solution sets of different levels are constructed, such as $F_1, F_2, \ldots F_n$. As the optimization objective of this paper is to minimize the maintenance cost per unit and maximize energy efficiency, when the target value of energy efficiency is the vertical coordinate, and the target value of maintenance cost per unit is the horizontal coordinate, the higher the rank of the points on the axis to the upper left.

Step 4: Crowding distance calculation. In order to select the better individuals of the population and prevent falling into local maxima and local minima, the crowding distance of individuals needs to be calculated. It is defined as the sum distance of the two points on either side of this point along each of the objectives, denoted by i_d. As shown in Figure 5, the crowding distance of the ith point is expressed as the sum of the variable lengths of the rectangular rectangle, that is, the sum of the distance along the direction of the first objective and the distance along the direction of the second objective. The formula is expressed as:

$$i_d = [f_1(i+1) - f_1(i-1)] + [f_2(i-1) - f_2(i+1)] \qquad (22)$$

Step 5: Elite retention strategy. In order to prevent the loss of outstanding individuals during the evolution of the population, an elite retention strategy for individuals is required. The total crowding distance for each individual is equal to the sum of the distances for each single target metric. According to the elite retention strategy, individuals are selected sequentially from the highest ranked nondominated solution set to the lower ranked solution set. If two individuals are in the same rank, the crowding distance between them is compared, and the individual with the greater crowding is selected. N individuals are eventually selected to form a new parent population Q.

Step 6: Determine if the maximum number of iterations has been reached. If the maximum number of iterations is reached, the Pareto solution set is output; if not, the new parent population Q is crossed and mutated, the resulting child population Q' is merged with the parent population Q, and the operation in Step 2 is repeated.

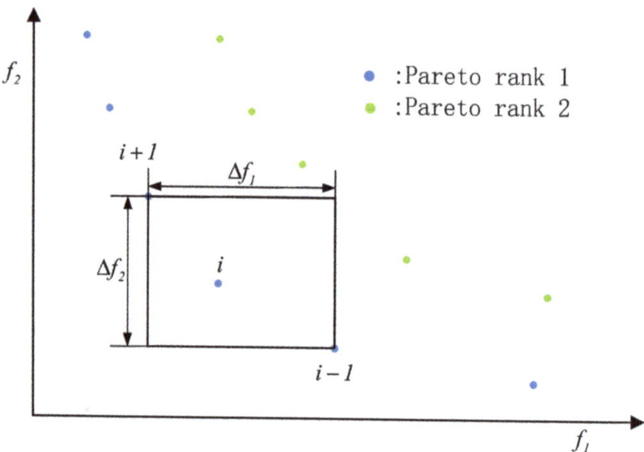

Figure 5. The crowding distance of the ith point.

5. Case Study

5.1. Data Preparation

The validity and adaptability of the multiobjective decision model are verified by a case study. In this paper, the production equipment of a manufacturing company is selected for the study. The production equipment produces products at a fixed rate every day. The equipment will produce a small number of defective products that can be recycled to a certain extent. Meanwhile, the defective product rate will increase with equipment degradation, and equipment operation and maintenance consume more energy. Referring to the historical data of the equipment, it can be found that the failure rate of the equipment obeys the Weibull distribution with the shape parameter of 3 and the size parameter of 110, and then referring to the general calculation of comprehensive energy consumption, the following parameters related to maintenance and energy consumption are obtained. As the types of products produced by the equipment will change with customer demand, the defective data of each product varies. In this paper, one of the products is selected, and the initial defective rate is obtained by analyzing the defective data. Other parameters of defective products are determined by referring to the literature [36]. Furthermore, due to the change in customer demand, the production rate of the equipment is not fixed. We assume that the production rate of the equipment is 100 pieces per day, thus obtaining the total parameter table, as shown in Table 1.

Table 1. Value of related parameters.

Maintenance Parameters		Energy Consumption Parameters		Production Parameters	
C_s^{pm}	400	x_0	180	P_0	0.008
C_{mv}^{cm}	150	ω	1.2	μ	0.075
C_{dv}	10	E_{iv}	25	σ	20
C_r	2000	E_{wv}	15	θ	1.1
t_m^{pm}	1	E_{pv}	50	τ	0.4
t_m^{cm}	0.5	E_{pr}	300	v	100
T_r	2				
γ	1.2				

5.2. Results Analysis

In order to narrow the search for multiobjective solution sets and ensure the accuracy and convenience of the solution, the maintenance cost per unit and energy efficiency of equipment under different combinations of preventive maintenance thresholds and maintenance times were obtained by numerical simulation.

The results can be seen in Figure 6. When the preventive maintenance threshold is in the range (0.6, 0.8), and the number of preventive maintenance is in the range (0, 6), there is a minimum value of maintenance cost per unit and a maximum value of energy efficiency. In addition, the graph of simulation results shows that the maintenance cost per unit tends to decrease and then increase as the number of maintenance increases. When preventive maintenance is less frequent, the time from the beginning of use to the replacement of equipment is shorter, and the equipment maintenance cost is mainly composed of replacement cost and breakdown maintenance cost, which makes the maintenance cost per unit higher. With the increase in maintenance times, the cycle time will gradually become longer, while the maintenance cost slowly increases, and the maintenance cost per unit shows a downward trend. When the number of maintenance exceeds a certain threshold due to frequent maintenance, the preventive maintenance cost of equipment significantly increases, and the cycle time slowly increases at this time, which makes the maintenance cost per unit show an upward trend.

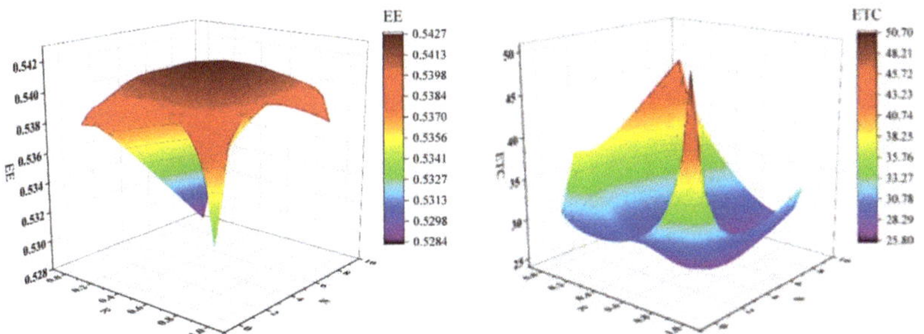

Figure 6. Maintenance cost per unit and energy efficiency under different threshold combinations.

Similarly, energy efficiency tends to increase and then decrease as the number of maintenance increases. When the number of preventive maintenance times is small, the energy consumption of equipment maintenance is mainly composed of replacement energy consumption and operation energy consumption, which makes the increase in equipment output exceed the increase in energy consumption, and the energy efficiency shows an upward trend. When the number of maintenance times exceeds a certain threshold, the cycle will gradually become longer as the number of maintenance increases. At this time, the effective output of the equipment slowly increases, but the energy consumption rapidly increases due to frequent preventive maintenance and breakdown maintenance, which makes the energy efficiency show a downward trend.

The optimal solution for the single objective can be obtained by conducting a simulation in the interval of the preventive maintenance thresholds of (0.6, 0.8) and the preventive maintenance times of (0, 6), as shown in Tables 2 and 3. The simulation results show that when the preventive maintenance threshold is 0.77 and the number of maintenance visits is 4, the lowest maintenance cost per unit is achieved at 25.81566. When the preventive maintenance threshold is 0.73 and the number of maintenance visits is 2, the highest equipment energy efficiency is achieved at 0.542660.

Thus, in this paper, we set the range of preventive maintenance threshold as (0.6, 0.8), the range of maintenance times as (2, 4), the number of individuals in the population as 100, the crossover rate as 0.9, the variation rate as 0.1, and the maximum number of iterations as 200. The above parameters were input into the algorithm of NSGAII, and the following results were obtained by a simulation using python, as shown in Figure 7. By comparing them, it is found that when $N = 2$, the energy efficiency of the equipment is the highest, but the maintenance cost per unit of equipment is also high. When $N = 3$, the energy efficiency of the equipment is lower than when $N = 2$, and the maintenance cost per unit of equipment is reduced more. When $N = 4$, the energy efficiency of the equipment is the lowest, but the maintenance cost per unit of equipment is not significantly reduced. Therefore, the comprehensive analysis yields that the energy efficiency and maintenance cost per unit of equipment is generally better for different maintenance thresholds at $N = 3$, so the Pareto curve at $N = 3$ is the final set of Pareto solutions for the model.

Table 2. Simulation results for maintenance costs per unit.

R	$N = 0$	$N = 1$	$N = 2$	$N = 3$	$N = 4$	$N = 5$	$N = 6$
0.6	30.733	28.83966	26.97263	26.50926	26.65826	27.13605	27.80899
0.61	30.85359	28.80936	26.91436	26.43532	26.57268	27.04048	27.7043
0.62	30.98098	28.78345	26.85985	26.36491	26.49058	26.9484	27.60315
0.63	31.11558	28.76212	26.80925	26.29816	26.41207	26.85993	27.50566
0.64	31.25781	28.74557	26.76272	26.23523	26.3373	26.7752	27.41196
0.65	31.40815	28.73406	26.72046	26.17628	26.26642	26.69437	27.3222
0.66	31.5671	28.72785	26.68267	26.1215	26.19961	26.6176	27.23655
0.67	31.73525	28.72725	26.6496	26.07111	26.13707	26.5451	27.15522
0.68	31.91322	28.73258	26.62151	26.02533	26.07903	26.47708	27.07842
0.69	32.10173	28.74424	26.59871	25.98445	26.02574	26.41381	27.00641
0.7	32.30155	28.76263	26.58153	25.94877	25.97751	26.35557	26.93948
0.71	32.51357	28.78824	26.57036	25.91863	25.93465	26.30268	26.87794
0.72	32.73875	28.82161	26.56564	25.89444	25.89755	26.25552	26.82218
0.73	32.97822	28.86334	26.56786	25.87662	25.86664	26.2145	26.77261
0.74	33.2332	28.91411	26.57757	25.86571	25.84239	26.1801	26.72972
0.75	33.50512	28.97472	26.59542	25.86226	25.82536	26.15288	26.69407
0.76	33.79558	29.04606	26.62213	25.86696	25.81621	26.13348	26.66628
0.77	34.10643	29.12916	26.65856	25.88057	25.81566	26.12261	26.6471
0.78	34.43977	29.22522	26.70568	25.90398	25.82457	26.12114	26.63738
0.79	34.79807	29.3356	26.76462	25.93823	25.84395	26.13004	26.63811
0.8	35.18419	29.46194	26.83671	25.98453	25.87495	26.15049	26.65046

Table 3. Simulation results for energy efficiency.

R	N = 0	N = 1	N = 2	N = 3	N = 4	N = 5	N = 6
0.6	0.540185	0.542253	0.542420	0.542160	0.541726	0.541198	0.540612
0.61	0.540153	0.542267	0.542452	0.542204	0.541779	0.541258	0.540678
0.62	0.540119	0.542279	0.542483	0.542245	0.541829	0.541315	0.540743
0.63	0.540081	0.542289	0.542511	0.542285	0.541877	0.541371	0.540805
0.64	0.540040	0.542296	0.542538	0.542323	0.541924	0.541425	0.540865
0.65	0.539995	0.542301	0.542562	0.542359	0.541968	0.541476	0.540923
0.66	0.539946	0.542303	0.542584	0.542392	0.542010	0.541525	0.540978
0.67	0.539894	0.542303	0.542603	0.542424	0.542050	0.541572	0.541032
0.68	0.539837	0.542299	0.542620	0.542452	0.542087	0.541617	0.541082
0.69	0.539775	0.542292	0.542634	0.542478	0.542122	0.541658	0.541130
0.7	0.539708	0.542282	0.542646	0.542502	0.542154	0.541697	0.541176
0.71	0.539636	0.542269	0.542654	0.542523	0.542183	0.541734	0.541218
0.72	0.539558	0.542251	0.542659	0.542540	0.542209	0.541767	0.541257
0.73	0.539474	0.542230	0.542660	0.542554	0.542232	0.541797	0.541293
0.74	0.539384	0.542204	0.542658	0.542565	0.542251	0.541823	0.541325
0.75	0.539286	0.542173	0.542652	0.542571	0.542267	0.541845	0.541353
0.76	0.539179	0.542137	0.542640	0.542574	0.542278	0.541863	0.541377
0.77	0.539064	0.542094	0.542624	0.542571	0.542285	0.541877	0.541396
0.78	0.538940	0.542046	0.542603	0.542564	0.542287	0.541885	0.541411
0.79	0.538804	0.541991	0.542576	0.542551	0.542283	0.541888	0.541419
0.8	0.538657	0.541927	0.542541	0.542532	0.542273	0.541885	0.541421

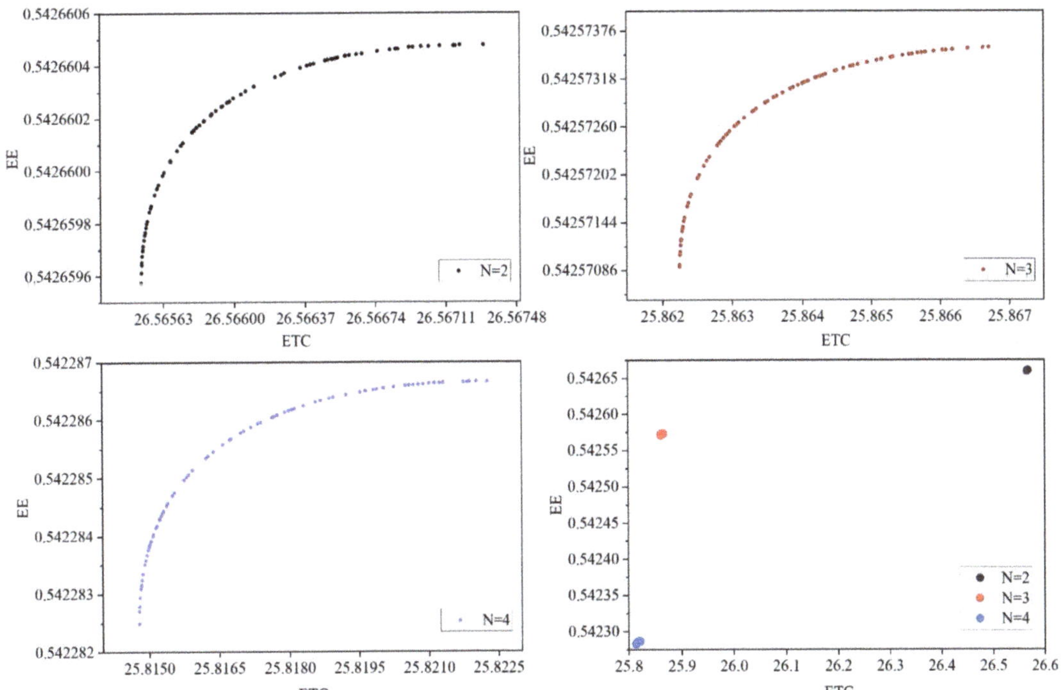

Figure 7. Pareto curve at different N.

5.3. Results Comparison

Then, in order to verify the superiority of the model, a comparison with a single target is required. By comparing the results of the single objective decision with the multiobjective decision, we can see that when only the cost of maintenance is considered, the optimal solution is obtained with an objective value of (25.8157, 0.5423). It is clear that improvements in energy efficiency need to be made. When only energy efficiency is considered, the optimal solution obtained corresponds to an objective value of (26.5679, 0.5427), a decision that is clearly not optimal in terms of maintenance costs. In comparison, the compromise solution is (25.8622, 0.5426), where the energy efficiency is not much different from the single objective and the maintenance cost per unit is better, thus showing that the integrated consideration of maintenance cost and energy efficiency can help enterprises to achieve the goals of energy conservation and emission reduction.

5.4. Sensitivity Analysis

Finally, in order to analyze the relationship between the objective function and the parameters, this paper selects the maintenance cost parameters for sensitivity analysis by changing one maintenance cost parameter, keeping the other parameters constant, and observing the sensitivity of the objective function to the maintenance cost parameter. In this paper, the fixed cost of preventive maintenance, breakdown maintenance, and replacement cost affect the objective function. The range of parameter variation is $-50\sim+50\%$. The analysis results obtained are shown in Table 4.

Table 4. Sensitivity analysis results.

Parameters	Range of Changes	ETC	Variation of ETC	EE	Variation of EE
\	\	25.8622	\	0.5426	\
C_s^{pm}	−50%	22.9814	−11.15%	0.5422	−0.7%
	−25%	24.6962	−4.6%	0.5425	−0.2%
	+25%	26.9908	4.3%	0.5425	−0.2%
	+50%	28.0859	9.3%	0.5425	−0.2%
C_m^{cm}	−50%	25.5247	−1.3%	0.5425	−0.2%
	−25%	25.6952	−0.7%	0.5425	−0.2%
	+25%	26.0260	0.6%	0.5425	−0.2%
	+50%	26.1867	1.2%	0.5425	−0.2%
C_r	−50%	21.7022	−16.1%	0.5425	−0.2%
	−25%	23.8954	−7.6%	0.5425	−0.2%
	+25%	27.4742	6.2%	0.5422	−0.7%
	+50%	29.0641	12.4%	0.5421	−0.9%

According to the sensitivity analysis results, the maintenance cost per unit is more sensitive to the fixed cost of preventive maintenance and the replacement cost, and it varies positively with both parameters. When the fixed cost of preventive maintenance and the replacement cost decrease, the change in cost is more obvious than when they increase, which means that the maintenance costs per unit can be reduced by reducing the fixed costs of preventive maintenance and replacement costs when making maintenance decisions. In addition, the maintenance costs per unit and energy efficiency are not sensitive to breakdown maintenance cost, and changes in the fixed cost of preventive maintenance, breakdown maintenance cost, and replacement cost do not have a significant impact on changes in energy efficiency.

6. Conclusions

A methodological framework and a new preventive maintenance model were proposed that make it possible to optimize maintenance strategies in manufacturing production equipment. More generally, in the context of reducing carbon emissions and mitigating global warming, this paper focuses on solving the problem of equipment maintenance and energy efficiency in production systems by modeling and calculating the costs and various energy consumptions in the process of equipment maintenance to achieve the goal of optimizing maintenance strategies. In addition, the difference between considering energy efficiency and not is shown in this paper. The main findings of the article are the following: (1) compared with a maintenance strategy that only considers maintenance costs, the integrated consideration of maintenance costs, energy efficiency, and product quality is more suitable for manufacturing systems; (2) the modeling of dynamic preventive maintenance costs as well as dynamic operational energy consumption makes the calculation of costs and energy consumption more accurate; (3) the recycling of defective products is consistent with the goal of energy saving and emission reduction, and the amount of recycling is closely related to the state of the equipment. The framework and methods presented in this paper can be applied to production, maintenance, quality, and architecture maintenance optimization in manufacturing, which makes it possible to support management decisions. The decision process regarding production, quality control, and maintenance will be influenced by the results of the contribution. For example, the energy efficiency in maintenance will influence the maintenance policy, and the manufacturing system will specify new solutions for recycling defective products.

However, there are also limitations of the study. In many cases, manufacturing systems often include much equipment, which may be connected in series, parallel, or groups. The limitations of this paper, which considers only single-device preventive maintenance, also indicate potential directions for further research. In further research, the model can be extended to more complex equipment models and the use of opportunistic maintenance.

Author Contributions: Formal analysis, J.L.; Funding acquisition, T.X. and C.Y.; Investigation, Q.L.; Methodology, L.Y. All authors have read and agreed to the published version of the manuscript.

Funding: This research was funded by grants from the National Key R&D Program of China (No. 2021YFF0900400), National Natural Science Foundation of China (Nos. 71840003), Natural Science Foundation of Shanghai (No. 19ZR1435600), Humanity and Social Science Planning Foundation of the Ministry of Education of China (No. 20YJAZH068), Action Plan for Scientific and Technological Innovation of Shanghai Science and Technology Commission (No. 21SQBS01404), and the Science and Technology Development Project of University of Shanghai for Technology and Science (No. 2020KJFZ038).

Data Availability Statement: Not applicable.

Acknowledgments: The authors are indebted to the reviewers and the editors for their constructive comments, which greatly improved the contents and exposition of this paper.

Conflicts of Interest: The authors declare no conflict of interest.

References

1. Arora, N.K.; Fatima, T.; Mishra, I.; Verma, M.; Mishra, J.; Mishra, V. Environmental sustainability: Challenges and viable solutions. *Environ. Sustain.* **2018**, *1*, 309–340.
2. Anderson, T.R.; Hawkins, E.; Jones, P.D. CO_2, the greenhouse effect and global warming: From the pioneering work of Arrhenius and Callendar to today's Earth System Models. *Endeavour* **2016**, *40*, 178–187. [CrossRef] [PubMed]
3. Liu, Z.; Deng, Z.; He, G.; Wang, H.; Zhang, X.; Lin, J.; Qi, Y.; Liang, X. Challenges and opportunities for carbon neutrality in China. *Nat. Rev. Earth Environ.* **2021**, *3*, 141–155. [CrossRef]
4. Zhang, S.; Chen, W. China's Energy Transition Pathway in a Carbon Neutral Vision. *Engineering* **2021**, *14*, 64–76. [CrossRef]
5. Beraud, J.-J.D.; Xicang, Z.; Jiying, W. Revitalization of Chinese's manufacturing industry under the carbon neutral goal. *Environ. Sci. Pollut. Res.* **2022**, *29*, 66462–66478. [CrossRef]
6. Jiang, Z.; Zhang, H.; Sutherland, J.W. Development of an environmental performance Assessment method for manufacturing process plans. *Int. J. Adv. Manufacturing. Technol.* **2012**, *58*, 783–790. [CrossRef]
7. Hassan, T.; Song, H.; Khan, Y.; Kirikkaleli, D. Energy efficiency a source of low carbon energy sources? Evidence from 16 high-income OECD economies. *Energy* **2022**, *243*, 123063. [CrossRef]
8. Howell, M.; Alshakhshir, F.S. *Energy Centered Maintenance—A Green Maintenance System*; River Publishers: Aalborg, Denmark, 2020.
9. Zhao, J.; Gao, C.; Tang, T. A Review of Sustainable Maintenance Strategies for Single Component and Multicomponent Equipment. *Sustainability* **2022**, *14*, 2992. [CrossRef]
10. Ben-Daya, M.; Duffuaa, S.O. Maintenance and quality: The missing link. *J. Qual. Maint. Eng.* **1995**, *1*, 20–26. [CrossRef]

11. Hajej, Z.; Rezg, N.; Gharbi, A. Quality issue in forecasting problem of production and maintenance policy for production unit. *Int. J. Prod. Res.* **2018**, *56*, 6147–6163. [CrossRef]

12. Mehdi, R.; Nidhal, R.; Anis, C. Integrated maintenance and control policy based on quality control. *Comput. Ind. Eng.* **2010**, *58*, 443–451. [CrossRef]

13. Lesage, A.; Dehombreux, P. Maintenance & quality control: A first methodological approach for maintenance policy optimization. *IFAC Proc. Vol.* **2012**, *45*, 1041–1046.

14. Jin, K.; Liu, X.; Liu, M.; Qian, K. Modified quality loss for the analysis of product quality characteristics considering maintenance cost. *Qual. Technol. Quant. Manag.* **2022**, *19*, 341–361. [CrossRef]

15. Jiang, A.; Dong, N.; Tam, K.L.; Lyu, C. Development and Optimization of a Condition-Based Maintenance Policy with Sustainability Requirements for Production System. *Math. Probl. Eng.* **2018**, *2018*, 4187575. [CrossRef]

16. Tlili, L.; Radhoui, M.; Chelbi, A. Condition-Based Maintenance Strategy for Production Systems Generating Environmental Damage. *Math. Probl. Eng.* **2015**, *2015*, 494162. [CrossRef]

17. Chouikhi, H.; Dellagi, S.; Rezg, N. Development and optimization of a maintenance policy under environmental constraints. *Int. J. Prod. Res.* **2012**, *50*, 3612–3620. [CrossRef]

18. Huang, J.; Chang, Q.; Arinez, J.; Xiao, G. A Maintenance and Energy Saving Joint Control Scheme for Sustainable Manufacturing Systems. *Procedia CIRP* **2019**, *80*, 263–268. [CrossRef]

19. Liu, Q.; Li, Z.; Xia, T.; Hsieh, M.; Li, J. Integrated Structural Dependence and Stochastic Dependence for Opportunistic Maintenance of Wind Turbines by Considering Carbon Emissions. *Energies* **2022**, *15*, 625. [CrossRef]

20. Van Horenbeek, A.; Kellens, K.; Pintelon, L.; Duflou, J.R. Economic and environmental aware maintenance optimization. *Procedia CIRP* **2014**, *15*, 343–348. [CrossRef]

21. Saez, M.; Barton, K.; Maturana, F.; Tilbury, D.M. Modeling framework to support decision making and control of manufacturing systems considering the relationship between productivity, reliability, quality, and energy consumption. *J. Manuf. Syst.* **2021**, *62*, 925–938. [CrossRef]

22. Yan, J.; Hua, D. Energy consumption modeling for machine tools after preventive maintenance. In Proceedings of the 2010 IEEE International Conference on Industrial Engineering and Engineering Management, Macao, China, 7–10 December 2010; IEEE: New York, NY, USA, 2010; pp. 2201–2205.

23. Zhou, L.; Li, J.; Li, F.; Meng, Q.; Li, J.; Xu, X. Energy consumption model and energy efficiency of machine tools: A comprehensive literature review. *J. Clean. Prod.* **2016**, *112*, 3721–3734. [CrossRef]

24. Shang, Z.; Gao, D.; Jiang, Z.; Lu, Y. Towards less energy intensive heavy-duty machine tools: Power consumption characteristics and energy-saving strategies. *Energy* **2019**, *178*, 263–276. [CrossRef]

25. Zhou, B.; Yi, Q. An energy-oriented maintenance policy under energy and quality constraints for a multielement-dependent degradation batch production system. *J. Manuf. Syst.* **2021**, *59*, 631–645. [CrossRef]

26. Mawson, V.J.; Hughes, B.R. The development of modelling tools to improve energy efficiency in manufacturing processes and systems. *J. Manuf. Syst.* **2019**, *51*, 95–105. [CrossRef]

27. Bermeo-Ayerbe, M.A.; Ocampo-Martinez, C.; Diaz-Rozo, J. Data-driven energy prediction modeling for both energy efficiency and maintenance in smart manufacturing systems. *Energy* **2022**, *238*, 121691. [CrossRef]

28. Xia, T.; Shi, G.; Si, G.; Du, S.; Xi, L. Energy-oriented joint optimization of machine maintenance and tool replacement in sustainable manufacturing. *J. Manuf. Syst.* **2021**, *59*, 261–271. [CrossRef]

29. Aramcharoen, A.; Mativenga, P.T. Critical factors in energy demand modelling for CNC milling and impact of toolpath strategy. *J. Clean. Prod.* **2014**, *78*, 63–74. [CrossRef]

30. Zhou, B.; Qi, Y.; Liu, Y. Proactive preventive maintenance policy for buffered serial production systems based on energy saving opportunistic windows. *J. Clean. Prod.* **2020**, *253*, 119791. [CrossRef]

31. Xia, T.; Xi, L.; Du, S.; Xiao, L.; Pan, E. Energy-Oriented Maintenance Decision-Making for Sustainable Manufacturing Based on Energy Saving Window. *J. Manuf. Sci. Eng.* **2018**, *140*, 051001. [CrossRef]

32. Brundage, M.P.; Chang, Q.; Zou, J.; Li, Y.; Arinez, J.; Xiao, G. Energy economics in the manufacturing industry: A return on investment strategy. *Energy* **2015**, *93*, 1426–1435. [CrossRef]

33. Xia, T.; Si, G.; Shi, G.; Zhang, K.; Xi, L. Optimal selective maintenance scheduling for series–parallel systems based on energy efficiency optimization. *Appl. Energy* **2022**, *314*, 118927. [CrossRef]

34. Hoang, A.; Do, P.; Iung, B. Energy efficiency performance-based prognostics for aided maintenance decision-making: Application to a manufacturing platform. *J. Clean. Prod.* **2017**, *142*, 2838–2857. [CrossRef]

35. Frigerio, N.; Matta, A. Energy-efficient control strategies for machine tools with stochastic arrivals. *IEEE Trans. Autom. Sci. Eng.* **2014**, *12*, 50–61. [CrossRef]

36. Bouslah, B.; Gharbi, A.; Pellerin, R. Joint economic design of production, continuous sampling inspection and preventive maintenance of a deteriorating production system. *Int. J. Prod. Econ.* **2016**, *173*, 184–198. [CrossRef]

37. Thiede, S. *Energy Efficiency in Manufacturing Systems*; Springer Science & Business Media: Berlin/Heidelberg, Germany, 2012.

MDPI

St. Alban-Anlage 66

4052 Basel

Switzerland

Tel. +41 61 683 77 34

Fax +41 61 302 89 18

www.mdpi.com

Energies Editorial Office

E-mail: energies@mdpi.com

www.mdpi.com/journal/energies